Lecture Notes in Computer Science 5800

Commenced Publication in 1973
Founding and Former Series Editors:
Gerhard Goos, Juris Hartmanis, and Jan van Leeuwen

T0223397

Kurt Jensen   Jonathan Billington
Maciej Koutny (Eds.)

Transactions on
Petri Nets
and Other Models
of Concurrency III

Editor-in-Chief

Kurt Jensen
University of Aarhus
Faculty of Science
Department of Computer Science
IT-parken, Aabogade 34, 8200 Aarhus N, Denmark
E-mail: kjensen@cs.au.dk

Guest Editors

Jonathan Billington
University of South Australia
School of Electrical and Information Engineering
Mawson Lakes Campus, Mawson Lakes, South Australia 5095, Australia
E-mail: jonathan.billington@unisa.edu.au

Maciej Koutny
Newcastle University
School of Computing Science
Newcastle upon Tyne, NE1 7RU, UK
E-mail: maciej.koutny@newcastle.ac.uk

Library of Congress Control Number: 2009938164

CR Subject Classification (1998): D.2.2, D.2.4, D.2.9, F.1.1, F.3.2, F.4.1, H.1, I.6

ISSN 0302-9743 (Lecture Notes in Computer Science)
ISSN 1867-7193 (Transactions on Petri Nets and Other Models of Concurrency)

ISBN 978-3-642-04854-8 Springer Berlin Heidelberg New York

Typesetting: Camera-ready by author, data conversion by Scientific Publishing Services, Chennai, India
Printed on acid-free paper   SPIN: 12769803   06/3180   5 4 3 2 1 0

Preface by Editor-in-Chief

The third issue of LNCS Transactions on Petri Nets and Other Models of Concurrency (ToPNoC) contains revised and extended versions of a selection of the best papers from the workshops associated with the 29th International Conference on Application and Theory of Petri Nets and Other Models of Concurrency, which took place in Xi'an, China, 23–27 June 2008, and from the 8th and 9th Workshop and Tutorial on Practical Use of Coloured Petri Nets and the CPN Tools, held in Aarhus, Denmark, in October 2007 and 2008. It also contains a paper that was submitted to ToPNoC directly.

I would like to thank the two guest editors of this special issue: Jonathan Billington and Maciej Koutny. Moreover, I would like to thank all authors, reviewers, and the organizers of the Petri net conference satellite workshops, without whom this issue of ToPNoC would not have been possible.

July 2009   Kurt Jensen
                                                                  Editor-in-Chief
LNCS Transactions on Petri Nets and Other Models of Concurrency (ToPNoC)

LNCS Transactions on Petri Nets and Other Models of Concurrency: Aims and Scope

ToPNoC aims to publish papers from all areas of Petri nets and other models of concurrency ranging from theoretical work to tool support and industrial applications. The foundation of Petri nets was laid by the pioneering work of Carl Adam Petri and his colleagues in the early 1960s. Since then, an enormous amount of material has been developed and published in journals and books and presented at workshops and conferences.

The annual International Conference on Application and Theory of Petri Nets and Other Models of Concurrency started in 1980. The International Petri Net Bibliography maintained by the Petri Net Newsletter contains close to 10,000 different entries, and the International Petri Net Mailing List has 1,500 subscribers. For more information on the International Petri Net community, see: http://www.informatik.uni-hamburg.de/TGI/PetriNets/

All issues of ToPNoC are LNCS volumes. Hence they appear in all large libraries and are also accessible in LNCS Online (electronically). It is possible to subscribe to ToPNoC without subscribing to the rest of LNCS.

ToPNoC contains:

- revised versions of a selection of the best papers from workshops and tutorials concerned with Petri nets and concurrency;
- special issues related to particular subareas (similar to those published in the *Advances in Petri Nets* series);
- other papers invited for publication in ToPNoC; and
- papers submitted directly to ToPNoC by their authors.

Like all other journals, ToPNoC has an Editorial Board, which is responsible for the quality of the journal. The members of the board assist in the reviewing of papers submitted or invited for publication in ToPNoC. Moreover, they may make recommendations concerning collections of papers for special issues. The Editorial Board consists of prominent researchers within the Petri net community and in related fields.

Topics

System design and verification using nets; analysis and synthesis, structure and behavior of nets; relationships between net theory and other approaches; causality/partial order theory of concurrency; net-based semantical, logical and algebraic calculi; symbolic net representation (graphical or textual); computer tools for nets; experience with using nets, case studies; educational issues related to nets; higher level net models; timed and stochastic nets; and standardization of nets.

Applications of nets to: biological systems, defence systems, e-commerce and trading, embedded systems, environmental systems, flexible manufacturing systems, hardware structures, health and medical systems, office automation, operations research, performance evaluation, programming languages, protocols and networks, railway networks, real-time systems, supervisory control, telecommunications, and workflow.

For more information about ToPNoC, please see: www.springer.com/lncs/topnoc

Submission of Manuscripts

Manuscripts should follow LNCS formatting guidelines, and should be submitted as PDF or zipped PostScript files to ToPNoC@cs.au.dk. All queries should be addressed to the same e-mail address.

Preface by Guest Editors

This issue of ToPNoC contains revised and extended versions of a selection of the best papers from the workshops held at the 29th International Conference on Application and Theory of Petri Nets and Other Models of Concurrency, Xi'an, China, 23–27 June 2008 and from the 8th and 9th Workshops and Tutorials on Practical Use of Coloured Petri Nets and the CPN Tools, Aarhus, Denmark, October 2007 and 2008. It also contains a paper that was submitted to ToPNoC directly.

We are indebted to the programme committees (PCs) of the workshops and in particular their chairs. Without their enthusiastic work this volume would not have been possible. Many members of the PCs participated in reviewing the revised and extended papers considered for this issue.

Papers from the following workshops were considered when selecting the best papers:

- PNDS 2008: International Workshop on Petri Nets and Distributed Systems, organized by Heiko Rölke (Germany), Natalia Sidorova (The Netherlands) and Daniel Moldt (Germany);
- CHINA 2008: Workshop on Concurrency Methods, Issues and Applications, organized by Jetty Kleijn (The Netherlands) and Maciej Koutny (UK);
- PNAM 2008: The 2008 Workshop on Petri Nets and Agile Manufacturing organized by MengChu Zhou (USA) and ZhiWu Li (China);
- CPN 2007: The 8th Workshop and Tutorial on Practical Use of Coloured Petri Nets and the CPN Tools, organized by Kurt Jensen and his group (Denmark); and
- CPN 2008: The 9th Workshop and Tutorial on Practical Use of Coloured Petri Nets and the CPN Tools, organized by Kurt Jensen and his group (Denmark).

The best papers of these workshops were selected in close cooperation with their chairs. The authors of these papers were invited to submit improved and extended versions. The papers needed to incorporate new results and to address comments made by the workshop's referees and those made during discussions at the workshop.

All invited papers were reviewed by three or four referees. We followed the principle of also asking for "fresh" reviews of the revised papers, i.e. from referees who had not been involved initially in reviewing the papers. After the first round of reviews, some papers were rejected while the authors of the others were asked to revise their papers in line with the reviewer comments and to include a response to each comment to indicate how changes had been incorporated as a result of the comment. The revised paper and the responses were then forwarded to the reviewers for a final recommendation and comment. We would like to thank all authors and the reviewers for their excellent cooperation and for their outstanding work which has led to a set of excellent papers in this issue.

After this rigorous review process, 9 papers were accepted out of the 18 initially considered as best papers. (Note that the workshops accepted about 60 papers in total and that the number of submissions to these workshops was considerably higher.)

The first three papers of this issue are concerned with workflow. The paper "Designing a Workflow System Using Coloured Petri Nets" by Nick Russell, Wil van der Aalst and Arthur ter Hofstede, addresses the design of a workflow system, called newYawl, the objective of which is to provide support for both data management and resources as well as the traditional control flow perspective. The language semantics are based on Coloured Petri Nets. The second paper, "From Requirements via Coloured Workflow Nets to an Implementation in Several Workflow Systems" by Ronny Mans et al, presents a large workflow case study concerning the process used for the diagnosis of cancer in a teaching hospital in The Netherlands. These first two papers are complemented by the final paper of this group, "Soundness of Workflow Nets with Reset Arcs" by Wil van der Aalst et al. There is a shift from Coloured Petri Nets to Place/Transition Nets, and a proof is provided that the important workflow property of soundness is undecidable for workflow nets with reset arcs. An excellent introduction to the problem is provided in this paper, set in the context of practical workflow systems.

The next 3 papers address various aspects of modelling and analysing computer network protocols. Jonathan Billington, Somsak Vanit-Anunchai and Guy Gallasch explore, in tutorial fashion, the problem of modelling bounded and unbounded communication channels that may or may not lose and reorder packets in their paper "Parameterised Coloured Petri Net Channel Models". Adhoc routing protocols are being developed and standardised for networks where there is no central administration, useful when there is a need to quickly establish a computer communications network in difficult terrain. The paper, "On Modelling and Analysing the Dynamic MANET On-Demand (DYMO) Routing Protocol" by Jonathan Billington and Cong Yuan, presents a modelling method that reduces state space explosion, and hence increases the range of analysis results possible for a mobile adhoc network (MANET) on-demand routing protocol being developed by the Internet Engineering Task Force. The mobility theme is continued in the last paper of this group, "Modelling Mobile IP with Mobile Petri Nets" by Charles Lakos. The paper presents a case study, the Mobile Internet Protocol (Mobile IP), to assess the utility of a hierarchical Petri net formalism, known as Mobile Petri Nets.

The last four papers have an increasingly theoretical flavour. In "A Discretization Method from Coloured to Symmetric Nets: Application to an Industrial Example", Fabien Bonnefoi, Christine Choppy and Fabrice Kordon propose a methodology for modelling and analysis of complex systems that involves translating an initial Coloured Petri Net model of the system to a Symmetric Net model for analysis. An important consideration in the translation is dealing with hybrid systems by discretization of continuous variables. The paper discusses the errors introduced by quantization and illustrates the method with a case study involving the emergency braking module of an intelligent transport system. "The ComBack Method Revisited: Caching Strategies and Extension with Delayed Duplicate Detection" by Sami Evangelista, Michael Westergaard and Lars Kristensen, tackles the state explosion problem which limits the extent to which model checking can be used for industrial systems. They improve the previously published ComBack technique for memory reduction, by proposing several strategies that reduce its time penalty. The algorithms are implemented and extensive experiments on various examples illustrate their utility. The third paper of this group also investigates the effectiveness of algorithms, this time in the setting of Petri net synthesis. In their paper, "Comparison of Different Algorithms to Synthesize a Petri Net from a Partial Language", Robin Bergenthum, Joerg Desel and Sebastian Mauser report on two new algorithms they have developed for synthesising a Place/Transition net from a finite partial language. They are implemented in the VipTool and compared with their predecessors. The new algorithms are also compared with respect to their theoretical complexity and by an extensive set of examples using the VipTool. The last paper of this issue is devoted to theoretical considerations of the π-calculus. In the paper, "On Bisimulation Theory in Linear Higher-Order π-Calculus", Xian Xu discusses several aspects of bisimulation in the higher-order π-calculus, and establishes two simpler bisimulation variants.

The 10 papers of this issue provide good coverage of a diverse range of topics including workflow systems, network protocols, mobility, intelligent transport systems, the state explosion problem, Petri net synthesis and the π-calculus. The volume offers a good mixture of theory, tools and practical applications related to concurrency and provides a useful snapshot of current research.

As guest editors we would like to thank Dorthe Haagen Nielsen of Aarhus University for providing administrative support and the Springer/ToPNoC team for the final production of this issue.

July 2009

Jonathan Billington
Maciej Koutny
Guest Editors, Third Issue of ToPNoC

Organization of This Issue

Guest Editors

Jonathan Billington, Australia
Maciej Koutny, UK

Co-chairs of the Workshops

Heiko Rölke (Germany)
Natalia Sidorova (The Netherlands)
Daniel Moldt (Germany)
Jetty Kleijn (The Netherlands)
Maciej Koutny (UK)
MengChu Zhou (USA)
ZhiWu Li (China)
Kurt Jensen (Denmark)

Referees

Eric Badouel
Kamel Barkaoui
Josep Carmona
Daniel Y. Chao
Thomas Chatain
Tom Chothia
Søren Christensen
Piotr
   Chrzastowski-Wachtel
Gianfranco Ciardo
José-Manuel Colom
Philippe Darondeau
Jörg Desel
Raymond Devillers
Susanna Donatelli
Sami Evangelista
Joaquin Ezpeleta

Berndt Farwer
Joern Freiheit
Guy Gallasch
Luis Gomes
Serge Haddad
Xudong He
Kees van Hee
Vladimír Janoušek
Hanna Klaudel
Jetty Kleijn
Michael Köhler
Lars M. Kristensen
Charles Lakos
Johan Lilius
Robert Lorenz
Agostino Mangini
Sebastian Mauser

Manuel Mazzara
Roland Meyer
Daniel Moldt
Jan Ortman
Chun Ouyang
Laure Petrucci
Lucia Pomello
Franck Pommereau
Heiko Rölke
Sol Shatz
Mark-Oliver Stehr
H.M.W. (Eric) Verbeek
Karsten Wolf
Jianli Xu
Wlodek Zuberek
Roberto Zunino

Table of Contents

Designing a Workflow System Using Coloured Petri Nets*,**

Nick C. Russell[1], Wil M.P. van der Aalst[1,2], and Arthur H.M. ter Hofstede[2]

[1] Eindhoven University of Technology,
P.O. Box 513, 5600MB, Eindhoven, The Netherlands
{n.c.russell,w.m.p.v.d.aalst}@tue.nl

[2] Queensland University of Technology,
P.O. Box 2434, QLD, 4001, Australia
a.terhofstede@qut.edu.au

Abstract. Traditional workflow systems focus on providing support for the control-flow perspective of a business process, with other aspects such as data management and work distribution receiving markedly less attention. A guide to desirable workflow characteristics is provided by the well-known workflow patterns which are derived from a comprehensive survey of contemporary tools and modelling formalisms. In this paper we describe the approach taken to designing the *new*YAWL workflow system, an offering that aims to provide comprehensive support for the control-flow, data and resource perspectives based on the workflow patterns. The semantics of the *new*YAWL workflow language are based on Coloured Petri Nets thus facilitating the direct enactment and analysis of processes described in terms of *new*YAWL language constructs. As part of this discussion, we explain how the operational semantics for each of the language elements are embodied in the *new*YAWL system and indicate the facilities required to support them in an operational environment. We also review the experiences associated with developing a complete operational design for an offering of this scale using formal techniques.

Keywords: *new*YAWL, workflow technology, workflow patterns, business process management, coloured Petri nets.

1 Introduction

There are a plethora of workflow systems on the market today providing organisations with various forms of automated support for their business processes. It is ironic however, that despite the rigour that workflow systems introduce

---

* An earlier version of this work was presented at PNDS'08, the International Workshop on Petri Nets and Distributed Systems [1].
** This research is conducted in the context of the *Patterns for Process-Aware Information Systems (P4PAIS)* project which is supported by the Netherlands Organisation for Scientific Research (NWO).

K. Jensen, J. Billington, and M. Koutny (Eds.): ToPNoC III, LNCS 5800, pp. 1–24, 2009.
© Springer-Verlag Berlin Heidelberg 2009

into the conduct of the processes that they coordinate, they themselves do not demonstrate the same rigour in the workflow languages that they enact. Indeed, it is a salient fact that, almost without exception, workflow languages are defined on an informal basis leaving their precise operation unclear to anyone other than the system developers. An additional shortcoming of existing workflow solutions is their focus on the control-flow aspects of business processes.

The *YAWL* Initiative sought to address the first of these issues. *YAWL* [2] is an acronym for *Yet Another Workflow Language*. It provides a comprehensive modelling language for business processes based on formal foundations. The content of the YAWL language is an adaptation of Petri Nets informed by the workflow patterns [3]. One of its major aims was to show that a relatively small set of constructs could be used to directly support most of the workflow patterns identified. It also sought to illustrate that they could coexist within a common framework. In order to validate that the language was capable of direct enactment, the *YAWL System*[1] was developed, which serves as a reference implementation of the language. Over time, the YAWL language and the YAWL System have increasingly become synonymous and have garnered widespread interest from both practitioners and the academic community alike[2].

Initial versions of the YAWL System focussed on the control-flow perspective and provided a complete implementation of 19 of the original 20 patterns. Subsequent releases incorporated limited support for selected data and resource aspects of processes, however this effort was hampered by the lack of a complete formal description of the requirements in these perspectives. Recent work conducted as part of the Workflow Patterns Initiative has identified the core elements in other process perspectives (data, resource, exception handling) and a recent review [4] of the control-flow perspective has identified 23 additional patterns which illustrate a number of commonly used control-flow constructs, many of which YAWL is unable to provide direct support for, including the partial join, transient and persistent triggers, iteration and recursion.

In an effort to manage the conceptual shortcomings of YAWL with respect to the range of workflow patterns that have now been identified, a substantial revision of the language — termed *new*YAWL is proposed — which aims to support the broadest range of the workflow patterns in the control-flow, data and resource perspectives. *new*YAWL synthesises this work to provide a fully formalised workflow language based on a comprehensive view of a business process. The validation of this proposal is to design (and ultimately build) the workflow system that embodies the workflow language. An interesting consequence of formalising the operational semantics for the language constructs in *new*YAWL, has been the establishment of the functional architecture for the system to be

---

[1] See http://www.yawl-system.com for further details of the YAWL System and to download the latest version of the software.

[2] Hereafter in this paper, we refer to the collective group of YAWL offerings developed to date — both the YAWL language as defined in [2] and also more recent *YAWL System* implementations of the language based on the original definition (up to and including release Beta 8.2) — as *YAWL*.

developed. This is based on a detailed consideration of the causal effects and data interactions required to support each of the language constructs and their behaviour in a broader operational environment. This paper outlines the approach taken to designing the *new*YAWL system. In this paper we not only describe the design of *new*YAWL using Coloured Petri Nets, but also reflect on the use of such a design approach from a software engineering standpoint.

The remainder of this paper proceeds as follows: Section 2 introduces the YAWL language from a functional perspective. Section 3 presents *new*YAWL, a significant extension to YAWL that encompasses the broad range of workflow patterns which identify desirable workflow functionality that have recently been identified. Section 4 describes the approach to designing a workflow system that can enact business processes described in terms of the *new*YAWL language. Section 5 overviews related work and Section 6 discusses the experiences of designing a workflow system using formal methods and concludes the paper.

2 YAWL: Yet Another Workflow Language

YAWL has its genesis in the workflow patterns which aimed to delineate desirable constructs in the control-flow perspective of workflow processes. Hence the initial version of the YAWL language focussed solely on control-flow aspects of processes. It aimed to show that the patterns could be operationalised in an integrated framework. Furthermore, it also aimed to show that this could be achieved in the context of a formal framework, providing both a syntax and semantics for language constructs to remove any potential for ambiguity or uncertainty in their interpretation.

The formal foundation for YAWL is based on hierarchical Petri nets and there is a striking similarity between the two graphical representations with tasks taking the place of classical Petri net transitions and conditions representing the various states between tasks in the same way that places typically serve as the inputs and outputs to transitions in Petri nets. However Petri nets only serves as a basis for the fundamental aspects of YAWL and it significantly extends its capabilities in a variety of ways.

1. YAWL allows for tasks to be directly connected by an arc in the situation where there would normally be a single condition between them (that was not connected to any other tasks);
2. YAWL provides for direct representation of AND-split, AND-join, XOR-split, XOR-join and OR-split constructs rather than requiring their explicit modelling in terms of fundamental language constructs. In conjunction with direct connections between tasks, this serves to significantly simplify process models;
3. The notion of the (inclusive) OR-join, which is frequently described in process modelling notations without any consideration of how it will actually be enacted, is directly available as a modelling construct in YAWL. Moreover, there is a complete formal treatment of its operationalisation described in [5];

4. Task concurrency, which is not a consideration in many process modelling formalisms, is directly represented via the notion of the multiple instance task. This denotes a task (or subprocess) which executes multiple times in parallel with some or all instances needing to be synchronised before the thread of control can pass to subsequent tasks;

5. Cancellation of individual tasks, portions of a process or even an entire process can be explicitly represented in YAWL process models through the notion of a cancellation region. Cancellation regions are attached to a specific task in a process and when it completes, any threads of control residing in conditions within the cancellation region are removed and any executing tasks within the cancellation region are terminated; and

6. A YAWL process model (whether it is the top-level net in a process or a subprocess definition) has a single start and endpoint denoted by specific input and output conditions. This provides a precise semantics for process enablement and termination and allows a range of verification techniques to be applied to YAWL process models.

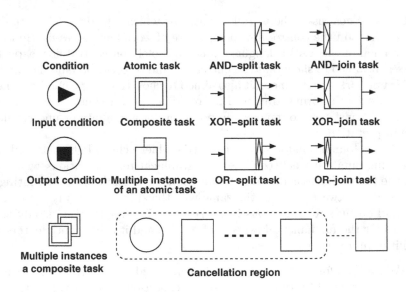

Fig. 1. YAWL symbology

The specific symbols used in a YAWL process are shown in Fig. 1. An example of a complete YAWL model using many of these symbols is depicted in Fig. 2. It denotes the sales fulfillment process for an ironmonger which sells and manufactures specialist metal fittings. An instance of the process is triggered when an order is received. This then initiates two distinct branches of activities (as signified by the outgoing AND-split). The first of these focusses on the financial aspects of the order. The other triggers the tasks associated with the actual assembly of the order for despatch, manufacturing and packing tasks. The first branch involves a *credit check*. If the customer has sufficient funds, then their order can be invoiced. It progresses to the *despatch* stage when the order handling is complete. If there is insufficient credit, then a reminder is sent and there is a waiting period for the required payment to be received from the customer. If it is not received within 10 days, the order is cancelled (denoted by the *timeout* task which is linked to a cancellation region which encompasses all tasks which might be active in the process). The branch associated with order assembly involves a series of tasks. First the order is prepared for picking. This is a composite task involving a sequence of three distinct activities: reviewing the order contents, producing a picking slip for items available from the warehouse and reviewing the requirements for any parts that need to be specially cast. Having done this, one or both of the *picking* and *custom cast* tasks are triggered via the OR-split operator from the *prepare picking* task. The *custom cast* task is a multiple instance task and a separate instance of it is triggered for each component that requires manufacturing. Once all of the *picking* and *custom cast* tasks that were initiated have been completed, the order is packed. After invoicing and packing have been completed, the *despatch* task can run, followed by the archiving of the order and completion of the case.

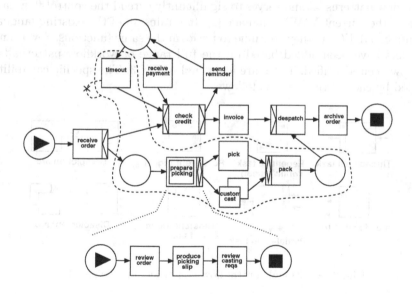

Fig. 2. Example of a YAWL process model: sales fulfillment

YAWL provides effective coverage of many commonly encountered control-flow constructs. However, a recent review [4] of the control-flow perspective identified a number of additional control-flow constructs that commonly arise in business processes. Moreover, there is also the need to consider other aspects of business processes, such as the requirements encountered in the data [6] and resource [7] perspectives as delineated by the data and resource patterns. In the next section, we introduce a comprehensive extension to the YAWL language that addresses these issues.

3   *new*YAWL: Extending YAWL to Multiple Perspectives

*new*YAWL is a multi-perspective business process modelling and enactment language founded on the workflow patterns. It provides a comprehensive and integrated formal description of the workflow patterns, which to date have only partially been formalised. It has a complete abstract syntax which identifies the characteristics of each of the language elements together with an executable semantic model in the form of a series of Coloured Petri Nets which define the runtime semantics of each of the language constructs. The following sections provide an overview of the features of *new*YAWL in the control-flow, data and resource perspectives.

3.1   Control-Flow Perspective

The control-flow perspective of *new*YAWL is based on the revised workflow control-flow patterns [4] and serves to significantly extend the control-flow capabilities of the current YAWL language [2]. It retains all of the existing language elements in *YAWL* and they continue to perform the same functions. Several new constructs have been added based on the full range of workflow patterns that have now been identified. These are identified in Fig. 3. The specific capabilities provided by each of them are as follows:

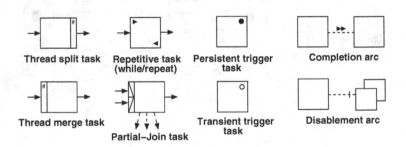

Fig. 3. Additional control-flow constructs in *new*YAWL

– the *Thread split* and *Thread merge* constructs, allow the thread of control to be split into multiple concurrent threads or several distinct threads to be merged into a single thread of control respectively. The number of threads being created/merged is specified for the construct in the design-time model. Fig. 4(a) illustrates these constructs. After the *make box* task, twelve threads of control are created ensuring that the *fill bottle* task runs 12 times before the *pack box* task can run (merging these threads before it commences);

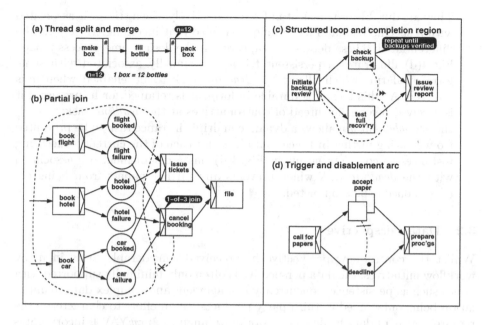

Fig. 4. Examples of *new*YAWL control-flow constructs

- the *Partial join* (also known as the m-out-of-n join) allows a series of incoming branches to be merged such that the thread of control is passed to
  the subsequent branch when m of the incoming n branches are enabled. The
  number of active threads required for the partial join to fire is specified in
  the design-time model. After firing, it resets (and can fire again) when all
  incoming branches have been enabled. In Fig. 4(b), the *cancel booking* task
  has a 1-out-of-3 join associated with it. If any of the incoming branches are
  enabled, then the *cancel booking* task is enabled (and any preceding tasks
  that are still executing in the associated cancellation region are withdrawn);
- the *Structured loop* (which supports while, repeat and combination loops)
  allows a task (or a sequence of tasks in the form of a subprocess) to execute repeatedly based on conditional tests at the beginning and/or end of
  each iteration. The loop is structured in form and it has a single entry and
  exit point. The entry and/or exit conditions are specified in the design-time
  process model. Fig. 4(c) illustrates a repeat loop for the *check backup* task
  which executes repeatedly until all backups have been verified (i.e. it is a
  post-tested loop);
- the *Completion region* supports the forced completion of tasks which it encompasses. In Fig. 4(c) the *test full recovery* task is forcibly completed once
  (all iterations of) the *check backup* task has finished. This allows the *issue
  review report* task to be immediately enabled;
- *Persistent triggers* and *Transient triggers* support the enablement of a task
  being contingent on a trigger being received from the operating environment.

They are durable or transient in form respectively. Each trigger is associated with a specific task and has a unique type so that incoming triggers can be differentiated. These details are captured in the design-time process model. Fig. 4(d) illustrates a persistent trigger (assumedly associated with some form of alarm) which allows the *deadline* task to be enabled when it is received. As this trigger is durable in form, it is retained for future use if it is received before the thread of control arrives at the *deadline* task;

- the *Disablement arc* allows a dynamic multiple instance task to be prevented from creating further instances but allows for each of the currently executing instances to complete normally. Fig. 4(d) has a disablement arc associated with the *deadline* task which prevents any further papers from being accepted once it has completed.

3.2 Data Perspective

Whilst the control-flow perspective has received considerable focus in many workflow initiatives, the data perspective is often only minimally supported with issues such as persistence, concurrency management and complex data manipulation being outsourced to third party products. In an effort to characterise the required range of data facilities in a workflow language, *new*YAWL incorporates a series of features derived from the data patterns. These include:

- Support for a variety of *distinct scopes* to which data elements can be bound. This allows the visibility and use of data elements to be restricted. The range of data scopes recognised include: *global* (available to all elements of all process instances), *folder* (available to the elements of process instances to which the folder is currently assigned), *case* (available to all elements in a given process instance), *block* (available to all elements of a specific process or subprocess definition for a given process instance), *scope* (available to a subset of the elements in a specific top-level process or subprocess definition for a given process instance), *task* (available to a given instance of a task) and *multiple-instance* (available to a specific instance of a multiple instance task);
- *Formal parameters* for specifying how data elements are transferred between process constructs (e.g. block to task, composite task to subprocess decomposition, block to multiple instance task). These parameters take a function-based approach to data transfer, thus providing the ability to support inline formatting of data elements and setting of default values. Parameters can be associated with tasks, blocks and processes;
- *Link conditions* for specifying conditions on outgoing arcs from OR-splits and XOR-splits that allow the determination of whether these branches should be activated;
- *Preconditions* and *postconditions* for tasks and processes. They are evaluated at the enablement or completion of the task or process with which they are associated. Unless they evaluate to true, the task or process instance with which they are associated cannot commence or complete execution; and

– *Locks* which allow tasks to specify data elements that they require exclusive
access to (within a given process instance) in order to commence. Once these
data elements are available, the associated task instance retains a lock on
them until it has completed execution preventing any other task instances
from using them concurrently. The lock is relinquished once the task instance
completes.

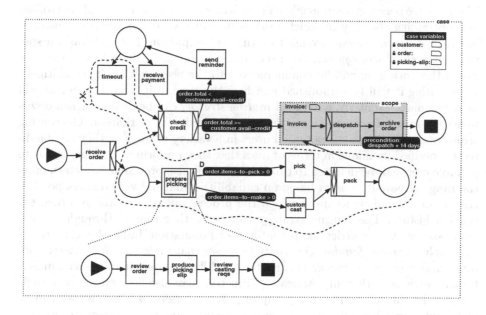

Fig. 5. Data perspective of sales fulfillment process

Figure 5 illustrates the main aspects of the data perspective for the sales fulfillment model (shown earlier in Fig. 2). The case variables *customer, order* and
picking-slip are used throughout the tasks in the model. They are record-based
in format and are passed on a reference basis between task instances. The lock
beside each variable name indicates that when it is passed to a task instance an
exclusive lock is applied to it whilst it is in use by that task instance to prevent problems arising from concurrent usage. In contrast to the case variables,
the *invoice* variable is bound to a specific scope involving only three tasks and,
although also passed by reference, does not require a lock as it is only updated
by the first task in the scope and cannot be used concurrently by several tasks
in the scope. The *credit check* task has an XOR-split associated with it and the
(disjoint) conditions on outgoing branches are illustrated. Similarly, the *prepare
picking* task has an OR-split associated with it and the two outgoing conditions are also shown although in this case, there is no requirement for them to
be disjoint since one or both outgoing branches can be enabled. For both split
constructs, the default branch is indicated with a D and this branch is enabled

if none of the conditions specified evaluate to true. The *archive order* task has a precondition associated with it which ensures it can only commence 14 days after the *despatch* task has completed. This is to allow for any returns or damage claims that might arise during transport.

3.3   Resource Perspective

The resource perspective in *new*YAWL provides a variety of means of controlling and optimising the way in which work is distributed to users and the manner in which it is progressed through to ultimate completion. For each task, a specific *interaction strategy* can be specified which precisely describes the way in which the work item will be communicated to the user, how their commitment to executing it will be established and how the time of its commencement will be determined. Similarly, a detailed *routing strategy* can be defined which determines the range of potential users that can undertake the work item. The routing strategy can nominate the potential users in a variety of ways — they can be directly specified by name, in terms of roles that they perform or the decision as to possible users can be deferred to runtime. There is also provision for determining the range of potential users based on capabilities that individual users possess, the organisational structure in which the process operates or the recorded execution history. The routing strategy can be further refined through the use of constraints that restrict the potential user population. Indicative constraints may include: *retain familiar* (i.e. route to a user that undertook a previous work item) and *four eyes principle* (i.e. route to a different user than one who undertook a previous work item). Allocation directives can also be used where a single user need to be selected from a group of potential users to whom a task can be allocated. Candidate allocation directives include *random allocation* (route to a user at random from the range of potential users), *round robin allocation* (route to a user from the potential population on an equitable basis such that all users receive a similar number of work items over time) and *shortest queue allocation* (route the work item to the user with the shortest work queue).

　*new*YAWL also supports two advanced operating modes that are designed to expedite the throughput of work by imposing a defined protocol on the way in which the user interacts with the system and work items are allocated to them. These modes are: *piled execution* where all work items corresponding to a given task are routed to the same user and *chained execution* where subsequent work items in a process instance are routed to the same user once they have completed a preceding work item. Finally, there is also provision for specifying a range of user privileges, both at process and individual task level, that restrict or augment the range of interactions that they can have with the workflow engine when they are undertaking work items.

　Figure 6 illustrates the resource perspective for the sales fulfillment process. Each task is annotated with the basic distribution strategy (DS) and interaction strategy (IS) for the task. The distribution strategy indicates which users and roles the task will be routed to at runtime. For the purposes of this model, routing is either to specific roles (e.g. A is the role for Administration users) or

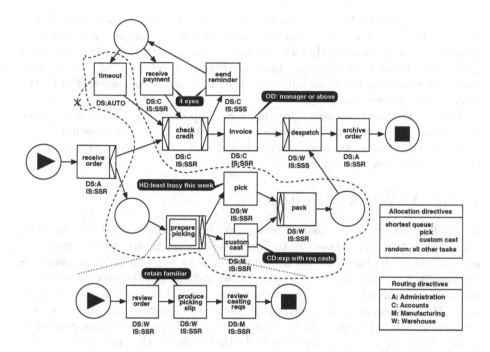

Fig. 6. Resource perspective of sales fulfillment process

AUTO where tasks are done automatically without requiring resource support. Extended routing directives apply to the *invoice, pick* and *custom cast* tasks. These operate in conjunction with the basic distribution strategy and further refine the specification of the user(s) to whom a task may be distributed. An organisational distribution directive applies to the *invoice* task requiring it be distributed to a member of the Accounts role that is at least a manager in organisational seniority. A historical distribution directive applies to the *pick* task requiring it be distributed to the member of the Warehouse role that has been least busy in the past week. A capability distribution directive applies to the *custom cast* task requiring it be distributed to the member of the Manufacturing role that has experience with the required casts that need to be manufactured for this order.

The interaction strategy indicates whether the system or an individual resource are responsible for triggering the offer, allocation and commencement of a task. For this model, the SSR and SSS interaction strategies are utilised. The former involves the system allocating the task to a specific user but the user being able to nominate the time at which they commence it, the latter (also known as "heads down processing") involves the system allocating a task to a user and indicating when they should start it. In both cases, allocation directives are employed to select the individual user to whom a task should be allocated, this is generally based on random selection of a user although the *pick* and *custom cast* tasks are allocated to user on the basis of who has the shortest current work queue. Distribution constraints exist between various pairs of tasks in the model. There is a four eyes constraint between the *receive payment* and *send reminder* tasks indicating they should not be allocated to the same user in a given case. There is also a retain familiar constraint between the *review order* and *produce picking slip* tasks indicating that they should be undertaken by the same user in a given case.

This section has focussed on providing a comprehensive introduction to the various language elements that make up *new*YAWL from a conceptual standpoint. In the next section, we discuss the design of a system that is able to operationalise these constructs.

4   *new*YAWL: The System

A workflow system encompasses a number of distinct functions as illustrated by the diagram in Fig. 7. Generally the business process that is to be automated is captured in the form of a *process model*. A *workflow management system* is responsible for coordinating the execution of instances of the process model. It comprises a number of discrete components. First the *workflow engine* is responsible for managing the control-flow and data elements that are associated with each process instance. As the thread of control flows through a process, it results in the triggering of individual tasks that make up the process model. The enabling of a task results in the creation of a new work item which needs to be executed by a human resource. However, in order for this to occur, the identity of one or more suitable resources needs to be determined. This activity is the responsibility of the *work item routing* component and is based on the interpretation of task routing information associated with each task in the context of the current state of the process instance.

Once a set of suitable resources have been determined for a work item, it is necessary to advise them of the pending work item. This function is undertaken by the *worklist management* component which places the work item on the worklist of each resource to whom it is to be routed. Thus the workflow management system retains a centralised view of the state of all work items and also

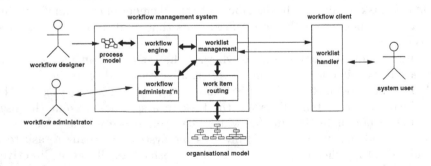

Fig. 7. Outline of major workflow system components

provides *workflow administration* facilities should it be necessary to intervene in the normal conduct of this process. However despite the consistent centralised view of pending work maintained by the workflow management system, there is another layer of complexity in managing the actual distribution and conduct of work items across the range of resources coordinated by the workflow system. This stems from the fact that resources typically operate independently of the workflow system. They retain a distinct view of the work that they are conducting which is accessed via a *worklist handler* (which typically takes the form of a software client running at a distinct location to that of the workflow management system). The worklist handler operates on a client-server basis with respect to the workflow management system. It is generally disconnected from the workflow system, connecting only when it wishes to refresh its view of the current work allocation or to advise the workflow system of a change of state in the work items it has been allocated.

Clearly a workflow system involves a relatively complex set of software components and interactions. In order to provide a precise definition of how a business process should actually be enacted in an operational environment, it is necessary not only to provide an operational semantics for the workflow language that describes the business process, but also to define the overall architecture and operation of the workflow system. This has been done for *new*YAWL using a series of interrelated Coloured Petri Nets developed using the CPN Tools environment [8]. This approach to formalising the system offers the dual benefits of establishing a precise definition of the operation of each of the language constructs which comprise *new*YAWL and also providing a means of describing exactly how an instance of a *new*YAWL specification should be executed. There are 55 distinct CPNs which make up the *new*YAWL system description. These are illustrated in Fig. 8 along with the relationships between them. The correspondence between the functional workflow system components identified in Fig. 7 and each of the CPNs is also delineated. An indication of the complexity of individual nets is illustrated by the p and t values included for each of them which indicate the number of places and transitions that they contain. Clearly it is not possible to discuss the operation of all of these nets in the confines of this paper, however some of them (indicated by the shaded boxes and cross-references) are discussed in further detail in subsequent sections. A comprehensive description of the 55 CPNs which comprise the *new*YAWL system can be found in [9]. In the following sections, we will outline the operation of three of these areas, illustrated by the shaded boxes in Fig. 8. These provide an overview of the *workflow engine*, *worklist management* and *worklist handler* components of the workflow system.

4.1 Workflow Engine

Figure 9, which is the topmost net in the *new*YAWL model, provides a useful summary of the operation of a workflow engine. The various aspects of control-flow, data management and work distribution information which make up a static *new*YAWL specification are encoded in the CPN model as tokens in individual places. The top level view of the lifecycle of a process instance is indicated by the

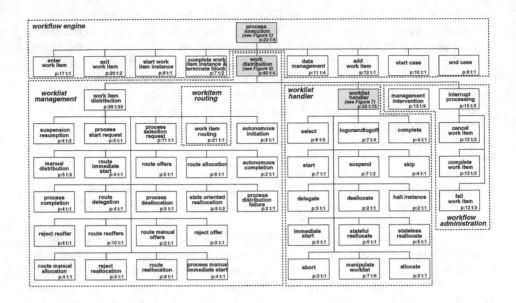

Fig. 8. *new*YAWL system CPN model hierarchy

transitions in this diagram connected by the thick black line. First a new process instance is started, then there is a succession of enter→start→complete →exit transitions which fire as individual task instances are enabled, the work items associated with them are started and completed and the task instances are finalised before triggering subsequent tasks in the process model. Each atomic work item needs to be routed to a suitable resource for execution, an act which occurs via the work distribution transition. This cycle repeats until the last task instance in the process is completed. At this point, the process instance is terminated via the end case transition. There is provision for data interchange between the process instance and the environment via the data management transition. Finally, where a process model supports task concurrency via multiple work item instances, there is provision for the dynamic addition of work items via the add transition.

The major data items shared between the activities which facilitate the process execution lifecycle are shown as shared places in this diagram. Not surprisingly, this includes both *static* elements which describe characteristics of individual processes such as the flow relation, task details, variable declarations, parameter mappings, preconditions, postconditions, scope mappings and the hierarchy of process and subprocess definitions which make up an overall process model, all of which remain unchanged during the execution of particular instances of the process. It also includes *dynamic* elements which describe how an individual process instance is being enacted at any given time. These elements are commonly known as the *state* of a process instance and include items such as the current marking of the place in the flow relation, variable instances and their

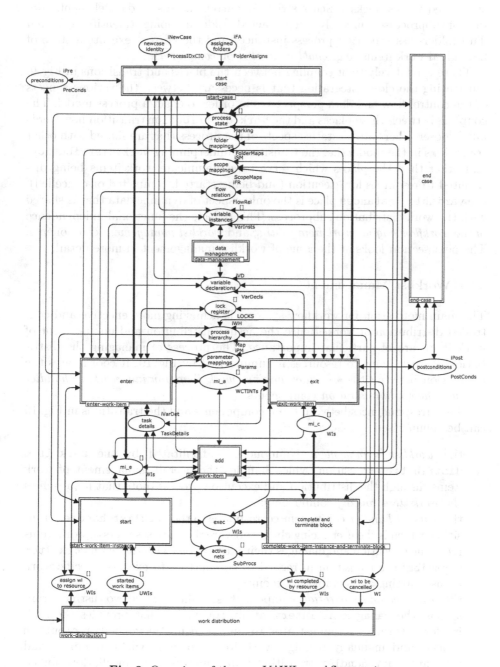

Fig. 9. Overview of the *new*YAWL *workflow engine*

associated values, locks which restrict concurrent access to data elements, details of subprocesses currently being enacted, folder mappings (identifying shared data folders assigned to a process instance) and the current execution state of individual work items (e.g. *enabled*, *started* or *completed*).

There is relatively tight coupling between the places and transitions in Fig. 9, illustrating the close integration that is necessary between the various aspects of the control-flow and data perspectives in order to enact a process model. The coupling between these places and the `work distribution` transition however is much looser. There are no static aspects of the process that are shared with other transitions in the model (i.e. the transitions underpinning `work distribution`) and other than the places which serve to communicate work items being distributed to resources for execution (and being started, completed or cancelled), the `variable instances` place is the only aspect of dynamic data that is shared with the work distribution subprocess. This reflects the functional independence of the *workflow engine*, *work item routing* and *worklist management* components. The next section looks at the issue of worklist management in more detail.

4.2   Worklist Management

The main motivation for workflow systems is achieving more effective and controlled distribution of work. Hence the actual distribution and management of work items are of particular importance. The process of managing the distribution of work items to resources is summarised by Fig. 10. It coordinates the interaction between the *workflow engine*, *work item routing*, *worklist handler* and *workflow administration* components.

The correspondences between these components and the transitions in Fig. 10 can be summarised as follows:

- the *worklist management* component is facilitated by the `work item distribution` transition, which handles the overall management of work items through the distribution and execution process (note that it subsumes the *work item routing* component);
- the *worklist handler* component corresponds to the `worklist handler` transition, which is the user-facing client software that advises users of work items requiring execution and manages their interactions with the main `work item distribution` transition in regard to committing to execute specific work items, starting and completing them;
- the *workflow administration* component is facilitated via two distinct transitions: the `management intervention` transition, that provides the ability for a workflow administrator to intervene in the `work distribution` process and manually reassign work items to users where required; and the `interrupt handler` transition that supports the cancellation, forced completion and forced failure of work items as may be triggered by other components of the workflow engine (e.g. the control-flow process, exception handlers).

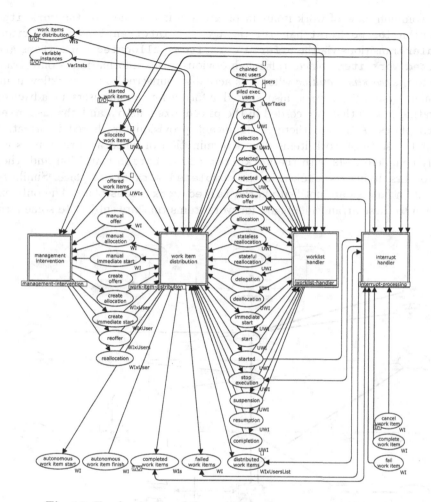

Fig. 10. Top level view of the *worklist management* component

Work items that are to be distributed are communicated between the *work-flow engine* and the *worklist management* components via the `work items for distribution` place. This then prompts the `work item distribution` transition to determine how they should be routed for execution. This may involve the services of the workflow administrator in which case they are sent to the `management intervention` transition or alternatively they may be sent directly to one or more resources via the `worklist handler` transition. The various places between these three transitions correspond to the range of requests that flow between them. In the situation where a work item corresponds to an *automatic* task, it is sent directly to the `autonomous work item start` place and no further distribution activities take place. An automatic task is considered complete when a token is inserted in the `autonomous work item finish` place.

A common view of work items in progress is maintained for the work item distribution, worklist handler, management intervention and interrupt handler transitions via the offered work items, allocated work items and started work items places (although obviously this information is only available to the *worklist handler* when it is actually connected to the workflow management system). There is also shared information about users in advanced operating modes that is recorded in the piled exec users and chained exec users places. Although there is significant provision for shared information about the state of work items, the determination of when a work item is actually complete rests with the work item distribution transition and when this occurs, it inserts a token in the completed work items place. Similarly, work item failures are notified via the failed work items place. The only exception to these arrangements are for work items that are subject to some form

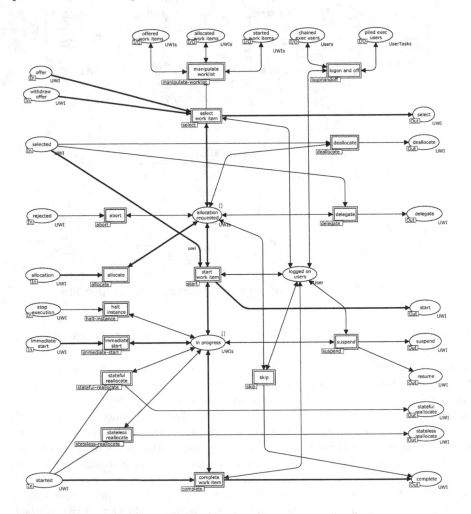

Fig. 11. *Worklist handler* component

of interrupt (e.g. an exception being detected and handled). The `interrupt handler` transition is responsible for managing these occurrences on the basis of cancellation, forced completion and failure requests received in the `cancel work item`, `complete work item` and `fail work item` places respectively. All of the activities in the *worklist management* component are illustrated by substitution transitions indicating that each of them are defined in terms of significantly more complex subprocesses. It is not possible to present each of them in this paper. Finally we focus on one other significant component: the *worklist handler*.

4.3   Worklist Handler

The *worklist handler* component is illustrated in Fig. 11 and describes how the user-facing workflow interface (typically a worklist handler software client) operates and interacts with the *worklist management* component. The main path through this process is indicated by the thick black arcs. There are various transitions that make up the process, these correspond to actions that individual users can request in order to alter the current state of a work item to more closely reflect their current handling of it. These actions may simply be requests to start or complete it or they may be "detour" requests to reroute it to other users e.g. via `delegation` or `deallocation`. The manner in which these requests operate is illustrated by the shared places in Fig. 10. Typically the inclusion of a request in one of these shared places results in a message flowing between the *worklist handler* and *worklist management* components which ultimately causes the relative states of the two components to be synchronised.

5   Related Work

There have been numerous papers advocating approaches to workflow and business process modelling based on Petri Nets (cf. [10,11,12,13]), however these tend to either focus on a single aspect of the domain (e.g. the control-flow perspective) or they are based on a relatively simplistic language. There have also been attempts to provide formal semantics using Petri Nets for many of the more widely used approaches to business process modelling including EPCs [14], UML 2.0 Activity Diagrams [15] and BPMN [16], although in each case arriving at a complete semantics has been hampered by inherent ambiguities in the informal descriptions for each of the formalisms. There has been minimal work on formalisation of the other workflow perspectives, one exception is [17] which investigates mechanisms for work distribution in workflows and presents CPN models for a number of the workflow resource patterns.

Historically, the modelling and enactment of processes have often been treated distinctly and it is not unusual for separate design and runtime models to be utilised by systems. Approaches to managing the potential disparities between these models have included the derivation of executable process descriptions from design-time models [18] and the direct animation of design-time models for requirements validation [19]. The latter of these approaches which uses a strategy based on Coloured Petri Nets [8] and CPN Tools [20] as an enablement vehicle is one of a number of initiatives that have successfully used the CPN Tools offering as a means of executing various design-time modelling formalisms including Protos models [21], sequence diagrams [22] and task descriptions [23].

There has been a significant body of work that describes software architectures for workflow management systems. Significant examples of such systems include MOBILE [24], WIDE [25], CrossFlow [26] and WAMO [27] amongst many others however none of these systems offer a fully formalised description both of their language elements and the overall operation of the workflow system.

6   Experiences and Conclusion

The selection of Coloured Petri Nets as the conceptual foundation for *new*YAWL proved to be a fortuitous choice. Being state-based and graphical, the formalism delivered immediate modelling benefits as a consequence of its commonalities with the domain it was used to represent. The availability of a means of integrating the handling of the data-related aspects of the *new*YAWL language into the model (i.e. using "colour") and partitioning the model on a hierarchical basis into units of related functionality meant that a more compact means of representing the overall *new*YAWL model was possible. The most significant advantage of this design choice however proved to be the availability of an interactive modelling and execution environment in the form of CPN Tools. Indeed, it is only with the aid of an interactive modelling environment such as CPN Tools that developing a formalisation of this scale actually becomes viable.

Although there are other candidates for developing large-scale system designs, none of them deliver the benefits inherent in the Coloured Petri Nets and CPN Tools combination. Conceptual foundations such as π-calculus and process algebra as well as software-oriented specification formalisms such as Z and VDM lack a graphical representation meaning that the visualisation and assessment of specific design choices is difficult. In contrast, lighter-weight approaches to business process modelling such as those embodied in offerings such as Protos, ARIS and other business process modelling tools do provide an intuitive approach to specifying business process that is both graphical and state-based, however they lack a complete formal semantics. Moreover, they only allow for the specification of a specific candidate model and do not provide a means of capturing an arbitrary range of business processes in a single model as is required for the *new*YAWL language. This shortcoming stems from the fact that the modelling formalisms employed in these tools are control-flow centric and lack a fully fledged data perspective. High-level CASE tools (e.g. Rational Rose) share similar shortcomings and their generalist nature means that they do not provide any specific support for business process modelling and enactment initiatives.

One of the major advantages of the approach pursued in developing *new*YAWL is that it provided a design that is executable. This allowed fundamental design decisions to be evaluated and tested much earlier than would ordinarily be the case during the development process. Where suboptimal design decisions were revealed, the cost of rectifying them was significantly less than it would have been later in the development lifecycle. There was also the opportunity to test alternate solutions to design issues with minimal overhead before a final decision was settled on. A particular benefit afforded by this approach to formalisation was that the CPN hierarchy established during the design process provided an excellent basis on which to make subsequent architectural and development decisions.

Whilst complete, the resultant model *new*YAWL system model[3] is extremely complex. *It incorporates 55 distinct pages of CPN diagrams and encompasses 480 places, 138 transitions and in excess of 1500 lines of ML code.* It took approximately six months to develop. The size of the model gives an indication of the relative complexity of formally specifying a comprehensive business process modelling language such as *new*YAWL. The original motivations for this research initiative were twofold: (1) to establish a fully formalised business process modelling language based on the synthesis of the workflow patterns and (2) to demonstrate that the language was not only suitable for conceptual modelling of business processes but that it also contained sufficient detail for candidate models to be directly enacted. *new*YAWL achieves both of these objectives and *directly supports 118 of the 126 workflow patterns* that have been identified. It is interesting to note however that whilst the development of a system model of this scale offers some extremely beneficial insights into the overall problem domain and provides a software design that can be readily utilised as the basis for subsequent programming activities, it also has its limitations. Perhaps the most significant of these is that the scale and complexity of the model obviates any serious attempts at verification. Even on a relatively capable machine (P4 2.1Ghz dual-core, 2Gb RAM), it takes almost 4 minutes just to load the model. Moreover the potentially infinite range of business process models that the *new*YAWL system can encode, rules out the use of techniques such as state space analysis. This raises the question as to how models of this scale can be comprehensively tested and verified.

Notwithstanding these considerations however, the development of the new-YAWL system model delivered some salient insights into areas of new YAWL that needed further consideration during the design activity. These included:

– the introduction of a deterministic mechanism for recording status changes in the work item execution lifecycle in order to ensure that the views of these details maintained by the *worklist management* and *worklist handler* components are consistent;
– the establishment of a coherence protocol to ensure that reallocation of work items to alternate resources either by resources themselves or the workflow administrator are handled in a consistent manner in order to ensure that potential race conditions arising during reallocation do not result in the workflow engine, workflow administrator or the initiating resource (i.e. *worklist handler*) having irreconcilable views of the current state of work item allocations;

---

[3] This model is available at www.yawl-system.com/newYAWL

- the introduction of a consistent approach for handling the evaluation of any functions associated with a *new*YAWL specification e.g. for outgoing links in an XOR-split, pre/postconditions, pre/post tests for iterative tasks etc. This issue was ultimately addressed by mapping any necessary function calls to ML functions and establishing a standard approach to encoding the invocation of these functions and the passing of any necessary parameters and the return of associated results;
- adoption of a standard strategy for characterising parameters to functions in order to ensure that they could be passed in a uniform way to the associated ML functions that evaluated them;
- the introduction of a locking strategy for data elements to prevent inadvertant side-effects of concurrent data usage; and
- recognition that when a self-cancelling task completes: (1) it should process the cancellation of itself last of all in order to prevent the situation where it cancels itself before all other cancellations have been completed and (2) it needs to establish whether it is cancelling itself before it can make the decision to put tokens in any relevant output places associated with the task.

The *new*YAWL system model provides a complete description of an operational environment for the *new*YAWL language. It is sufficiently detailed to be directly useful for system design and development activities. It will serve as the design blueprint for upcoming versions of the open-source YAWL System. In fact the resource management component of the *new*YAWL language has already been incorporated in the YAWL System.

Acknowledgement. The authors would like to thank the anonymous reviewers for their constructive comments and suggestions.

References

1. Russell, N., ter Hofstede, A.H.M., van der Aalst, W.M.P.: *new*YAWL: Specifying a workflow reference language using Coloured Petri Nets. In: Proceedings of the Eighth Workshop and Tutorial on Practical Use of Coloured Petri Nets and the CPN Tools. Number DAIMI PB-584, Department of Computer Science, pp. 107–126. University of Aarhus, Denmark (2007)
2. van der Aalst, W.M.P., ter Hofstede, A.H.M.: YAWL: Yet another workflow language. Information Systems 30(4), 245–275 (2005)
3. van der Aalst, W.M.P., ter Hofstede, A.H.M., Kiepuszewski, B., Barros, A.: Workflow patterns. Distributed and Parallel Databases 14(3), 5–51 (2003)
4. Russell, N., ter Hofstede, A.H.M., van der Aalst, W.M.P., Mulyar, N.: Workflow control-flow patterns: A revised view. Technical Report BPM-06-22 (2006), http://www.BPMcenter.org
5. Wynn, M., Edmond, D., van der Aalst, W.M.P., ter Hofstede, A.H.M.: Achieving a general, formal and decidable approach to the OR-join in workflow using Reset nets. In: Ciardo, G., Darondeau, P. (eds.) ICATPN 2005. LNCS, vol. 3536, pp. 423–443. Springer, Heidelberg (2005)

6. Russell, N., ter Hofstede, A.H.M., Edmond, D., van der Aalst, W.M.P.: Workflow data patterns: Identification, representation and tool support. In: Delcambre, L.M.L., Kop, C., Mayr, H.C., Mylopoulos, J., Pastor, Ó. (eds.) ER 2005. LNCS, vol. 3716, pp. 353–368. Springer, Heidelberg (2005)
7. Russell, N., van der Aalst, W.M.P., ter Hofstede, A.H.M., Edmond, D.: Workflow resource patterns: Identification, representation and tool support. In: Pastor, Ó., Falcão e Cunha, J. (eds.) CAiSE 2005. LNCS, vol. 3520, pp. 216–232. Springer, Heidelberg (2005)
8. Jensen, K.: Coloured Petri Nets. Basic Concepts, Analysis Methods and Practical Use. Basic Concepts. Monographs in Theoretical Computer Science, vol. 1. Springer, Heidelberg (1997)
9. Russell, N., ter Hofstede, A.H.M., Edmond, D., van der Aalst, W.M.P.: newYAWL: achieving comprehensive patterns support in workflow for the control-flow, data and resource perspectives. Technical Report BPM-07-05 (2007), http://www.BPMcenter.org
10. van der Aalst, W.M.P.: The application of Petri nets to workflow management. Journal of Circuits, Systems and Computers 8(1), 21–66 (1998)
11. Ellis, C., Nutt, G.: Modelling and enactment of workflow systems. In: Ajmone Marsan, M. (ed.) ICATPN 1993. LNCS, vol. 691, pp. 1–16. Springer, Heidelberg (1993)
12. Adam, N., Atluri, V., Huang, W.: Modeling and analysis of workflows using Petri nets. Journal of Intelligent Information Systems 10(2), 131–158 (1998)
13. Moldt, D., Rölke, H.: Pattern based workflow design using reference nets. In: van der Aalst, W.M.P., ter Hofstede, A.H.M., Weske, M. (eds.) BPM 2003. LNCS, vol. 2678, pp. 246–260. Springer, Heidelberg (2003)
14. van der Aalst, W.M.P.: Formalization and verification of event-driven process chains. Information and Software Technology 41(10), 639–650 (1999)
15. Störrle, H., Hausmann, J.: Towards a formal semantics of UML 2.0 activities. In: Liggesmeyer, P., Pohl, K., Goedicke, M. (eds.) Proceedings of the Software Engineering 2005, Fachtagung des GI-Fachbereichs Softwaretechnik, Essen, Germany, Gesellschaft fur Informatik. LNI, vol. 64, pp. 117–128 (2005)
16. Dijkman, R., Dumas, M., Ouyang, C.: Semantics and analysis of business process models in BPMN. Information and Software Technology 50(12), 1281–1294 (2008)
17. Pesic, M., van der Aalst, W.M.P.: Modelling work distribution mechanisms using colored Petri nets. International Journal on Software Tools for Technology Transfer 9(3), 327–352 (2007)
18. Di Nitto, E., Lavazza, L., Schiavoni, M., Tracanella, E., Trombetta, M.: Deriving executable process descriptions from UML. In: ICSE 2002: Proceedings of the 24th International Conference on Software Engineering, pp. 155–165. ACM Press, New York (2002)
19. Machado, R., Lassen, K., Oliveira, S., Couto, M., Pinto, P.: Requirements validation: Execution of UML models with CPN tools. International Journal on Software Tools for Technology Transfer 9(3), 353–369 (2007)
20. Jensen, K., Kristensen, L., Wells, L.: Coloured Petri nets and CPN tools for modelling and validation of concurrent systems. International Journal of Software Tools for Technology Transfer 9(3), 213–254 (2007)
21. Gottschalk, F., van der Aalst, W., Jansen-Vullers, M., Verbeek, H.: Protos2CPN: Using colored Petri nets for configuring and testing business processes. In: Jensen, K. (ed.) Proceedings of the 7th Workshop and Tutorial on Practical Use of Coloured Petri Nets and the CPN Tools. PB-579 of DAIMI Reports, Aarhus, Denmark, pp. 137–155 (2006)

22. Ribeiro, O., Fernandes, J.: Some rules to transform sequence diagrams into coloured Petri nets. In: Jensen, K. (ed.) Proceedings of the 7th Workshop and Tutorial on Practical Use of Coloured Petri Nets and the CPN Tools. PB-579 of DAIMI Reports, Aarhus, Denmark, pp. 137–155 (2006)
23. Jørgensen, J.B., Lassen, K.B., van der Aalst, W.M.P.: From task descriptions via coloured Petri nets towards an implementation of a new electronic patient record. In: Jensen, K. (ed.) Proceedings of the 7th Workshop and Tutorial on Practical Use of Coloured Petri Nets and the CPN Tools, Aarhus, Denmark. PB-579 of DAIMI Reports, pp. 137–155 (2006)
24. Jablonski, S., Bussler, C.: Workflow Management: Modeling Concepts, Architecture and Implementation. Thomson Computer Press, London (1996)
25. Ceri, S., Grefen, P., Sanchez, G.: WIDE: a distributed architecture for workflow management. In: Proceedings of the Seventh International Workshop on Research Issues in Data Engineering (RIDE 1997), Birmingham, England. IEEE Computer Society Press, Los Alamitos (1997)
26. Ludwig, H., Hoffner, Y.: Contract-based cross-organisational workflows - the Cross-Flow project. In: Grefen, P., Bussler, C., Ludwig, H., Shan, M. (eds.) Proceedings of the WACC Workshop on Cross-Organisational Workflow Management and Co-Ordination, San Francisco (1999)
27. Eder, J., Liebhart, W.: The workflow activity model (WAMO). In: Laufmann, S., Spaccapietra, S., Yokoi, T. (eds.) Proceedings of the Third International Conference on Cooperative Information Systems (CoopIS 1995), pp. 87–98. University of Toronto Press, Vienna (1995)

From Requirements via Colored Workflow Nets to an Implementation in Several Workflow Systems

Ronny S. Mans[1,2], Wil M.P. van der Aalst[1], Nick C. Russell[1],
Piet J.M. Bakker[2], Arnold J. Moleman[2], Kristian B. Lassen[3],
and Jens B. Jørgensen[3]

[1] Department of Information Systems, Eindhoven University of Technology,
P.O. Box 513, NL-5600 MB, Eindhoven, The Netherlands
{r.s.mans,w.m.p.v.d.aalst,n.c.russell}@tue.nl
[2] Academic Medical Center, University of Amsterdam, Department of Quality
Assurance and Process Innovation, Amsterdam, The Netherlands
{p.j.bakker,a.j.moleman}@amc.uva.nl
[3] Department of Computer Science, University of Aarhus, IT-parken, Aabogade 34,
DK-8200 Aarhus N, Denmark
{krell,jbj}@daimi.au.dk

Abstract. Hospitals and other healthcare organizations need to support complex and dynamic workflows. Moreover, these processes typically invoke a number of medical disciplines. This makes it important to avoid the typical disconnect between requirements and the actual implementation of the system. In this paper we apply a development approach where an Executable Use Case (EUC) and a Colored Workflow Net (CWN) are used to close the gap between a given requirements specification and the realization of these requirements based on workflow technology. In order to do so, we describe a large case study where the diagnostic process of the gynecological oncology care process of the Academic Medical Center (AMC) hospital is used as a candidate process. The process consists of hundreds of activities. These have been modeled and analyzed using an EUC and a CWN. Moreover, based on the CWN, the process has been implemented using four different workflow systems. In this way, we demonstrate the general application of the approach and its applicability to distinct technology systems.

Keywords: Workflow Management, Executable Use Cases, Colored Petri Nets, healthcare.

1 Introduction

For some time, particularly in academic hospitals, there has been the need for better support in controlling and monitoring health care processes for patients [25]. One of the objectives of hospitals is to increase the quality of care for patients [15], while in the future, an increase in the demand for care is expected.

K. Jensen, J. Billington, and M. Koutny (Eds.): ToPNoC III, LNCS 5800, pp. 25–49, 2009.

Workflow technology presents an interesting vehicle for the support and monitoring of health care processes as it facilitates process automation by managing the flow of work such that constituent activities are done at the right time by the proper person [4]. The advantages of successfully applying workflow technology are faster and more efficient process execution [34,22,14].

Typically, there is a large gap between an actual hospital process and its implementation in a workflow system. One approach to bridging this gap is to go from a real-world process, via a requirements model and a design model to an implementation in a workflow system as is described in [8,20]. The different steps in this development approach are shown in Figure 1. First a requirements model is developed, based on a real-life case. The next phase in the process is the construction of a design model, followed by its implementation in a workflow system. The construction of the requirements and the design model is done using *Colored Petri Nets (CPNs)* [17]. CPNs provide a well-established and well-proven language with formal semantics. CPNs are particularly suitable for describing the behavior of systems requiring support for concurrency, resource sharing, and synchronization, and are therefore well suited for modeling business processes.

To be more precise, in the requirements phase, a so-called Executable Use Case (EUC) is created, a CPN model augmented with a graphical animation. The next step in the development process is to build a design model. Here a so-called Colored Workflow Net (CWN), which is also a CPN model, and which is closer to an actual implementation in a workflow system, is used. In an EUC any concepts and entities deemed relevant may be included, whereas for the CWN we are restricted to the workflow domain as only concepts and entities which are common in workflow languages may be used. Ultimately, we have the implementation in several workflow systems. The main advantages of this development process are that concerns, as shown in Figure 1, are dealt with at the right time and that the models constructed in the EUC and CWN phase are

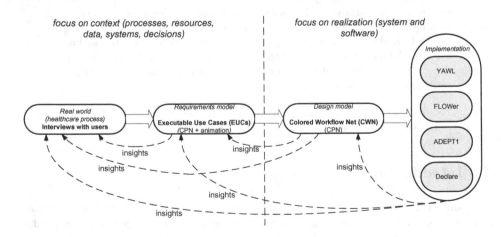

Fig. 1. Overall approach

based on a formal foundation. In that way, rigor is added to the development process but is also allows for a seamless integration between the requirements and design phase.

In [8,20] rather small cases are used, whereas real world processes typically consist of hundreds of activities and are far from trivial. Therefore, in order to investigate the *general applicability of the approach*, shown in Figure 1, it is applied to a large real world healthcare process which is non-trivial. In this way, we can also investigate whether the approach is *repeatable*.

The healthcare process that will be considered in this paper is the diagnostic process of patients visiting the gynaecological oncology outpatient clinic in the AMC hospital, a large academic hospital in the Netherlands. To give an idea about the size of the healthcare process, it should be emphasized that the EUC consists of *689 transitions, 601 places and 1648 arcs* and that the CWN consists of *324 transitions, 522 places and 1221 arcs*. This shows that the healthcare process is far from trivial.

Typical difficulties that hospitals have to cope with when they want to apply workflow technology stem from the fact that healthcare processes are *diverse, flexible* and that *several medical departments* can be involved in the treatment process. For example, as a consequence of the way a patient reacts to the treatment offered, and the condition of the patient themself, it may be necessary to continuously adapt the care process an individual patient [13]. This shows that flexibility is a key requirement in the healthcare domain which consequently needs to be provided by the workflow system.

The following workflow management systems were selected to implement (parts) of this process: YAWL [6], FLOWer [9], ADEPT1 [33], and Declare [29]. These systems were selected because they all provide a certain kind of flexibility, which in this context is deemed relevant[1]. In this way, it allows us to demonstrate the applicability of the development approach to *distinct technologies*. The selected systems cover various flexibility paradigms: adaptive workflow (ADEPT1), case handling (FLOWer), and declarative workflow (Declare). An additional reason for implementing a hospital process in different workflow systems is that we wanted to identify the requirements that need to be fulfilled by workflow systems, in order to be successfully applied in a hospital environment. These requirements have been discussed in [27].

There are some notable differences with the work presented in this paper and that presented in [8,20] which also went from an informal description of a real world process, via an EUC and CWN, to an implementation in a workflow system. As already indicated, we studied an existing healthcare process of a hospital in detail, whereas in the earlier work rather small cases are used. Moreover, we developed an implementation in four different workflow systems instead of just one and systematically collected feedback from the host care organization (AMC). In [8] there was only user involvement in the EUC phase, and in [20]

---

[1] Of course other workflow systems could have been selected. An additional reason for choosing these systems is that they were already available to us, or could easily be obtained.

there was no user involvement at all. Note, that an earlier version of this paper has already appeared as an (informal) workshop paper [26].

This paper is structured as follows: Section 2 introduces the approach followed. Section 3 introduces the EUC and the healthcare process we studied. Section 4 discusses the CWN, which is followed by an analysis of the model in Section 5. In Section 6, the implementations in the different workflow systems are discussed. Related work is presented in Section 7. The paper finishes with conclusions in Section 8.

2   Approach

In this section, we first elaborate on the approach that has been followed, as shown in Figure 1, to go from a real-life process to its implementation in several workflow systems via an EUC and CWN. Our approach commences by interviewing users who are involved in the diagnostic part of the gynecological oncology healthcare process. In these interviews we focused on identifying the work processes that constituted the requirements for the system to be built. These requirements were captured during the requirements phase. In this phase we developed a requirements model using the EUC method. EUCs are formal and executable representations of work processes to be supported by a new IT system and can be used in a prototyping fashion to specify, validate, and elicit requirements [19]. EUCs have the ability to "talk back to the user" and can be used in a trial-and-error fashion. Moreover, they have the ability to spur communication between stakeholders [19] about issues relevant to process definition.

In our case, the EUC consists of a CPN model, which describes the real-life process, and an animation layer on top of it, which can be shown to the people involved in the process. Our main reason for using EUCs, was that we did not expect the people to understand the underlying CPN model directly but anticipated that they would be able to derive an understanding from the animations as to how we modeled their work processes. This provided a suitable opportunity for validating our model. Moreover, the use and importance of animations in the very early stages of the development process has been stressed in [30].

After validation of the requirements model, we move on to the next phase, the design phase, in which the CWN method is used. We took the underlying CPN model of the EUC and translated it into a CWN. By developing this workflow model we restricted ourselves to the workflow domain. More specifically, by creating the workflow model, we restricted ourselves to concepts and entities which are common in workflow languages. *In comparison to the EUC model, we now only use a fixed library of concepts and entities, whereas in the EUC any concept or entity deemed relevant may be used.* In addition, the CWN only contains *actions that will be supported by the new system* whereas actions that are not going to be supported by the new system, are left out. Finally, once the CWN model was finished, we used it as a basis to configure each of the four workflow systems.

As is indicated in Figure 1, during the construction of the EUC and the CWN and the associated implementations in the different workflow systems, additional insights were obtained about previous phases. So, there are often iterations involving the repetition of earlier stages of the process as shown in Figure 1. When repeating an earlier phase, the model of that phase and subsequent phases are updated until the current phase is reached again. As we only had feedback from the people involved in the process during the interview, requirements and design phases. This means that we did not receive any user feedback during the implementation phase as we only developed prototypes that were not used as fully operational production systems. Although feedback during the implementation phase could have given us further information about the process, we felt that we had already sufficiently identified the process during the construction of the EUC and CWN.

In Figure 1, there is a dashed line between the first two blocks and the last two blocks indicating a shift of focus. On the left side of the dashed line the focus is on *context*, whereas on the right side the focus is on *realization*. With *context* we mean that the focus is on processes, resources, data, systems and decisions and with *realization* we mean that the focus is on the system and software itself. To support this shift of focus, the CPN modeling language, which has been used for the EUC and the CWN, provides a smooth transition between the two foci. In addition, we believe this addresses the classical "disconnect" which exists between business processes and IT. Moreover, building an EUC and CWN allows for a *separation of concerns*. EUCs are good for capturing the requirements of a process without thinking about how it is realized and this information serves as input for a CWN. CWNs define the control-flow, resource, data and operation perspective while at the same time abstracting from implementation details and language/application specific issues[2]. Using this development process, we are sure that these concerns are dealt with at the right time as we have to deal with them anyway.

Figure 1 should not be read as if we are proposing a waterfall development process. Instead, it should illustrate that the phases are partially ordered. During the development process we can go back to a preceding phase and make changes. Consequently, these changes will result in changes in subsequent phases.

Note that in principle other process languages can also be used with this approach. We used CPN as we are familiar with the language and because they provide a well-established and well-proven formalism for describing and analyzing the behavior of systems with characteristics such as concurrency, resource sharing, and synchronization. In this way, they are well-suited for modeling workflows or work processes [4]. It is worth noting that the process-oriented nature of both the EUC and CWN limits our approach to process-oriented systems, like workflow and document handling systems. Nevertheless, it is important to mention that the development approach used in this paper is in no way limited

---

[2] The control-flow perspective specifies the ordering of activities in a process, the resource perspective deals with the resource allocation required for the execution of the activities within a process, the data perspective deals with data transfer between activities, and the operation perspective describes the elementary operations performed by resources and applications.

to the healthcare domain. In [8] the work process of bank advisers is considered using similar techniques.

It is important to mention that all the models and translations between models have been done manually. In the end, this leads to a *full* implementation of a *complete* healthcare process in four different workflow systems. Because of the specific nature of the EUC model and the CWN, the CWN cannot be generated automatically from the EUC. This is because certain parts of the EUC need to be refined in order to allow for system support. Depending on what is specified in the EUC, parts can be reused and refined in the CWN model. However, in principle, a (semi-) automatic translation from the CWN to each of the workflow systems is possible and examples of this are shown in [8,20].

We already indicated that we studied a large healthcare process; for creating the EUC and the CWN more than 100 man hours were needed to develop each model. As we also needed to get acquainted with the workflow systems used, their configuration took around 240 man hours in total. Additionally, around 60 man hours were needed for interviewing and obtaining feedback from the people involved in the process.

3   Executable Use Case for the Gynecological Oncology Healthcare Process

In this section, we first introduce the gynecological oncology healthcare process which we studied. After that, we will consider one part of the healthcare process in more detail and for this part we will elaborate on how the animations have been set-up within the EUC. Moreover, we will elaborate on the experiences associated with this activity. Note that given the size of the CPN model of the EUC (689 transitions, 601 places, 1648 arcs and 2 color sets) it is only possible to show a small fragment of the overall model.

In Figure 2, the topmost page of the CPN model of the EUC is shown, which gives a general overview of the diagnostic process of the gynecologic oncology healthcare process in the AMC hospital. In the remainder of this paper, we will simply refer to the gynecological oncology healthcare process itself, instead of the diagnostic process of the gynecological oncology healthcare process.

As can be seen in Figure 2, the gynecological oncology process consists of two different processes. The *first* process, which is modeled in the upper part of the picture, deals with the diagnostic process that is followed by a patient who is referred to the AMC hospital, up until they are diagnosed. In this process, the patient can have several consultations with a doctor, either via visiting the outpatient clinic or via telephone.

During such a consultation, the status of the patient is discussed and a decision is made about whether diagnostic tests and/or further consultations need to be scheduled, canceled or rescheduled. Moreover, during the course of the process, several administrative activities such as brochure recommendation and patient registration can also occur.

A doctor can request a series of different diagnostic tests, undertaken at different medical departments. The interactions with these medical departments

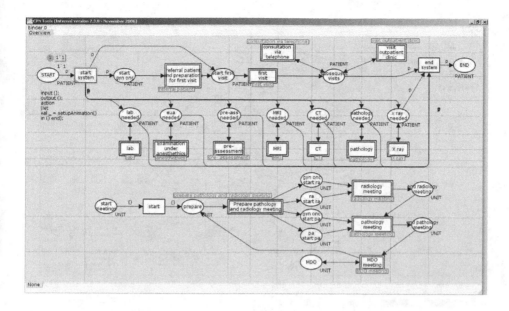

Fig. 2. General overview of the gynecological oncology healthcare process

and also the processes within these departments are modeled in the middle of Figure 2. The interactions with these medical departments are considered to be a 'black box' where a request or a cancelation of a diagnostic test is delivered and ends when a result is known or the test is canceled.

The *second* part of the process, which can be found at the lower part of Figure 2, deals with the weekly organized meetings, on Monday afternoon, for discussing the status of patients and what needs to be done in the future treatment of these patients.

Note that some connections exist between the two processes. However, as we only focus on the content and ordering of activities within one process, we did not put any effort in making these connections more explicit.

Figure 3 shows a part of the subnet for the substitution transition "referral patient and preparation for first visit" illustrating the very beginning of the process. At this time, a doctor of a referring hospital calls a nurse or doctor at the AMC followed by the necessary preparations for the first visit of the patient, like planning an MRI (transition "plan MRI").

In Figure 3, we see how the CPN model and the animation layer are related within the EUC. At the top, we see the CPN model that is executed in CPN Tools. At the bottom we see the animation that is provided within the BRITNeY tool[3], the animation facility for CPN Tools. The CPN model and the animation layer are connected by adding animation drawing primitives to transitions in the CPN model, which update the animation. The animation layer shows for

---

[3] http://wiki.daimi.au.dk/britney/britney.wiki

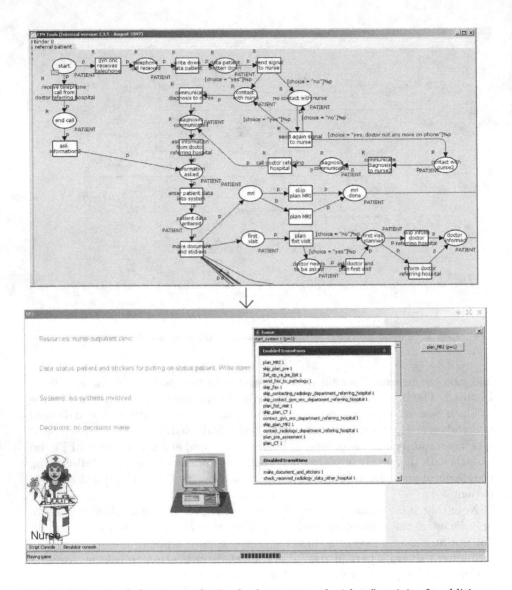

Fig. 3. Animation belonging to the "make document and stickers" activity. In addition, the panel at the top right side shows which activities are enabled now.

the last executed activity in the CPN model, which *resources*, *data* and *systems* are involved in executing the activity and it also shows which *decisions* are made during the activity. This ensures the focus is on what happens in the *context* of the process which will also be of help when constructing the CWN in the design phase. In addition, a separate panel is shown which indicates which activities are currently enabled and may be executed. One of the enabled activities in the panel can be selected and executed, which changes the state of the process and in this way, we can directly influence the routing within a process. When an activity is executed in the CPN model it is reflected by updates to the animation layer. Consequently, the CPN model and the animation layer remain synchronized.

In Figure 3, the animation visualizes the "make document and stickers" activity. The panel at the top right side of the snapshot in Figure 3, shows which activities can be executed after the "make document and stickers" activity has been executed. It shows that the "plan MRI" activity may be executed but the "make document and stickers" activity may not. Moreover, we see that a nurse of the outpatient clinic is responsible for executing the "make document and stickers" activity and that no decisions need to be made.

We have shown the animations to the people that were involved in the gynecological oncology process. The people were very positive and indicated that through their use, they were able to check whether the process modeled in the EUC corresponded with their work process. Moreover, they provided valuable feedback for improvements, like activities that needed to be reordered. Consequently, we can say that the EUC method was helpful in validating the model and we believe that better results have been obtained than if we had simply shown the plain CPN models or process definitions of a workflow management system.

4   Colored Workflow Net for the Gynecological Oncology Healthcare Process

In this section, we elaborate on CWNs in general. After this, we consider the same part of the CWN as we did for the EUC in more detail and explain the differences. Note that also in this case, given the size of the CWN model (324 transitions, 522 places and 1221 arcs and 53 color sets) it is only possible to show a small fragment of the overall model.

As can be seen in Figure 1, both the EUC and CWN are CPN models. Remember that a CWN is a *workflow* model in which we restricted ourselves to concepts and entities which are common in workflow languages, whereas in the EUC any concept or entity deemed relevant may be used. So, the CWN needs to cover the control-flow, resource, data and operation perspective.

The syntactical and semantic requirements for a CWN have already been defined in [8]. However, when using CWNs in this paper, we have introduced some subtle changes so that they suit our needs. According to [8], a CWN should be a CPN with only places of type Case, Resource or CxR. Tokens in a place of type Case refer only to a case and the corresponding attributes (e.g. patient name, patient id), and tokens in a place of type Resource refer only to resources. Finally, tokens in a place of type CxR refer to both a case and a resource. However, we have decided to separate the case data from the case, so that case data can be accessed anywhere in the model and is always up-to-date. For example, in Figure 4, the "plan first visit" and "plan MRI" activities might be executed concurrently. In the "plan first visit" activity, some data may be written which is interesting for the "plan MRI" activity, and vice-versa. To this end, instead of

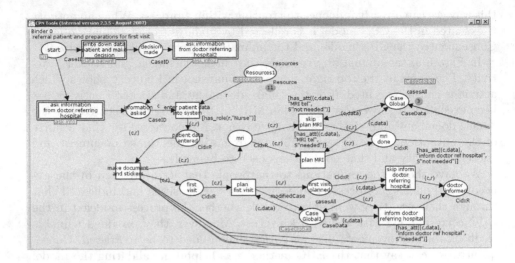

Fig. 4. CWN for the EUC shown in Figure 3

places of type Case and CxR, we have places of type CaseID for storing the case identifier, CidxR to refer to both a case identifier and a resource, and CaseData for accessing the data attributes of the case. Furthermore, instead of having data attributes of which the data values may only be integers or strings, we also allow them to have a list data type.

In Figure 4, we see the CWN for the EUC CPN which has been shown in Figure 3. If we compare the CWN of Figure 4 with the EUC CPN of Figure 3, we see that there are some differences. First of all, some activities which are shown in the EUC CPN do not appear in the CWN, as they are not supported by the workflow system. The activities that will not be supported are indicated by the character "R" at the top left of each individual activity that will be removed. On the other hand, the activities of the EUC that will be supported do not have the character "R" at the top left of them. For example, the "plan MRI" activity will be supported by the workflow, whereas the "send signal to nurse" activity is not.

Moreover, the place "Resources1" is only present in the CWN as it contains information about the availability of resources which are needed for the organizational perspective of the CWN. Also, the places "Case global" and "Case global1" are only present in the CWN as they contain the corresponding data attributes for each case instance which are needed for the data perspective of the CWN. Furthermore, the guards belonging to transitions explicitly reference the resource and data perspective.

For example, the guard of the "enter patient data into system" activity indicates that this activity may only be performed by a nurse, whereas the guard of the "plan MRI" activity indicates that this activity may only be performed if data attribute "MRI tel" has value "needed".

When we compare the EUC and CWN with each other it is clear that the CWN provides a complete specification for the control flow, resource, data, and operation perspectives. As a consequence, it is more closely aligned with the way workflow systems are described because similar concepts are used. When we look at the EUC, its goal is to support specification, validation and elicitation of requirements [19]. Its use narrows the gap between informal requirements and the formalization required for supporting the process by a workflow system which on its turn is provided by the CWN. In contrast to the EUC, in the CWN certain parts are refined in order to allow for system support. For example, in the EUC a sequence of activities may be possible which is prohibited in the CWN. Clearly, when defining the control flow, resource, data, and operation perspective of the CWN, parts of the EUC may be reused and refined depending on what is already specified.

A considerable amount of time has been spent in defining both the EUC and CWN (100 man hours for each) in which several iterations, back and forth, have been made. When defining the resource, data, and operation perspective in the CWN, questions arose about the process that had been modeled in the requirements phase. This was due to the fact, that not all process details were obtained when interviewing people involved in the process and showing the animations to them. For example, in the EUC we modeled that after the appointment with the doctor appointments can only be booked with an internist and radiotherapist for treatment of the patient. However, in principle, it should be possible to make an appointment with any specialist if needed. This issue was identified during the design phase. After showing the updated animation to the doctor, it was agreed that this was a correct observation and the CWN was updated accordingly.

Because of the aforementioned reason, it is important that we can jump back to the requirements phase, adapt the model and move on to the design phase again. We believe that this leads to a more complete description of the process itself (EUC) and its implementation in a chosen workflow system (CWN). Consequently, we can say that the CWN provides a useful step towards the implementation of a process in a chosen workflow system.

5 Analysis

In this section, we will focus on the analysis of the CWN model. Within CPN Tools there are two possibilities for the analysis of a CPN model, namely, simulation/animation and state space analysis [17]. Simulation can be used to investigate different scenarios and explore the behavior of the model but does not guarantee that it is free from errors. To this end, we used state space analysis which computes the full state space of a CPN model and makes it possible to *verify*, in the mathematical sense of the word, that the model possesses a certain formally specified property. The state space analysis of CPN Tools can handle state spaces of up to 200,000 nodes and 2,000,000 arcs [18] and provides, amongst other features, visual inspection and query functions for investigating (behavioral) properties of a model.

The primary task of a workflow management system is to enact case-driven business processes by integrating several perspectives, such as the control-flow, resource, and data perspective. One of the properties deemed to be most relevant in this context, and which can be easily checked for, is the so-called soundness property. We feel this property provides the most insights into the correct operation of a model. Moreover, the soundness property is generic and does not require process or domain specific knowledge. In [1] it is motivated why it is reasonable to abstract from the other perspectives when verifying a workflow. Verification with resources is possible (cf. [16]) but provides few surprising insights as there is little "locking of resources" in this domain. Verification with data is more problematic. As shown in [36] and other papers some verification is possible. However, analysis then only checks for the presence of data rather than the actual values. A full analysis involving data is impossible. If data is infinite, analysis is impossible (cf. halting problem). If data is finite but large, analysis is intractable because of the state-space explosion problem. Moreover, it is often impossible to model human behavior and applications completely. Therefore, we think it is justified to use the classical soundness notion. In [5] various alternative soundness notions are compared and discussed. Moreover, this paper provides a survey of analysis approaches for verifying soundness in the presence of advanced workflow patterns (e.g. cancelation).

Soundness for workflow nets is defined in [2] as: *for any case, the procedure will terminate eventually and the moment the procedure terminates there is a token in the sink place (i.e. a place with no outgoing arcs) and all the other places are empty. Moreover, there should be no dead transitions.* To check for soundness of the CWN, we need to abstract from resources. Moreover, as the CWN is exceptionally large we also need to simplify the color sets and verify things in a hierarchical manner (i.e. in a modular fashion). To be more precise, for each transition on the top page of the CWN which is linked to a subnet, we check the soundness of the subnet, including the net on the top page. Note that such a subnet can also contain subnets.

Checking the soundness of such a subnet has been done according to the following procedure. First, we focused on the individual subnet, removing all nodes and subnets from the total model which did not belong to the subnet being considered. Second, we removed all color sets and all data attributes which were not relevant for the subnet being considered. A data attribute was considered not to be relevant when there was no function in the subnet which actually accessed the data attribute. After finishing the two preceding steps, we were able to check whether the subnet was sound. For the actual check for soundness we added an extra transition $t*$ in the subnet which connected the only output place with the only input place. This is often called a *short-circuited net* [2]. According to [2], if the short-circuited net is live and bounded, then the original net is sound[4]. Moreover, as we are dealing with a hierarchical structure of subnets, another important requirement that needs to be satisfied for the whole net to be sound,

---

[4] More details about liveness, boundedness, and short-circuited nets can be found in [2].

Liveness Properties

Dead Markings
    96 [6392,5864,5863,5861,5860,...]

Dead Transition Instances
    None

Live Transition Instances
    None

Fig. 5. Liveness Properties section of the state space report generated for the short-circuited net of the erroneous CWN. The report shows that there are multiple dead markings. Moreover, there are no dead transitions and there are no live transitions.

is that all the nets need to be safe (i.e. for each place the maximum number of tokens must not exceed one). If an error was found, the subnet was adapted and again checked for its soundness. This last step was been repeated until the subnet was sound. A limitation of the approach is that a subnet which is checked for its soundness should not be too large and/or have many color sets.

For example, for the CWN, shown in Figure 4, we found that for the "skip plan MRI" activity a double-headed arc had not been used between this transition and place "CaseGlobal". The fact that there was an error could be derived from the liveness properties of the state space report of the short-circuited net which is shown in Figure 5. Informally, liveness means that all transitions in the net have always the opportunity to become enabled no matter which transitions are fired in the net. This means that there may not be any dead markings and all transitions instances in the net should be live. This requirement is not satisfied as the report indicates the existence of multiple dead markings and that no live transition instances exists. After solving the error we obtained a sound CWN consisting of 39808 nodes, 156800 arcs for which all transitions were live and there were no dead transitions.

In general, for each subnet we found one or two errors, but we also had subnets which were error-free. Although there are additional structural properties which could be checked for, we only checked for soundness. Other relevant workflow correctness criteria considering the resource and data perspective, and which can be checked for, are discussed in [36,16]. Note that the analysis of the above mentioned properties does not remove the possibility of semantic errors.

6 Realization of the System

In this section, we discuss how the four selected workflow systems (YAWL, FLOWer, ADEPT1 and Declare) have been configured in order to support the gynecological oncology healthcare process. For each system, we will first briefly introduce the system. Then the mapping from the CWN to the modeling language used in the workflow system itself, which is done in a manual way, is

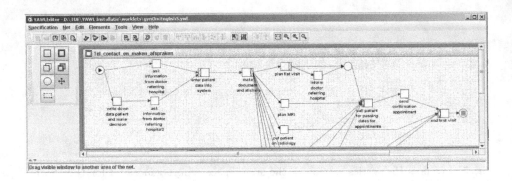

Fig. 6. Screenshot of the YAWL editor

exemplified. Finally, we elaborate on the applicability of a CWN for implementation of a process in a workflow system.

6.1   YAWL / Worklets

YAWL (Yet Another Workflow Language) [6] is an open source workflow management system[5], based on the well-known workflow patterns[6] which is more expressive than any other workflow languages available today. Moreover, in addition to supporting the control-flow and data perspectives, YAWL also supports the resource perspective.

YAWL supports the modeling, analysis and enactment of flexible processes by, so called, worklets [10] which can be seen as a type of dynamically configurable process fragment. Specific activities in a process are linked to a repertoire of possible actions. Based on the properties of the case at runtime and other associated context information, an appropriate execution option is chosen. The selection process is based on a set of rules. This can be extended at runtime and it is possible to dynamically add new actions to the repertoire.

In Figure 6, we can see how the CWN of Figure 4 is mapped to the YAWL language. Given the fact that YAWL can be seen as a superset of CWNs, it was easy to translate the CWN of Figure 4 to YAWL. It was possible to directly translate the transitions into YAWL tasks. Furthermore, the places of the CWN model can also be directly translated to YAWL conditions, but due to the syntactical sugaring inherent in YAWL there is no need to add all places as conditions in the YAWL model. For example, the "make document and stickers" activity in YAWL has an associated OR-split, after which it is possible to either plan an MRI or CT. In addition to this transformation, the worklet approach has been used as a selection mechanism for the most appropriate diagnostic test.

---

[5] YAWL can be freely downloaded from www.yawl-system.com

[6] More information about the workflow patterns can be found at www.workflowpatterns.com

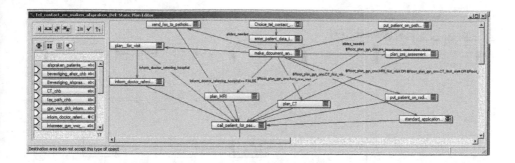

Fig. 7. Screenshot of the FLOWer editor

The resultant YAWL model consists of 231 nodes and 282 arcs and it took around 120 hours to construct a model that could be executed by the YAWL workflow engine.

6.2 FLOWer

FLOWer is a commercial workflow management system developed by Pallas Athena, the Netherlands[7]. FLOWer is a case-handling product [9]. Case-handling promotes flexibility in process enactment by focusing on the data rather than the control-flow aspects of a process.

In Figure 7, we see how the CWN of Figure 4 is mapped to the FLOWer language. In this case, it was quite easy to translate the CWN in FLOWer. In particular, it was possible to directly translate the transitions into FLOWer activities and the causal relationships could also be taken into account. In Figure 7, all nodes are activities except the "choice tel contact" node. This node represents a deferred choice which needs to be made at the beginning of the process, as can be seen in Figure 4. So, at the commencement of the process, either the "ask information from doctor referring hospital" activity can be chosen or the "write down data patient and make decision" activity can be chosen, but not both.

The resultant FLOWer model consists of 236 nodes and 190 arcs and it took around 100 hours to construct a model that could be executed by the FLOWer workflow engine.

6.3 ADEPT1

ADEPT1 is an academic prototype workflow system [33], developed at the University of Ulm, Germany. ADEPT1 supports *dynamic change* which means that the process of an individual case can be dynamically adapted during execution. So, it is possible to deviate from the static process model (skipping steps, going back to previous steps, inserting new steps, etc.) in a safe and secure way.

---

[7] http://www.pallas-athena.com/

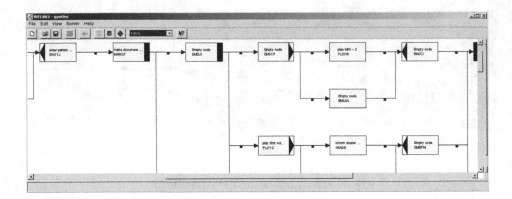

Fig. 8. Screenshot of the ADEPT1 editor. AND-splits/joins are represented by a black rectangle in a node and XOR-splits/joins are represented by a black triangle in a node.

That is, the system guarantees that all consistency constraints (e.g., no cycles, no missing input data when a task program will be invoked) which have been enforced prior to the dynamic (ad hoc) modification of the process instance are also enforced after the modification.

In Figure 8, we see how the first part of the CWN of Figure 4 is mapped to the ADEPT language. In the ADEPT language, the activities are represented by rectangles. However, as the ADEPT language only has an XOR and an AND split/join, we need to introduce dummy activities, i.e. they are not executed by users. For example, the "make stickers and document" activity in ADEPT is an AND-split which is followed by several dummy XOR-splits such that several activities, like "plan MRI" or "plan CT" can be performed or skipped.

Note that due to the restriction of data elements to simple data types, we only modeled 10% of the whole process[8]. The resultant ADEPT model consists of 40 nodes and 53 arcs and it took around 8 hours to construct a model that could be executed by the ADEPT1 workflow engine.

6.4   Declare

Declare is an academic prototype workflow system [29], developed at Eindhoven University of Technology in the Netherlands[9]. In Declare the language used for specifying processes, which is called *ConDec*, is a *declarative* process modeling language, which means that it specifies *what* should be done. *Imperative* process modeling languages, like YAWL and FLOWer, specify *how* it should

---

[8] For ADEPT1 only the part of the CWN associated with the referral of the patient and preparation for the first visit subprocess has been mapped and is shown in Figure 4. The number of activities contained in this subprocess constitutes around 10% of the activities present in the whole CWN. This selection can be considered as representative for the whole CWN.

[9] http://www.win.tue.nl/declare/

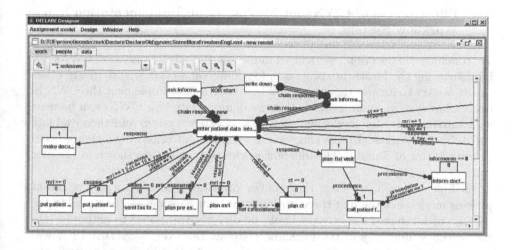

Fig. 9. Screenshot of the Declare editor

be done, which often leads to over-specified processes. By using a declarative rather than an imperative / procedural approach, ConDec aims at providing an under-specification of the process where workers have room to "manoeuver". This allows users to execute activities in any order and as often as they want, but they are bound to certain rules. These rules are based on *Linear Temporal Logic (LTL)*. Moreover, Declare also supports *dynamic change*.

In Figure 9, we see how the first part of the CWN of Figure 4 is mapped to the ConDec language. For the same reason as with the ADEPT1 implementation, here we also only modeled 10% of the whole process due to the restriction of data elements to simple data types.

In the ConDec language, the activities are represented by rectangles. Moreover, each different LTL formula, which can be used in the model, is represented by a different *template*, and can be applied to one or more activities, depending on the arity of the LTL formula which is used. Note that the language is extensible, i.e., it is easy to add a new construct by selecting its graphical representation and specifying its semantics in terms of LTL.

In Figure 9, we see that after the "enter patient data into system" activity a lot of subsequent activities need to be done, which is indicated by a *response* arc going from the "enter patient data into system" activity to these activities, e.g. "plan CT" and "plan MRI". However, it is only indicated that these activities need to be done afterwards, and no ordering is specified. At runtime, it is indicated to the user which activities need to done. The user can then decide in which order the activities will be executed and how often each one will be done.

The Declare model consists of 23 nodes and 44 LTL formulae and it took around 12 hours to construct a model that could be executed by the Declare workflow engine.

6.5 Effectiveness of CWNs for Moving to an Implementation in a Workflow System

To conclude this section, we will examine on the effectiveness of CWNs for implementing the healthcare process in four different workflow systems. In other words, we try to answer the question of how easy it is to implement the CWN in a workflow system. In order to do this, we demonstrate how CWNs can be used as a starting point for implementation in a workflow system and then evaluate the different systems. To this end, we use five different criteria, which are listed in the top row of Table 1. The different workflow systems are shown in the first column.

As first criterion, we have the number of nodes and arcs in the resultant process model. For the first three workflow systems considered, the nodes in the CWN which referred to activities could be directly translated to activities into the workflow language of the workflow systems. However, for ADEPT1 it was necessary to use dummy nodes as only XOR and AND split/joins are available.

As a CWN covers the *control flow*, *resource*, *data*, and *operation* perspective, we indicate how much effort was required to specify each perspective in each workflow system. This effort corresponded to the time taken by a single person having no experience in configuring processes using these systems. Furthermore, within Declare and ADEPT1 we defined a part of the CWN which constituted 10 percent of the model whereas for YAWL and FLOWer the whole model was implemented. The results are shown in Table 1. The numbers in brackets in the last four columns show the normalized value for the man hours required, so that each of the systems can be compared. However, for the Declare and ADEPT1 system these numbers should only be seen as an indication of the effort required as gradually the level of experience increases when configuring a workflow system. Moreover, some parts of the model might be more difficult to implement than others in different workflow systems.

The main conclusions that can be drawn from the table are discussed below (see [26] for full details). First of all, it can be seen that the control flow perspective can easily be translated for imperative languages (YAWL, FLOWer, ADEPT1). For declarative languages (Declare) more effort is needed. For the resource perspective it can be seen that the resource related parts can be easily translated as in all of the systems roles can be defined which can be linked to activities. Although not immediately visible from the table, the data perspective also could be easily translated for each of the systems. This stems from the fact that for YAWL and FLOWer, although the complex data types that needed to be converted were complex, there is great similarity between data types in these two systems and the CWN, and therefore the translation is still an easy process. Unfortunately, the operation perspective of the CWN could not be easily translated. This can easily be derived from the table as for each system half or more of the time required for configuring the system is spent on defining this perspective only. This is a consequence of the fact that every system has its own specific way/language for defining the operation perspective and that a big difference exists between these languages and the language used in the CWN. It is

Table 1. Translation of the CWN into the workflow languages of YAWL, FLOWer, ADEPT1, and Declare. Note that for Declare and ADEPT1 we only defined a part of the CWN which constitutes 10 percent of that model. To this end, to obtain normalized values for these systems, we multiplied the hours spent by 10. The hours enclosed in brackets indicate the normalized time. However, for the Declare and ADEPT1 system these figures should be seen as indicative figures only.

	number of nodes / arcs	effort control flow perspective (hours)	effort organizational perspective (hours)	effort data perspective (hours)	effort operation perspective (hours)
YAWL	231 / 282	20 (20)	5 (5)	30 (30)	65 (65)
FLOWer	236 / 190	20 (20)	5 (5)	20 (30)	55 (65)
ADEPT1	40 / 53	2 (20)	1 (10)	1 (10)	4 (40)
Declare	23 / 44	6 (60)	1 (10)	1 (10)	4 (40)

worth commenting that in Declare and ADEPT we only had to deal with simple rather than complex data types explaining the difference in effort required. For example, in YAWL and FLOWer, multiple instance tasks needed to be specified which is far from trivial.

During the implementation in the different workflow systems no further insights were gained about the healthcare process itself. So, all the time has been spent on the implementation in the workflow systems. This is due to the fact that we did not receive any user feedback during the implementation phase and the users did not evaluate the resultant systems themselves. If we had received further user feedback during the implementation phase, it is possible that further insights would have been gained about the process captured during the requirements phase (EUC). In essence, the approach focuses on identifying the real world process to be supported during the requirements phase. In the implementation phase the focus is solely on implementation aspects. If new requirements are identified during this phase, in principle, one should go back to the requirements phase and restart the development process from there.

In conclusion, we can say that a CWN is of help when implementing a given process in a workflow system, with regard to the control flow, resource and data perspective. These perspectives have to be defined anyway and the CWN allows for an implementation independent specification of these perspectives. Moreover, it is a useful step towards the implementation of the process in a chosen workflow system.

Furthermore, in principle, a (semi-) automatic translation from the CWN to each of the workflow systems is possible. The control flow and resource perspective are capable of automatic translation. With regard to the data perspective, data attributes which are name-value pairs for a simple data type, e.g. a string or an integer, are capable of automatic translation. However, for the operation perspective and for name-value pairs for which a list type has been used, more work is needed. For both of them, some form of semi-automatic translation would be necessary. However, as in this paper the focus is on the general application of the approach and its applicability to distinct technology systems, we do not focus on developing translation from the CWN to each of the workflow systems. In [8,20] it has been shown that (semi-) automatic translations from the CWN to a nominated workflow language are possible.

An issue related to the expressiveness of the CWN is the fact that CPNs cannot capture with three categories of workflow patterns, namely advanced synchronization (e.g. the synchronizing merge), multiple instances, and cancelation [6]. When constructing the CWN model already knowledge is gained about how the process exactly needs be supported by a certain workflow system. So, it can already be identified that such pattern might be more appropriate in the final implementation. In that way, if a workflow system offers support for such a pattern it can be selected instead.

As we manually configured each workflow system, there is the risk of making mistakes. Some of the workflow systems offer functionality for checking the soundness property of the configured model. Only in case that such functionality is available we can guarantee the soundness property for the translated model. For example, YAWL offers the opportunity to check for soundness which we subsequently used for checking whether the model was sound. Although such checks can prevent some types of mistakes, they do not prevent mistakes made at the semantic level. For the latter, interactions with users are of vital importance. Note that mistakes made at the semantic level cannot be prevented when making a (semi-) automatic translation. One hopes that these kinds of mistakes have already been detected during the requirements and design phase.

7   Related Work

A workflow application development process is a workflow development methodology aiming at improving the planning and construction of workflow projects [37]. So far, surprisingly little research has been performed on this topic. In [37,23,12,21,22] different approaches regarding this topic are described. These approaches differ in the number of phases that need to be followed, the steps in these phases, and the scope of the development methodology. Without focusing on specific modeling techniques, [37,23,22] specify the steps which needs to be taken in order to end with a workflow implementation of a certain process. These steps are not limited to model construction only, but also address issues like the selection of a workflow system. However, whilst [23,22] provide a rather high-level description of a workflow development methodology, [37] provides a detailed overview of the required phases and the steps in these phases, based on experiences obtained during real-world workflow processes.

In contrast, [12,21] propose specific modeling techniques that have to be used during the development process. In [12], UML descriptions are proposed for identifying business process requirements. Then the WIDE meta-model is used to design an implementation independent workflow schema and associated exception handling which can then subsequently mapped to the required workflow language. In [21], UML use cases and UML interaction diagrams are used to develop a so-called multi-level workflow schema which provides an abstraction of the workflow at different levels.

Compared to these approaches, the development methodology described in Section 2 brings more rigor to the development process as the models constructed in the EUC and CWN phase are based on a formal foundation. This provides various possibilities for verification whereas in [12,21] different modeling mechanisms are mentioned but nothing is said about the verification of these models. Moreover, as both the EUC and CWN use the same language, this allows for a seamless integration between the requirements and design phase. None of the approaches described in [37,23,12,21,22] report on obtained experiences when applying the proposed approach.

From the literature, it can be deduced that workflow systems are not applicable to the healthcare domain [11,24]. The current generation of workflow systems adequately supports administrative and production workflows but they are less suited for healthcare processes which have more complex requirements [11]. In addition, in [31,32], it has been indicated that so called "careflow systems", i.e. systems for supporting care processes in hospitals, have special demands with regard to workflow technology. One of these requirements is that flexibility needs to be provided by the workflow system [28,35]. Unfortunately, current workflow systems fall short in this area, a fact which is recognized in the literature [7,9]. Once a workflow-based application has been configured on the basis of explicit process models, the execution of related process instances tends to be rather inflexible [3,33]. Consequently, the lack of demonstrated flexibility has significantly limited the application of workflow technology. The workflow systems that we chose in this paper were specifically selected because they allow for more flexibility than classical workflow systems.

This paper uses the approach initially proposed in [8,20] where an EUC and CWN have been used to progress from an informal inscription of a real world process to an implementation of the same process in a particular workflow system. In [20], an electronic patient record system has been implemented using the YAWL system. However, in both papers, only small examples were used and an implementation was only completed in one workflow system. Furthermore, in [8] direct user involvement was limited to the requirements phase, whereas in [20] there was no user involvement at all. In this paper, we modeled a much larger, representative healthcare process from a hospital using four different workflow systems. Moreover, the approach was evaluated by the people involved in the process.

8   Conclusions

In this paper, we have focused on the implementation of a large hospital process consisting of hundreds of activities, in different workflow systems. To support the implementation process, we first developed an EUC, followed by a CWN. This CWN was used as the input for configuring the different workflow systems. The approach, shown in Figure 1, effectively bridges the gap between the modeling of a real-world process and the implementation of the process in a workflow system. We have successfully shown that the approach works for a large real-world healthcare process and for distinct workflow technologies. However, some important observations can be drawn from this research effort.

The first observation is that the combination of animations and EUCs are of great help when validating the modeled process. Based on the feedback of users we believe that better results have been obtained than when using the plain CPN models.

The second observation is that the CWN helps in elaborating how the candidate process, needs to be made ready for implementation in a workflow system. The CWN covers the control-flow, resource, data and operation perspective, which necessitates the specification of these perspectives during the construction of the CWN. Another advantage is that the CWN abstracts from implementation details and language/application specific issues.

The third observation is that a CWN is helpful during the configuration of the control-flow, resource and data perspective in a workflow system. However, this does not hold for the operation perspective of the CWN.

Furthermore, EUCs and CWNs are useful as they allow for a separation of concerns. The approach adopted enforces that the issues we have to deal during workflow development are tackled at the right time and in the right sequence.

A possible direction for future research is to develop animations specific for CWN. In this way, before a process is supported by a workflow system, the associated staff can already become acquainted with how their process will be facilitated and start experimenting with it. The case study has provided valuable insights into the requirements for workflow technology in care organizations. Besides the need for flexibility, it also revealed the need for a better integration of patient flow with the scheduling of appointments and peripheral systems supporting small and loosely coupled workflows (e.g. lab tests).

Acknowledgements. We would like to thank Lisa Wells and Kurt Jensen for their support with CPN Tools. Also, we would like to thank Michael Westergaard for his support in using the BRITnEY tool. Furthermore, we also want to thank the people of the gynecological oncology, radiology, pathology and anesthetics department, and especially Prof. Mathé Burger of the gynecological oncology department for their comprehensive explanation of the gynecological oncology healthcare process.

Moreover, we would like to thank the reviewers of ToPNoC, as well as the editors, for their fruitful comments to improve this paper.

References

1. van der Aalst, W.M.P.: Workflow Verification: Finding Control-Flow Errors using Petri-net-based Techniques. In: van der Aalst, W.M.P., Desel, J., Oberweis, A. (eds.) Business Process Management. LNCS, vol. 1806, pp. 161–183. Springer, Heidelberg (2000)

2. van der Aalst, W.M.P.: Business Process Management Demystified: A Tutorial on Models, Systems and Standards for Workflow Management. In: Desel, J., Reisig, W., Rozenberg, G. (eds.) Lectures on Concurrency and Petri Nets. LNCS, vol. 3098, pp. 1–65. Springer, Heidelberg (2004)
3. van der Aalst, W.M.P., Barthelmess, P., Ellis, C.A., Wainer, J.: Workflow Modeling using Proclets. In: Scheuermann, P., Etzion, O. (eds.) CoopIS 2000. LNCS, vol. 1901, pp. 198–209. Springer, Heidelberg (2000)
4. van der Aalst, W.M.P., van Hee, K.M.: Workflow Management: Models, Methods, and Systems. MIT Press, Cambridge (2002)
5. van der Aalst, W.M.P., van Hee, K.M., ter Hofstede, A.H.M., Sidorova, N., Verbeek, H.M.W., Voorhoeve, M., Wynn, M.T.: Soundness of Workflow Nets: Classification, Decidability, and Analysis. Computer Science Report No. 08-13, Technische Universiteit Eindhoven, The Netherlands (2008)
6. van der Aalst, W.M.P., ter Hofstede, A.H.M.: YAWL: Yet Another Workflow Language. Information Systems 30(4), 245–275 (2005)
7. van der Aalst, W.M.P., Jablonski, S.: Dealing with Workflow Change: Identification of Issues and Solutions. International Journal of Computer Systems, Science, and Engineering 15(5), 267–276 (2000)
8. van der Aalst, W.M.P., Jørgensen, J.B., Lassen, K.B.: Let's Go All the Way: From Requirements via Colored Workflow Nets to a BPEL Implementation of a New Bank System Paper. In: Meersman, R., Tari, Z. (eds.) OTM 2005. LNCS, vol. 3760, pp. 22–39. Springer, Heidelberg (2005)
9. van der Aalst, W.M.P., Weske, M., Grünbauer, D.: Case Handling: A New Paradigm for Business Process Support. Data and Knowledge Engineering 53(2), 129–162 (2005)
10. Adams, M., ter Hofstede, A.H.M., ter Edmond, D., van der Aalst, W.M.P.: Facilitating Flexibility and Dynamic Exception Handling in Workflows. In: Belo, O., Eder, J., Pastor, O., Falcao e Cunha, J. (eds.) Proceedings of the CAiSE 2005 Forum, FEUP, Porto, Portugal, pp. 45–50 (2005)
11. Anyanwu, K., Sheth, A., Cardoso, J., Miller, J., Kochut, K.: Healthcare Enterprise Process Development and Integration. Journal of Research and Practice in Information Technology 35(2), 83–98 (2003)
12. Baresi, L., Casati, F., Castano, S., Fugini, M.G., Mirbel, I., Pernici, B.: WIDE Workflow Development Metholodogy. In: Proceedings of International Joint Conference on Work Activities Coordination and Collaboration, pp. 19–28 (1999)
13. Dadam, P., Reichert, M., Khuhn, K.: Clinical Workflows - The Killer Application for Process-oriented Information Systems? In: Abramowicz, W., Orlowska, M.E. (eds.) BIS 2000, pp. 36–59. Springer, Heidelberg (2000)
14. Graeber, S.: The impact of workflow management systems on the design of hospital information systems. In: AMIA 2001 Symposium Proceedings (2001)
15. Greiner, U., Ramsch, J., Heller, B., Löffler, M., Müller, R., Rahm, E.: Adaptive Guideline-based Treatment Workflows with AdaptFlow. In: Kaiser, K., Misch, S., Tu, S.W. (eds.) CGP 2004. Computer-based Support for Clinical Guidelines and Protocols, pp. 113–117. IOS Press, Amsterdam (2004)
16. van Hee, K.M., Serebrenik, A., Sidorova, N., Voorhoeve, M.: Resource-Constrained Workflow Nets. In: Ciardo, G., Darondeau, P. (eds.) ICATPN 2005. LNCS, vol. 3536, pp. 250–267. Springer, Heidelberg (2005)

17. Jensen, K., Kristensen, L.M., Wells, L.: Coloured Petri Nets and CPN Tools for Modelling and Validation of Concurrent Systems. STTT 9(3–4), 213–254 (2007)
18. Jensen, K., Christensen, S., Kristensen, L.M.: CPN Tools State Space Manual. Department of Computer Science, Univerisity of Aarhus (2006)
19. Jørgensen, J.B., Bossen, C.: Executable Use Cases: Requirements for a Pervasive Health Care System. IEEE Software 21, 34–41 (2004)
20. Jørgensen, J.B., Lassen, K.B., van der Aalst, W.M.P.: From task descriptions via colored Petri nets towards an implementation of a new electronic patient record workflow system. STTT 10(1), 15–28 (2006)
21. Kim, J., Robert Karlson, C.: A Design Methodology for Workflow System Development. In: Bhalla, S. (ed.) DNIS 2002. LNCS, vol. 2544, pp. 15–28. Springer, Heidelberg (2002)
22. Kobielus, J.G.: Workflow Strategies. IDG Books (1997)
23. Kwan, M., Balasubramanian, P.R.: Adding Workflow Analysis Techniques to the IS Development Toolkit. In: HICSS 1998, vol. 4, pp. 312–321. IEEE Computer Society Press, Los Alamitos (1998)
24. Lenz, R., Elstner, T., Siegele, H., Kuhn, K.: A Practical Approach to Process Support in Health Information Systems. JAMIA 9(6), 571–585 (2002)
25. Lenz, R., Reichert, M.: IT Support for Healthcare Processes - Premises, Challenges, Perspectives. DKE 61, 49–58 (2007)
26. Mans, R.S., van der Aalst, W.M.P., Bakker, P.J.M., Moleman, A.J., Lassen, K.B., Jørgensen, J.B.: From Requirements via Colored Workflow Nets to an Implementation in Several Workflow Systems. In: Jensen, K. (ed.) Proceedings of the Eighth Workshop and Tutorial on Practical Use of Coloured Petri Nets and the CPN Tools, pp. 187–206 (2007)
27. Mans, R.S., van der Aalst, W.M.P., Russell, N.C., Bakker, P.J.M.: Flexibility Schemes for Workflow Management Systems. In: Pre-Proceedings of ProHealth 2008, pp. 50–61 (2008)
28. Maruster, L., van der Aalst, W.M.P., Weijters, A.J.M.M., van den Bosch, A., Daelemans, W.: Automated Discovery of Workflow Models from Hospital Data. In: Dousson, C., Höppner, F., Quiniou, R. (eds.) Proceedings of the ECAI Workshop on Knowledge Discovery and Spatial Data, pp. 32–36 (2002)
29. Pesic, M., van der Aalst, W.M.P.: A Declarative Approach for Flexible Business Processes Management. In: Eder, J., Dustdar, S. (eds.) BPM Workshops 2006. LNCS, vol. 4103, pp. 169–180. Springer, Heidelberg (2006)
30. Philippi, S., Hill, H.J.: Communication Support for Systems Engineering - Process Modelling and Animation with APRIL. The Journal of Systems and Software 80(8), 1305–1316 (2007)
31. Quaglini, S., Stefanelli, M., Cavallini, A., Micieli, G., Fassino, C., Mossa, C.: Guideline-based Careflow Systems. Artificial Intelligence in Medicine 20(1), 5–22 (2000)
32. Quaglini, S., Stefanelli, M., Lanzola, G., Caporusso, V., Panzarasa, S.: Flexible Guideline-based Patient Careflow Systems. Artificial Intelligence in Medicine 22(1), 65–80 (2001)
33. Reichert, M., Dadam, P.: ADEPTflex: Supporting Dynamic Changes of Workflow without Loosing Control. Journal of Intelligent Information Systems 10(2), 93–129 (1998)

34. Reijers, H.A., van der Aalst, W.M.P.: The Effectiveness of Workflow Management Systems: Predictions and Lessons Learned. International Journal of Information Management 56(5), 457–471 (2005)
35. Stefanelli, M.: Knowledge and Process Management in Health Care Organizations. Methods Inf. Med. 43, 525–535 (2004)
36. Trcka, N., van der Aalst, W.M.P., Sidorova, N.: Analyzing control-flow and data-flow in workflow processes in a unified way. Computer Science Report No. 08-31, Technische Universiteit Eindhoven, The Netherlands (2008)
37. Weske, M., Goesmann, T., Holten, R., Striemer, R.: Analysing, modelling and improving workflow application development processes. Software Process Improvement and Practice 6, 35–46 (2001)

Soundness of Workflow Nets with Reset Arcs

Wil M.P. van der Aalst[1,2], Kees M. van Hee[1], Arthur H.M. ter Hofstede[2],
Natalia Sidorova[1], H.M.W. Verbeek[1], Marc Voorhoeve[1], and Moe T. Wynn[2]

[1] Department of Mathematics and Computer Science,
Eindhoven University of Technology
P.O. Box 513, NL-5600 MB, Eindhoven, The Netherlands
w.m.p.v.d.aalst@tue.nl
[2] Business Process Management Group, Queensland University of Technology
P.O. Box 2434, Brisbane Qld 4001, Australia

Abstract. Petri nets are often used to model and analyze workflows.
Many workflow languages have been mapped onto Petri nets in order to
provide formal semantics or to verify correctness properties. Typically,
the so-called *Workflow nets* are used to model and analyze workflows
and variants of the classical *soundness property* are used as a correctness
notion. Since many workflow languages have *cancelation features*, a mapping to workflow nets is not always possible. Therefore, it is interesting
to consider workflow nets with *reset arcs*. Unfortunately, soundness is
undecidable for workflow nets with reset arcs. In this paper, we provide
a proof and insights into the theoretical limits of workflow verification.

Keywords: Petri Nets, Decidability, Workflow Nets, Reset Nets, Soundness, and Verification.

1 Introduction

Information systems have become "process-aware", i.e., they are driven by process models [13]. Often the goal is to automatically configure systems based
on process models rather than to code the control-flow logic using some conventional programming language. Early examples of process-aware information
systems were called WorkFlow Management (WFM) systems [4,24,34]. In more
recent years, vendors prefer the term Business Process Management (BPM) systems. BPM systems have a wider scope than the classical WFM systems and
are not just focusing on process automation. BPM systems tend to provide more
support for various forms of analysis and management support. Both WFM and
BPM aim to support operational processes that we refer to as "workflow processes" or simply "workflows".

The flow-oriented nature of workflow processes makes the Petri net formalism
a natural candidate for the modeling and analysis of workflows. This paper
focuses on the so-called *workflow nets* (WF-nets) introduced in [1,2]. A WF-net
is a Petri net with a start place i and an end place o such that all nodes are on
paths from i to o. A case, i.e., process instance, is initiated by putting a token
in the source place i. The completion of a case is signalled when a token appears
in the sink place o.

K. Jensen, J. Billington, and M. Koutny (Eds.): ToPNoC III, LNCS 5800, pp. 50–70, 2009.

In the context of WF-nets a correctness criterion called *soundness* has been defined [1,2]. A WF-net with source place *i* and sink place *o* is *sound* if and only if the following three requirements are satisfied: (1) *option to complete*: for each case starting in source place *i* it is always still possible to reach the state which just marks sink place *o*, (2) *proper completion*: if sink place *o* is marked, then all other places are unmarked for a given case, and (3) *no dead transitions*: it should be possible to execute an arbitrary activity by following an appropriate route through the WF-net. In [1,2] it was shown that soundness is decidable and that it can be translated into a liveness and boundedness problem, i.e., a WF-net is sound if and only if an extension of the net (the so called "short-circuited net") is live and bounded. In the last decade, the soundness property has become the most widely used correctness notion for workflow. This is illustrated by the fact that, according to Google Scholar, [2] is among the most cited papers both in the workflow/BPM community and Petri net community.

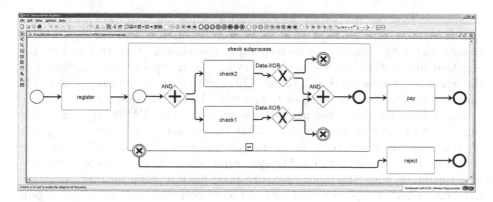

Fig. 1. A BPMN diagram constructed using ILOG JViews. The subprocess is canceled if one of the two checks is negative.

Since the mid-nineties many people have been looking at the verification of workflows. These papers all assume some underlying model (e.g., WF-nets) and some correctness criterion (e.g., soundness). However, in many cases a rather simple model is used (WF-nets or even less expressive) and practical features such as *cancelation* are missing. Many practical languages have a cancelation feature, e.g., Staffware has a withdraw construct, YAWL has a cancelation region, BPMN has cancel, compensate, and error events, etc. To illustrate this, consider the BPMN diagram shown in Fig. 1. The process describes the handling of some claim that requires two checks. The process starts with a *registration* step, followed by the parallel execution of *check1* and *check2*. The outcome of each of these checks may be negative or positive. If both are positive, activity *pay* follows and concludes the process. If one of the checks is negative, no further processing is needed and the process ends with activity *reject*. Fig. 1 uses four BPMN gateways depicted using a diamond shape. The gateways with a "+"

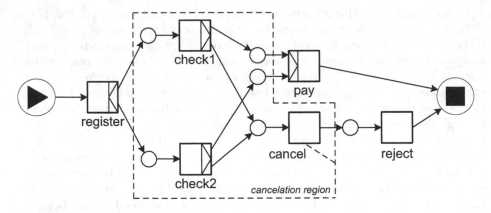

Fig. 2. A YAWL model using cancelation

annotation have an AND-split/join behavior, while the gateways with a "×" annotation have an XOR-split/join behavior. Events are depicted by a circle. Events with a "×" annotation correspond to cancelation. Note that in Fig. 1 a negative outcome triggers a cancelation event which triggers the cancelation of the whole subprocess *check subprocess*. This means that the first negative result cancels any activities scheduled or taking place inside this subprocess.

Fig. 2 shows the same example, but now modeled using YAWL. By comparing Fig. 1 and Fig. 2 it becomes obvious what the semantics of the various YAWL notations are. Task *register* is an AND-split and task *pay* is an AND-join. The two check tasks are XOR-splits. The two input places of *pay* correspond to positive outcomes, while the input place of *cancel* holds a token for every negative outcome. YAWL supports the concept of a "cancelation region", i.e., a region from which everything is removed for a particular instance. In Fig. 2 the cancelation region is depicted using dashed lines. This region is activated by the execution of task *cancel*, i.e., after executing *cancel* there is only a token in the input place of task *reject*.

BPMN and YAWL are two of many process modeling languages that support cancelation. Table 1 provides some more examples. This table illustrates that many systems and languages support cancelation functionality.

Since we are interested in verification of real-life problems, we need to support cancelation. Therefore, it is interesting to investigate the notion of soundness in the context of *WF-nets* with *reset arcs*. Petri nets with reset arcs are called *reset nets* [6,11,12,16,31]. WF-nets form a subclass of Petri nets tailored towards workflow modeling and analysis [1,2]. A reset arc connects a place to a transition. For the enabling of this transition the reset arc plays no role. However, whenever this transition fires, the place is emptied. Clearly, this concept can be used to model various cancelation concepts encountered in modern workflow languages. To illustrate this, consider the reset net shown in Fig. 3. Note that the two XOR-splits have both been replaced by a small network of transitions. For example, *check1* is followed by two transitions (*OK* and *NOK*) modeling the different

Table 1. Examples of languages supporting cancelation (see also [35])

BPMN	Cancelation is supported by adding some intermediate event trigger attached to the boundary of the activity to be canceled.
YAWL	Cancelation is supported by the cancelation region which "empties" a selected part of the process.
Staffware	Cancelation is supported using the so-called "withdraw construct", i.e., a connection entering the top of a workflow step.
UML ADs	Cancelation is supported by incorporating the activity in an interruptible region triggered either by a signal or execution of another activity.
SAP Workflow	Cancelation is supported through the use of the "process control" step that can be used to "logically delete" activities by specifying the node number of the corresponding step.
FileNet	Cancelation is supported via a so-called "Terminate Branch" step.
BPEL	Cancelation is supported by fault and compensation handlers.
XPDL	Cancelation is supported via an error type trigger attached to the boundary of the activity to be canceled.

outcomes of this check. Moreover, the cancelation region has been replaced by seven reset arcs. The double-headed arcs in Fig. 3 correspond to reset arcs. These arcs empty all places where tokens may remain after firing *cancel* for the first time. It is easy to see that Fig. 3 has indeed the behavior described earlier using BPMN and YAWL.[1]

It is far from trivial to express the desired behavior without reset arcs. In Fig. 3, transition *cancel* is the only transition having reset arcs. To remove these reset arcs, we would need to consider all possible markings before this point. In this case, there are (only) 7 possible markings when *cancel* fires, i.e., *cancel* would need to be replaced by 7 transitions. In general there is an exponential number of possible states. If there are n checks in the example (rather than 2), then $4^n - 3^n$ transitions are needed to replace *cancel* and its reset arcs (e.g., for 10 checks, 1048576-59049=989527 transitions are needed). This illustrates the relevance of reset arcs from a modeling point of view. Therefore, it is interesting to investigate the verification of WF-nets with reset arcs.

This paper will prove that *soundness is undecidable for reset WF-nets*. This result is not trivial since other properties such as e.g. coverability are decidable for reset nets. However, we managed to develop a construction that maps the soundness problem onto a reachability problem. Reachability is known to be undecidable for reset nets [6,11]. The construction shown in this paper suggests that there is not a simple mapping between soundness and reachability, thus making the result interesting.

---

[1] Note that here we assume activities to be atomic. It is also possible to describe the more refined behavior using reset nets. However, this would only complicate the presentation.

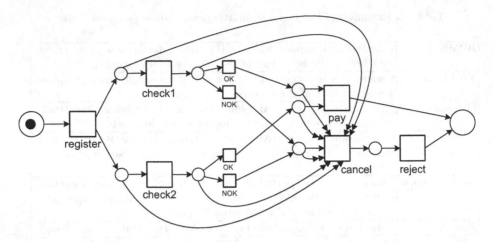

Fig. 3. A WF-net with reset arcs to model cancelation

The remainder of this paper is organized as follows. First, we briefly present an overview of related work (Section 2). Then, Section 3 presents some of the preliminaries (mathematical notations and Petri net basics). Section 4 presents the basic notion of reset WF-nets. In Section 5 the classical notion of soundness is introduced. Section 6 presents the main result: undecidability of soundness for reset WF-nets. Moreover, we will show that soundness is also undecidable for weaker notions such as relaxed soundness [8,9]. Section 7 concludes the paper.

2   Related Work

This paper builds on classical decidability results for (variants of) Petri nets. In [18], it is shown that the equality problem for Petri nets is undecidable (i.e., it is impossible to provide an algorithm that checks whether the sets of reachable markings of two nets are equivalent). In [26], it is shown that the reachability problem is decidable (i.e., it is possible to provide an algorithm that checks whether a particular marking is reachable). However, almost any extension of the basic Petri net formalism makes the reachability problem undecidable. In [6], Araki and Kasami show that the reachability problem for Petri nets with priorities or reset arcs is undecidable. In [31], Valk shows various undecidability results for so-called "self-modifying nets". (Reset nets can be seen as a subclass of self-modifying nets.) In [11], Dufourd et al. show that coverability and termination are decidable for reset nets, but that boundedness and reachability are undecidable. In [12] it is shown that boundedness is undecidable for nets with three reset arcs but still decidable for reset nets with two resetable places. In [33] it is shown that reduction rules can be applied to reset nets (and even to inhibitor nets) to speed-up analysis and improve diagnostics. In [16] several results are given for well-structured transition systems. In this paper these results are related to analysis and decidability issues for various Petri nets extensions (e.g., reset arcs). For a survey of decidability results for ordinary Petri nets, we refer to [14,15].

Since the mid nineties, many researchers have been working on workflow verification techniques. It is impossible to give a complete overview here. Moreover, most of the papers on workflow verification focus on rather simple languages, e.g., AND/XOR-graphs which are even less expressive than Petri nets. Therefore, we only mention the work directly relevant to this paper.

The use of Petri nets in workflow verification has been studied extensively. In [1,2] the foundational notions of WF-nets and soundness are introduced. In [19,20] two alterative notions of soundness are introduced: k-soundness and generalized soundness. These notions allow for dead parts in the workflow but address problems related to multiple instantiation. In [25] the notion of weak soundness is proposed. This notion allows for dead transitions. The notion of relaxed soundness is introduced in [8,9]. This notion allows for potential deadlocks and livelocks, however, for each transition there should be at least one proper execution. Lazy soundness [28] is another variant that only focuses on the end place and allows for excess tokens in the rest of the net. Finally, the notions of up-to-k-soundness and easy soundness are introduced in [30]. More details on these notions proposed in the literature are given in Section 5.

Most soundness notions (except generalized soundness [19,20]) can be investigated using classical model checking techniques that explore the state space. However, such approaches can be intractable or even impossible because the state space may be infinite. Therefore, alternative approaches that avoid constructing the (full) state space have been proposed. [3] describes how structural properties of a workflow net can be used to detect the soundness property. [32] presents an alternative approach for deciding relaxed soundness in the presence of OR-joins using invariants. The sketched approach results in the approximation of OR-join semantics and the transformation of YAWL nets into Petri nets with inhibitor arcs. The cancelation regions in YAWL are closely related to the notion of reset arcs. In [36] it is shown that the backward reachability graph can be used to determine the enabling of OR-joins in the context of cancelation. In [38] it is shown how reduction rules can be used to improve performance. The techniques proposed in the latter two papers heavily rely on reset nets and their applicability is demonstrated in [37].

3 Preliminaries

This section introduces some of the basic mathematical and Petri-net related concepts used in the remainder of this paper.

3.1 Multi-sets, Sequences, and Matrices

Let A be a set. $\mathbb{B}(A) = A \to \mathbb{N}$ is the set of multi-sets (bags) over A, i.e., $X \in \mathbb{B}(A)$ is a multi-set where for each $a \in A$: $X(a)$ denotes the number of times a is included in the multi-set. The sum of two multi-sets $(X + Y)$, the difference $(X - Y)$, the presence of an element in a multi-set $(x \in X)$, and the notion of sub-multi-set $(X \leq Y)$ are defined in a straightforward way. When appropriate, these operators can be applied to sets as well, considering them as multi-sets with multiplicities equal to one. $|X| = \sum_{a \in A} X(a)$ is the size of the multi-set. $\pi_{A'}(X)$ is the projection of X onto $A' \subseteq A$, i.e., $(\pi_{A'}(X))(a) = X(a)$ if $a \in A'$ and $(\pi_{A'}(X))(a) = 0$ if $a \notin A'$.

To represent a concrete multi-set we use square brackets, e.g., $[a, a, b, a, b, c]$, $[a^3, b^2, c]$, and $3[a] + 2[b] + [c]$ all refer to the same multi-set with six elements: 3 a's, 2 b's, and one c. $[\,]$ refers to the empty bag, i.e., $|[\,]| = 0$.

For a given set A, A^* is the set of all finite sequences over A (including the empty sequence $\langle \rangle$). A finite sequence over A of length n is a mapping $\sigma \in \{1, \dots, n\} \to A$. Such a sequence is represented by a string, i.e., $\sigma = \langle a_1, a_2, \dots, a_n \rangle$ where $a_i = \sigma(i)$ for $1 \leq i \leq n$.

For a relation R on A, i.e., $R \subseteq A \times A$, we define R^* as the reflexive transitive closure of R.

3.2   Reset Petri Nets

This subsection briefly introduces some basic *Petri net* terminology [10,21,29] and notations used in the remainder of this paper. Our starting point is a Petri net with reset arcs and arc weights. Such a Petri net is called a *reset net*.

Definition 1 (Reset net). *A reset net is a tuple (P, T, F, W, R), where:*

- *(P, T, F) is a Petri net with a finite set of places P, a finite set of transitions T (such that $P \cap T = \emptyset$), and a flow relation $F \subseteq (P \times T) \cup (T \times P)$,*
- *$W \in F \to \mathbb{N} \setminus \{0\}$ is an (arc) weight function, and*
- *$R \in T \to 2^P$ is a function defining reset arcs.*

A reset net extends the basic Petri-net notion with arc weights and reset arcs. The arc weights specify the number of tokens to be consumed or produced and the reset arcs are used to remove all tokens from the reset places independent of the number of tokens. To illustrate these concepts we use Fig. 4. This figure shows a reset net with seven places and six transitions. The arc from $t1$ to $p3$ has weight 6, i.e., $W(t1, p3) = 6$. Moreover, $W(p5, t5) = 6$, $W(p3, t4) = 2$, and $W(t4, p5) = 2$. All other arcs have weight 1, e.g., $W(p1, t1) = 1$. Transition tr has four reset arcs, i.e., $R(tr) = \{p2, p3, p4, p5\}$, and $R(t) = \emptyset$ for all other transitions t.

Because we allow for arc weights, the preset and postset operators return bags rather than sets: $\bullet a = [x^{W(x,y)} \mid (x, y) \in F \wedge a = y]$ and $a \bullet = [y^{W(x,y)} \mid (x, y) \in F \wedge a = x]$. For example, $\bullet t5 = [p4, p5^6, pr]$ is the bag of input places of $t5$ and $t1 \bullet = [p2, p3^6, pr]$ is the bag of output places of $t1$.

Places may contain *tokens*. The *marking* of a reset net is a distribution of tokens over places. A marking M is represented as a bag over the set of places, i.e., $M \in \mathbb{B}(P)$. The initial marking shown in Fig. 4 is $[p1]$. For a marked reset net we define standard notions such as *enabling* and *firing*.

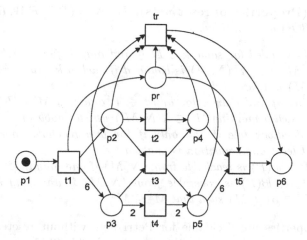

Fig. 4. A reset net. Transition tr is enabled if pr is marked and removes all tokens from $p2$, $p3$, $p4$, and $p5$.

Definition 2 (Firing rule). *Let $N = (P, T, F, W, R)$ be a reset net and $M \in \mathbb{B}(P)$ be a marking.*

- *A transition $t \in T$ is enabled at M, denoted by $(N, M)[t\rangle$, if and only if, $M \geq \bullet t$.*
- *An enabled transition t can fire while changing the marking to M', denoted by $(N, M)[t\rangle(N, M')$, if and only if $M' = \pi_{P \setminus R(t)}(M - \bullet t) + t\bullet$.*

The resulting marking $M' = \pi_{P \setminus R(t)}(M - \bullet t) + t\bullet$ is obtained by first removing the tokens required for enabling: $M - \bullet t$. Then all tokens are removed from the reset places of t using projection. Note that $\pi_{P \setminus R(t)}$ removes all tokens except the ones in the non-reset places $P \setminus R(t)$. Finally, the specified numbers of tokens are added to the output places, i.e., $t\bullet$ is a *bag* of places corresponding to the tokens to be added.

In Fig. 4, transition tr is enabled if and only if there is a token in place pr, i.e., reset arcs do not influence enabling. However, after the firing of tr all tokens are removed from the four places $p2$, $p3$, $p4$, and $p5$.

$(N, M)[t\rangle(N, M')$ defines how a Petri net can move from one marking to another by firing a transition. We can extend this notion to firing sequences. Suppose $\sigma = \langle t_1, t_2, \ldots, t_n \rangle$ is a sequence of transitions present in some Petri net N with initial marking M. $(N, M)[\sigma\rangle(N, M')$ means that there is also a sequence of markings $\langle M_0, M_1, \ldots, M_n \rangle$ where $M_0 = M$, $M_n = M'$, and for any $0 \leq i < n$: $(N, M_i)[t_{i+1}\rangle(N, M_{i+1})$. Using this notation we define the set of reachable markings $R(N, M)$ as follows: $R(N, M) = \{M' \in \mathbb{B}(P) \mid \exists_{\sigma \in T^*}(N, M)[\sigma\rangle(N, M')\}$. Observe that, by definition, $M \in R(N, M)$ because the initial marking M is trivially reachable via the empty sequence ($n = 0$).

Definition 3 (Properties of reset nets). *Let $N = (P, T, F, W, R)$ be a reset net and $M \in \mathbb{B}(P)$ be a marking.*

- *(N, M) is k-bounded for some $k \in \mathbb{N}$ if and only if for all $M' \in R(N, M)$ and $p \in P$: $M'(p) \leq k$. (N, M) is bounded if such a k exists and (N, M) is safe if (N, M) is 1-bounded.*
- *(N, M) is live if and only if for any $t \in T$ and any $M' \in R(N, M)$ it is possible to reach a marking $M'' \in R(N, M')$ which enables t.*
- *(N, M) is deadlock free if and only if for any reachable marking $M' \in R(N, M)$ at least one transition is enabled.*
- *Marking $M' \in \mathbb{B}(P)$ is reachable from (N, M) if and only if $M' \in R(N, M)$.*
- *Marking $M' \in \mathbb{B}(P)$ is coverable from (N, M) if and only if there is a marking $M'' \in R(N, M)$ such that $M' \leq M''$.*

The above properties are decidable for Petri nets without reset arcs [14,15]. However, for reset nets most properties are undecidable [6,11], e.g., there is no algorithm to check the reachability of a particular marking. A notable exception is coverability. Using a so-called "backward coverability algorithm" one can decide whether some marking M' is coverable from (N, M) [11].

Theorem 1 (Undecidability of reachability [6,11]). *The reachability problem ("Is a particular marking reachable?") is undecidable for reset nets.*

Theorem 2 (Decidability of coverability [11]). *The coverability problem ("Is a particular marking coverable?") is decidable for reset nets.*

Any reset net with arc weights can be transformed into a reset net without arc weights, i.e., all arcs have weight 1. Therefore, in proofs we can assume arc weights of 1. Fig. 5 illustrates how a Petri net with arc weights of 2 can be transformed into a Petri net without arc weights. If k is the maximum arc weight, the construction illustrated by Fig. 5 requires the splitting of place p into k places $(p_1, \ldots, p_k)$. See [5] for details about this construction. Note that the reset nets

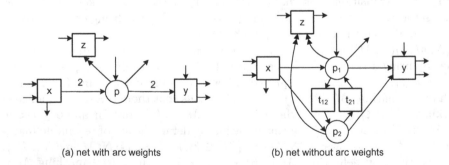

(a) net with arc weights   (b) net without arc weights

Fig. 5. Construction illustrating that it is possible to transform any reset net with arc weights into an equivalent reset net without arc weights

in Fig. 5(a) and Fig. 5(b) are branching bisimilar [17,7]. This means that after renaming the added transitions to τ (i.e. so-called silent steps), any (non-τ) action by one net can be simulated by the other. Any marking M in Fig. 5(a) corresponds to a set of markings X_M in Fig. 5(b). Moreover, any marking in X_M is reachable from any other marking in X_M. Therefore, it is possible to enable the same set of non-silent transitions in both nets. It is easy to see that the number of tokens in p equals the sum of tokens in $p_1, \ldots, p_k$ if one net follows the other. This property is invariant, showing that there is a tight relation between both nets. Based on this invariant, it is easy to see that the construction shown in Fig. 5 preserves boundedness. Liveness is also preserved since any firing sequence in one net can be mimicked in the other (abstracting from silent steps). Hence transitions that are live in Fig. 5(a) remain live in Fig. 5(b) while transitions that are not live are also not live in the new net. If the transitions in Fig. 5(a) are live, the newly added silent transitions are also live. The construction can remove deadlocks, as silent transitions may be enabled in an originally dead state. However, in such a state, non-silent transitions are blocked forever. It is possible to establish an explicit bisimulation relation between both nets [17,7]. However, since the relationship between the two nets is obvious we refrain from doing so.

4   Reset Workflow Nets

In the previous section, we considered arbitrary Petri nets without having an application in mind. However, when looking at workflows, we can make some assumptions about the structure of the Petri net. The idea of a workflow process is that many *cases* (also called *process instances*) are handled in a uniform manner. The workflow definition describes the ordering of *activities* to be executed for each case including a clear *start state* and *end state*. These basic assumptions lead to the notion of a *WorkFlow net* (WF-net) [1,2]. In the introduction, we already informally introduced the notion of WF-nets and now it is time to formalize this notion in the presence of reset arcs.

Definition 4 (RWF-net). *A reset net $N = (P, T, F, W, R)$ is a Reset Work-Flow net (RWF-net) if and only if*

- *There is a single source place i, i.e., $\{p \in P \mid \bullet p = [\ ]\} = \{i\}$.*
- *There is a single sink place o, i.e., $\{p \in P \mid p\bullet = [\ ]\} = \{o\}$.*
- *Every node is on a path from i to o, i.e., for any $n \in P \cup T$: $(i, n) \in F^*$ and $(n, o) \in F^*$ (where F^* is the transitive closure of F).*
- *There is no reset arc connected to the sink place, i.e., $\forall_{t \in T} \ o \notin R(t)$.*

Fig. 4 shows an RWF-net. Also the example used in the introduction (Fig. 3) is an RWF-net. The requirement that $\forall_{t \in T} \ o \notin R(t)$ has been added to emphasize that termination should be irreversible, i.e., it is not allowed to complete (put a token in o) and then undo this completion (remove the token from o).

5   Soundness

Based on the notion of RWF-nets, we now investigate the fundamental question: "Is the workflow correct?". If one has domain knowledge, this question can be answered in many different ways. However, without domain knowledge one can only resort to generic questions such as: "Does the workflow terminate?", "Are there any deadlocks?", "Is it possible to execute activity A?", etc. Such kinds of generic questions triggered the definition of *soundness* [1,2].

Definition 5 (Classical soundness [1,2]). *Let* $N = (P, T, F, W, R)$ *be an RWF-net. N is sound if and only if the following three requirements are satisfied:*

- *Option to complete:* $\forall_{M \in R(N,[i])} [o] \in R(N, M)$.
- *Proper completion:* $\forall_{M \in R(N,[i])} (M \geq [o]) \Rightarrow (M = [o])$.
- *No dead transitions:* $\forall_{t \in T} \exists_{M \in R(N,[i])} (N, M)[t\rangle$.

The RWF-nets depicted in figures 3 and 4 are sound.

The first requirement in Definition 5 states that starting from the initial state (just a token in place i), it is always possible to reach the state with one token in place o (state $[o]$). If we assume a strong notion of fairness, then the first requirement implies that eventually state $[o]$ is reached. Strong fairness, sometimes also referred to as "impartial" [22] or "recurrent" [23], means that in every infinite firing sequence, each transition fires infinitely often. Note that weaker notions of fairness are not sufficient, see for example Fig. 2 in [23]. However, such a fairness assumption is reasonable in the context of workflow management since all choices are made (implicitly or explicitly) by applications, humans or external actors. If we required termination without this assumption, all nets allowing loops in their execution sequences would be considered erroneous, which is clearly not desirable. The reason for mentioning fairness in this context is that for workflow verification we are forced to abstract from data and application logic by introducing non-deterministic choices as discussed in [2,3].

The second requirement states that at the moment a token is put in place o, all the other places should be unmarked. The last requirement states that there are no dead transitions (tasks) in the initial state $[i]$. By carefully looking at Definition 5 one can see that the second requirement is implied by the first one. Hence while analyzing we can ignore the second requirement in Definition 5. The reason that we include it anyway is because it represents an intuitive behavioral requirement.

As pointed out in [1,2], classical soundness of a WF-net without reset arcs corresponds to liveness and boundedness of an extension: the so-called short-circuited net. The short-circuited net is the Petri net obtained by connecting o to i, thus making the net cyclic. This result also holds for WF-net with arc weights. In fact, the construction shown in Fig. 5 preserves liveness and boundedness. Hence, also soundness is preserved by removing arc weights as shown in Fig. 5.

After the initial papers on soundness of WF-nets [1,2,3], many other papers have followed. Some extend the results while others explore alternative notions of soundness. These notions strengthen or weaken some of the requirements mentioned in Definition 5. Some examples are: k-soundness [19,20], weak soundness [25], up-to-k-soundness [30], generalized soundness [19,20], relaxed soundness [8,9], lazy soundness [28], and easy soundness [30].

A detailed discussion of these soundness notions is beyond the scope of this paper, see [5] for a complete overview. Nevertheless, we would like to define *relaxed soundness* as an example of an alternative soundness notion.

Definition 6 (Relaxed soundness [8,9]). *Let N be an RWF-net. N is relaxed sound if and only if for each transition $t \in T$:*
$$\exists_{M,M' \in R(N,[i])} \ (N,M)[t\rangle(N,M') \ \wedge \ [o] \in R(N,M').$$

Classical soundness considers all possible execution paths and if for one path the desired end state is not reachable, the net is not sound. In a way this implies that the workflow is "monkey proof", e.g., the user cannot select a path that will deadlock. The notion of relaxed soundness assumes a responsible user or environment, i.e., the net does not have to be "lunacy proof" as long as there exist "good" execution paths, i.e., for each transition there has to be at least one execution from the initial state to the desired final state that executes this transition.

6 Decidability of Soundness

In this section we explore the decidability of soundness in the presence of reset arcs. First, we show that classical soundness is undecidable, then we show that relaxed soundness is also undecidable for RWF-nets.

6.1 Classical Soundness Is Undecidable for RWF-Nets

In this subsection, we explore the decidability of soundness for RWF-nets. If a WF-net has no reset arcs, soundness is decidable. Such a WF-net $N = (P, T, F)$ (without reset arcs) is sound if and only if the short-circuited net $(\overline{N}, [i])$ with $\overline{N} = (P, T \cup \{t^*\}, F \cup \{(o, t^*), (t^*, i)\})$ and $t^* \notin T$ is live and bounded. Since liveness and boundedness are both decidable [14,15], soundness is also decidable. For some subclasses (e.g., free-choice nets), this is even decidable in polynomial time [1,2].

Unfortunately, soundness is not decidable for RWF-nets *with reset arcs* as is shown by the following theorem.

Theorem 3 (Undecidability of soundness). *Soundness is undecidable for RWF-nets.*

Proof. To prove undecidability of soundness, we use the fact that reachability is undecidable for reset nets [6,11].

Let (N, M_I) be an arbitrary marked reset net. Without loss of generality we can assume that N is connected and that every transition has input and output places. Note that by adding dummy self-loop places the behavior is not influenced. Moreover, since coverability is decidable for reset nets [6,11], we can assume that all dead transitions and dead places (i.e., places not connected to non-dead transitions) have been removed. (Because we can check whether $\bullet t$ is coverable from the initial marking, we can test whether transition t is dead for any $t \in T$.) Hence we may assume that (N, M_I) is connected, every transition has input and output places, and there are no dead transitions and no dead places.

Let M_X be a marking of N. To show that soundness is undecidable, we construct a new net $(N', [i])$ which embeds (N, M_I) such that N' is sound if and only if marking M_X is *NOT* reachable from (N, M_I). By doing so, we show that reachability in an arbitrary reset net can be analyzed through soundness, making soundness also undecidable.

The construction is shown in Fig. 6. To explain it we first need to introduce some notation. P is the set of places in N and T is the set of transitions in N. Assume $\{i, o, u, s, v, w\} \cap P = \emptyset$ and $(\{a, b, c, z\} \cup \{z_p \mid p \in P\}) \cap T = \emptyset$. These are the "fresh" identifiers corresponding to the places and transitions added to N to form N'. $I \subseteq P$ are all the places that are initially marked in (N, M_I) and $X \subseteq P$ are the places that are marked in (N, M_X). As Fig. 6 shows, transition c initializes the places in I, i.e., for $p \in I$: $W(c, p) = M_I(p)$.[2] Similarly, transition b can fire and consume all tokens from X if marking M_X is reached, i.e., for $p \in X$: $W(p, b) = M_X(p)$, and transition a marks the places in X appropriately, i.e., for $p \in X$: $W(a, p) = M_X(p)$. The transitions z and z_p ($p \in P$) have reset arcs from all places in N' except the new sink place o. Any transition in the original net has a bidirectional arc with s, i.e., a self-loop. All other connections are as shown in Fig. 6.

The constructed net $(N', [i])$ has the following behavior. First a fires, marking u, v and the places in X. No transition $t \in T$ can fire because s is still unmarked and c is also blocked because w is unmarked. The only two transitions that can fire are b and z. If z occurs, the net ends in marking $[o]$. If b fires, it will be followed by c. The firing of c brings the net into marking $M_I + [s, v]$. Note that in marking $M_I + [s, v]$ the original transitions are not constrained in any way and the embedded subnet can evolve as in (N, M_I) until one of the newly added transitions fires. Transitions $\{z_p \mid p \in P\}$ can fire as long as there is at least one token in a place in P and z can fire as long as there is a token in v. The firing of such a transition always leads to $[o]$, i.e., firing a transition in $\{z\} \cup \{z_p \mid p \in P\}$ always leads to the proper end state. Transition b can fire as soon as the embedded subnet has a marking which covers M_X.[3]

It is obvious that net N' shown in Fig. 6 is an RWF-net, i.e., there is one source place i, one sink place o, all nodes are on a path from i to o, and there

---

[2] Note that we are assuming weighted arcs here. However, as shown before these can be removed using the construction in Fig. 5.

[3] It is important to note that a first needs to enable b so that b occurs before the original net is initialized. Otherwise, the net is not sound even if M_X is not coverable (because b would be dead).

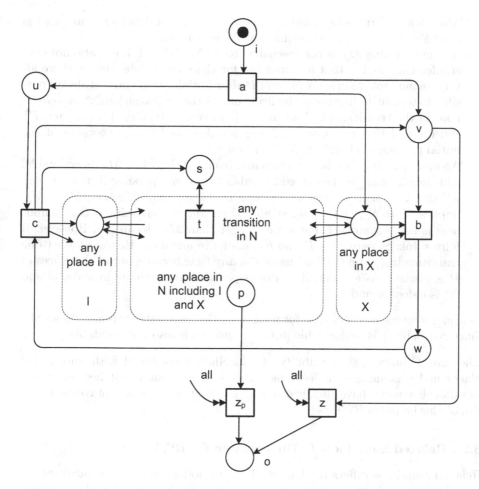

Fig. 6. Construction showing that soundness is undecidable for WF-nets with reset arcs. The original net comprises the three dashed areas: I is the set of places of N initially marked, X is the set of places that are marked in M_X, and all other nodes of N are shown in the dashed area in the middle. Note that I and X may overlap. The "all" annotations refer to all places including the newly added ones.

is no reset on o. Note that here we rely on the preprocessing of (N, M_I) making sure that the net is connected, that every transition has input and output places, and that there are no dead transitions and no dead places.

Now we can show that N' is sound if and only if the specified marking M_X is NOT reachable from (N, M_I):

– Assume marking M_X is reachable from (N, M_I). This implies that from $(N', [i])$ the marking $M_X + [s, v]$ is reachable. Hence b can fire for the second time resulting in a state $[s, w]$. In this state all transitions in T are blocked because transitions have input places and all input places in P are unmarked.

Also all added transitions are dead in $[s, w]$. Hence a deadlock state $[s, w]$ is reachable from $(N', [i])$ implying that N' is not sound.

- Assume marking M_X is not reachable from (N, M_I) and M_X is also not coverable. This implies that b cannot fire for the second time. Hence, there always remain tokens in some place of P after initialization and it is always possible to terminate in state $[o]$ by firing one of the "z transitions".[4] Moreover, none of the transitions is dead in $(N', [i])$ because $\{a, b, c, z\} \cup \{z_p \mid p \in P\}$ can fire and the transitions in T are not dead in (N, M_I) (because of the initial cleaning). Therefore, N' is indeed sound.
- Assume marking M_X is not reachable from (N, M_I) but M_X is coverable. This implies that in the embedded subnet it is only possible to reach states M' that are not covering M_X or that are bigger than M_X, i.e., $M' \geq M_X$ implies $M' \neq M_X$. For states smaller than M_X we have shown that soundness is not jeopardized. For states bigger than M_X, b can fire. However, if b fires, tokens remain in P and b cannot fire anymore. Hence, at least one transition in $\{z_p \mid p \in P\}$ is enabled at any time because one of the places in P is marked. As a result, it is always possible to terminate in state $[o]$ and N' is indeed sound.

Hence, if soundness is decidable for reset nets, then reachability is also decidable. Since reachability is undecidable [6,11], soundness is also not decidable.   □

Theorem 3 shows that the ability of cancellation combined with unbounded places makes soundness undecidable. This is a relevant result because many workflow languages have such features. (Note that soundness is of course undecidable for bounded RWF-nets.)

6.2   Relaxed Soundness Is Undecidable for RWF-Nets

Relaxed soundness differs fundamentally from notions such as classical soundness, because it allows for deadlocks, etc. as long as there is a "good execution" possible for each transition. Like classical soundness, relaxed soundness is decidable for WF-nets without reset arcs. Unfortunately, relaxed soundness is also undecidable for RWF-nets.

Theorem 4 (Undecidability of relaxed soundness). *Relaxed soundness is undecidable for RWF-nets.*

Proof. Let (N, M_I) be an arbitrary marked reset net. Without loss of generality we can again assume that N is connected and that every transition has input and output places. Add dummy self-loop places to ensure this. This does not change the behavior and there is a one-to-one correspondence between the states of the net with such places and the net without. Similarly, we can remove unconnected places.

---

[4] Note that we assume that I is not empty and that (by adding self loops) all transitions produce and consume tokens. If I is empty, then the net does not contain any non-dead transitions, so we can exclude this.

Let M_X be a marking of N. To show that relaxed soundness is undecidable, we construct an RWF-net $(N', [i])$ which embeds (N, M_I) such that N' is relaxed sound if and only if M_X is reachable from (N, M_I). By doing so, we show that reachability in an arbitrary reset net can be analyzed through relaxed soundness, making relaxed soundness undecidable because reachability is undecidable for reset nets [6,11].

Here we choose a fundamentally different strategy than in Theorem 3 where soundness corresponds to the *non*-reachability of a given marking M_X. Here, we make a construction such that relaxed soundness of N' corresponds to the reachability of M_X in (N, M_I).

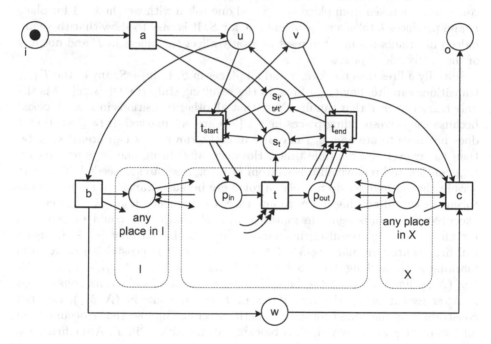

Fig. 7. Construction showing that reachability can be expressed in terms of relaxed soundness for WF-nets with reset arcs. (Note that I and X may overlap.)

Fig. 7 shows the basic idea underlying the construction of N' from N. P is the set of places in N and T is the set of transitions in N. $I \subseteq P$ is the set of places marked in M_I and $X \subseteq P$ is the set of places marked in M_X. Although it is not shown in Fig. 7, I and X may overlap. Let $T_{start} = \{t_{start} \mid t \in T\}$ and $T_{end} = \{t_{end} \mid t \in T\}$ be new transitions and let $S = \{s_t \mid t \in T\}$ be new places, i.e., for each $t \in T$ we add a self-loop place s_t and transitions t_{start} and t_{end}. Assume $(\{i, o, u, v, w\} \cup S) \cap P = \emptyset$ and $(\{a, b, c\} \cup T_{start} \cup T_{end}) \cap T = \emptyset$. For any t: $\bullet t_{start} = [u] + S$, $t_{start} \bullet = (\bullet t) + [s_t, v]$, $\bullet t_{end} = (t \bullet) + [s_t, v]$, and $t_{end} \bullet = [u] + S$. Also note that t_{end} resets all the places in N and that $s \in \bullet t \cap t \bullet$. As Fig. 7 shows, transition b initializes the places in I, i.e., for $p \in I$: $W(b, p) = M_I(p)$.

Similarly, transition c consumes all tokens from X if marking M_X is reached, i.e., for $p \in X$: $W(p, c) = M_X(p)$.

To better understand the structure of N' it is useful to observe the following place invariants: $i + u + v + w + o$ and $k.i + \sum_{t \in T} s_t + (k-1).v + k.o$ where $k = |T|$. The first invariant indicates that there will always be one token in exactly one of the places i, u, v, w, and o. The second invariant shows that there is a token in i (weight k), or there is a token in o (weight k), or there are tokens in $S \cup \{v\}$. In the latter case, there may be one token in v with weight $k-1$ and one token in one of the places in S with weight 1. So the sum of these two tokens is also k. Note that t_{start} consumes k tokens with weight one from S, returns one token to place $s_t \in S$, and puts a token with weight $k-1$ in place v. Transition t_{end} consumes one token from place $s_t \in S$ and one token with weight $k-1$ for place v, and produces k tokens with weight one for S. It is easy to show that these are indeed invariants because the reset arcs only affect the places in P and not any of the newly added places.

Initially a fires thus marking u and all places in S. In $[u] + S$, any of the T_{start} transitions can fire. Say t_{start} fires. In the resulting state $((\bullet t) + [s_t, v])$, t is the only transition in T that can fire. Note that all other transitions in T are blocked because the corresponding places in $S \setminus \{s_t\}$ are not marked. If $t\bullet \subseteq \bullet t$, then t does not have to fire and t_{end} may fire directly. However, t can fire. If $\bullet t \subseteq t\bullet$, then t may even fire multiple times. However, after firing one or more times t, t_{end} can fire and remove all tokens from $t\bullet$ using reset arcs if needed. The reset arcs in the original net do not play a role here because transition t removes the tokens in $\bullet t$ and nothing more. In any case, the sequence $\langle t_{start}, t, t_{end} \rangle$ can be executed and results again in marking $[u] + S$. Hence this could be repeated for all $t \in T$, still resulting in marking $[u] + S$. In marking $[u] + S$, also b can fire resulting in marking $M_I + S + [w]$. Hence, it is possible to move from marking $[i]$ to marking $M_I + S + [w]$ by firing $\sigma_b = \langle a, \dots, t_{start}, t, t_{end}, \dots, b \rangle$, i.e., $(N', [i])[\sigma_b\rangle(N', M_I + S + [w])$. Note that σ_b contains all transitions except c. After executing σ_b, the transitions in T can fire like in (N, M_I), i.e., not constrained by the added constructs, until c occurs. Suppose that c occurs, then all tokens in S are removed, thus blocking all transitions in T. After firing c a token is put into o and no transition can fire anymore.

Now we can show that N' is relaxed sound if and only if the specified marking M_X is reachable in (N, M_I):

- Assume marking M_X is reachable from (N, M_I). There exists a firing sequence σ_N such that $(N, M_I)[\sigma_N\rangle(N, M_X)$. This sequence is also enabled in the state after executing σ_b: $(N', M_I + S + [w])[\sigma_N\rangle(N', M_X + S + [w])$. Hence, $(N', [i])[\sigma_b\sigma_N c\rangle(N', [o])$ and it becomes clear that N' is indeed relaxed sound.
- Assume N' is relaxed sound. Hence there is a sequence σ: $(N', [i])[\sigma\rangle(N', [o])$. σ needs to have the following structure $\sigma_b = \langle a, \dots, b, \dots, c \rangle$ because in order to mark o, c must have been the last step and must have been preceded by b which in turn must have been preceded by a. Recall that $i + u + v + w + o$ is a place invariant illustrating the main control-flow in the net and the linear

dependencies between a, b and c. It is also clear that a, b, and c can fire only once. Just before firing c the marking must have been precisely $M_X + S + [w]$ because c does not have any reset arcs. Just after firing b the marking must have been $M_I + S + [w]$. Hence, there exists a firing sequence σ_N such that $(N', M_I + S + [w])[\sigma_N\rangle(N', M_X + S + [w])$. Note that in σ_N only transitions of T can be present ($T_{start} \cup T_{end}$ are dead after removing the token from u). Hence, σ_N is also enabled in the original net, i.e., $(N, M_I)[\sigma_N\rangle(N, M_X)$. Therefore, M_X must be reachable in (N, M_I) thus completing the proof.

$\square$

Theorems 3 and 4 show that two of the most obvious soundness properties are undecidable for RWF-nets. This illustrates that the cancelation feature present in the more powerful workflow notations may lead to analysis problems.

7   Conclusion

In this paper we explored decidability of soundness notions in the presence of cancelation. As a basic model, we used RWF-nets, i.e., workflow nets with reset arcs. As shown in Theorem 3, the classical notion of soundness becomes undecidable by adding reset arcs. Moreover, the weaker notion of relaxed soundness is also undecidable for RWF-nets (cf. Theorem 4). Interestingly, the strategies used to prove undecidability are very different for both soundness notions. Theorem 3 uses an ingenious construction where soundness corresponds to *non*-reachability while Theorem 4 uses a completely different construction where soundness corresponds to reachability.

In a technical report [5] we also show that most other notions of soundness are undecidable for RWF-nets. Of the many soundness notions described in literature only generalized soundness *may* be decidable (this is still an open problem). All other notions are shown to be undecidable. Besides the open problem of deciding generalized soundness for RWF-nets, there are several other research questions:

- *Is soundness decidable for RWF-nets with certain restrictions?* Here one could use insights such as the fact that boundedness is undecidable for reset nets with three reset arcs but still decidable for reset nets with two resetable places [12].
- *How are the various soundness notions related and can one notion be translated the other?* Some initial answers to this question have already been given in [5].
- *What is the exact relationship between soundness and reachability?* Results so far suggest that soundness is decidable if reachability is decidable. However, it is not clear whether this holds for all kinds of workflow models (e.g., restricted forms of inhibitors, resets, or priorities). Moreover, even if both are decidable there is the question of computational complexity.

We expect that our decidability results are useful for researchers working on workflow verification. The results provide insights into the boundaries of workflow verification. We would like to stress that undecidability does not make things hopeless. Soundness can be checked for RWF-nets with a finite state space and even if the state space is infinite a partial exploration of the state space may reveal various errors. Moreover, many errors can be discovered using techniques such as invariants and reduction rules [27,32,33,36]. Motivated by the findings in [27], we are planning more empirical studies on workflow verification.

Acknowledgments. We thank the reviewers for their detailed comments and suggestions which helped to improve the quality and readability of the paper.

References

1. van der Aalst, W.M.P.: Verification of Workflow Nets. In: Azéma, P., Balbo, G. (eds.) ICATPN 1997. LNCS, vol. 1248, pp. 407–426. Springer, Heidelberg (1997)
2. van der Aalst, W.M.P.: The Application of Petri Nets to Workflow Management. The Journal of Circuits, Systems and Computers 8(1), 21–66 (1998)
3. van der Aalst, W.M.P.: Workflow Verification: Finding Control-Flow Errors using Petri-net-based Techniques. In: van der Aalst, W.M.P., Desel, J., Oberweis, A. (eds.) Business Process Management. LNCS, vol. 1806, pp. 161–183. Springer, Heidelberg (2000)
4. van der Aalst, W.M.P., van Hee, K.M.: Workflow Management: Models, Methods, and Systems. MIT Press, Cambridge (2004)
5. van der Aalst, W.M.P., van Hee, K.M., ter Hofstede, A.H.M., Sidorova, N., Verbeek, H.M.W., Voorhoeve, M., Wynn, M.T.: Soundness of Workflow Nets: Classification, Decidability, and Analysis. Computer Science Report No. 08-13, Technische Universiteit Eindhoven, The Netherlands (2008)
6. Araki, T., Kasami, T.: Some Decision Problems Related to the Reachability Problem for Petri Nets. Theoretical Computer Science 3(1), 85–104 (1977)
7. Basten, T.: Branching Bisimilarity is an Equivalence indeed! Information Processing Letters 58(3), 141–147 (1996)
8. Dehnert, J., van der Aalst, W.M.P.: Bridging the Gap Between Business Models and Workflow Specifications. International Journal of Cooperative Information Systems 13(3), 289–332 (2004)
9. Dehnert, J., Rittgen, P.: Relaxed Soundness of Business Processes. In: Dittrich, K.R., Geppert, A., Norrie, M.C. (eds.) CAiSE 2001. LNCS, vol. 2068, pp. 157–170. Springer, Heidelberg (2001)
10. Desel, J., Esparza, J.: Free Choice Petri Nets. Cambridge Tracts in Theoretical Computer Science, vol. 40. Cambridge University Press, Cambridge (1995)
11. Dufourd, C., Finkel, A., Schnoebelen, P.: Reset Nets Between Decidability and Undecidability. In: Larsen, K.G., Skyum, S., Winskel, G. (eds.) ICALP 1998. LNCS, vol. 1443, pp. 103–115. Springer, Heidelberg (1998)
12. Dufourd, C., Jančar, P., Schnoebelen, P.: Boundedness of Reset P/T Nets. In: Wiedermann, J., Van Emde Boas, P., Nielsen, M. (eds.) ICALP 1999. LNCS, vol. 1644, pp. 301–310. Springer, Heidelberg (1999)
13. Dumas, M., van der Aalst, W.M.P., ter Hofstede, A.H.M.: Process-Aware Information Systems: Bridging People and Software through Process Technology. Wiley & Sons, Chichester (2005)

14. Esparza, J.: Decidability and Complexity of Petri Net Problems: An Introduction. In: Reisig, W., Rozenberg, G. (eds.) APN 1998. LNCS, vol. 1491, pp. 374–428. Springer, Heidelberg (1998)

15. Esparza, J., Nielsen, M.: Decidability Issues for Petri Nets: A Survey. Journal of Information Processing and Cybernetics 30, 143–160 (1994)

16. Finkel, A., Schnoebelen, P.: Well-structured Transition Systems everywhere! Theoretical Computer Science 256(1–2), 63–92 (2001)

17. van Glabbeek, R.J., Weijland, W.P.: Branching Time and Abstraction in Bisimulation Semantics. Journal of the ACM 43(3), 555–600 (1996)

18. Hack, M.H.T.: The Equality Problem for Vector Addition Systems is Undecidable. Theoretical Computer Science 2, 77–95 (1976)

19. van Hee, K.M., Sidorova, N., Voorhoeve, M.: Soundness and Separability of Workflow Nets in the Stepwise Refinement Approach. In: van der Aalst, W.M.P., Best, E. (eds.) ICATPN 2003. LNCS, vol. 2679, pp. 337–356. Springer, Heidelberg (2003)

20. van Hee, K.M., Sidorova, N., Voorhoeve, M.: Generalised Soundness of Workflow Nets Is Decidable. In: Cortadella, J., Reisig, W. (eds.) ICATPN 2004. LNCS, vol. 3099, pp. 197–215. Springer, Heidelberg (2004)

21. Jensen, K.: Coloured Petri Nets. Basic Concepts, Analysis Methods and Practical Use. EATCS monographs on Theoretical Computer Science, vol. 1. Springer, Berlin (1997)

22. Jensen, K.: Coloured Petri Nets. Basic Concepts, Analysis Methods and Practical Use. Analysis Methods. Monographs in Theoretical Computer Science, vol. 2. Springer, Berlin (1997)

23. Kindler, E., van der Aalst, W.M.P.: Liveness, Fairness, and Recurrence. Information Processing Letters 70(6), 269–274 (1999)

24. Leymann, F., Roller, D.: Production Workflow: Concepts and Techniques. Prentice-Hall PTR, Upper Saddle River (1999)

25. Martens, A.: Analyzing Web Service Based Business Processes. In: Cerioli, M. (ed.) FASE 2005. LNCS, vol. 3442, pp. 19–33. Springer, Heidelberg (2005)

26. Mayr, E.W.: An Algorithm for the General Petri Net Reachability Problem. In: STOC 1981: Proceedings of the thirteenth annual ACM symposium on Theory of computing, pp. 238–246. ACM, New York (1981)

27. Mendling, J., Neumann, G., van der Aalst, W.M.P.: Understanding the Occurrence of Errors in Process Models Based on Metrics. In: Meersman, R., Tari, Z. (eds.) OTM 2007, Part I. LNCS, vol. 4803, pp. 113–130. Springer, Heidelberg (2007)

28. Puhlmann, F., Weske, M.: Investigations on Soundness Regarding Lazy Activities. In: Dustdar, S., Fiadeiro, J.L., Sheth, A.P. (eds.) BPM 2006. LNCS, vol. 4102, pp. 145–160. Springer, Heidelberg (2006)

29. Reisig, W., Rozenberg, G. (eds.): Lectures on Petri Nets I: Basic Models. LNCS, vol. 1491. Springer, Heidelberg (1998)

30. van der Toorn, R.: Component-Based Software Design with Petri nets: An Approach Based on Inheritance of Behavior. PhD thesis, Eindhoven University of Technology, Eindhoven, The Netherlands (2004)

31. Valk, R.: Self-Modifying Nets: A Natural Extension of Petri Nets. In: Ausiello, G., Böhm, C. (eds.) ICALP 1978. LNCS, vol. 62, pp. 464–476. Springer, Heidelberg (1978)

32. Verbeek, H.M.W., van der Aalst, W.M.P., ter Hofstede, A.H.M.: Verifying Workflows with Cancellation Regions and OR-joins: An Approach Based on Relaxed Soundness and Invariants. The Computer Journal 50(3), 294–314 (2007)

33. Verbeek, H.M.W., Wynn, M.T., van der Aalst, W.M.P., ter Hofstede, A.H.M.: Reduction Rules for Reset/Inhibitor Nets. BPM Center Report BPM-07-13, BPM-center.org (2007)
34. Weske, M.: Business Process Management: Concepts, Languages, Architectures. Springer, Berlin (2007)
35. Workflow Patterns Home Page, http://www.workflowpatterns.com
36. Wynn, M.T., van der Aalst, W.M.P., ter Hofstede, A.H.M., Edmond, D.: Verifying Workflows with Cancellation Regions and OR-joins: An Approach Based on Reset Nets and Reachability Analysis. In: Dustdar, S., Fiadeiro, J.L., Sheth, A.P. (eds.) BPM 2006. LNCS, vol. 4102, pp. 389–394. Springer, Heidelberg (2006)
37. Wynn, M.T., Verbeek, H.M.W., van der Aalst, W.M.P., ter Hofstede, A.H.M., Edmond, D.: Business Process Verification: Finally a Reality! Business Process Management Journal 15(1), 74–92 (2009)
38. Wynn, M.T., Verbeek, H.M.W., van der Aalst, W.M.P., ter Hofstede, A.H.M., Edmond, D.: Soundness-preserving Reduction Rules for Reset Workflow Nets. Information Sciences 179(6), 769–790 (2009)

Parameterised Coloured Petri Net Channel Models*

Jonathan Billington[1], Somsak Vanit-Anunchai[2], and Guy E. Gallasch[1]

[1] Computer Systems Engineering Centre
University of South Australia
Mawson Lakes Campus, SA 5095, Australia
{jonathan.billington,guy.gallasch}@unisa.edu.au
[2] School of Telecommunication Engineering
Institute of Engineering, Suranaree University of Technology
Muang, Nakhon Ratchasima 30000, Thailand
somsav@sut.ac.th

Abstract. Computer protocols operate over communication channels with varying properties. Channel characteristics include order of delivery (e.g. first-in first-out (FIFO)), whether it is lossy or not, and capacity. Important Internet protocols, such as the Transmission Control Protocol (TCP) and the Datagram Congestion Control Protocol (DCCP), are designed to operate over channels which can duplicate, reorder and lose packets. It is important to be able to analyse protocol behaviour incrementally, to ensure that channel imperfections, such as reordering or loss, do not hide protocol errors such as unspecified receptions. In order to analyse protocols progressively using FIFO channels without (and then with) loss and then reordering channels without (and then with) loss, this paper proposes Coloured Petri Net (CPN) models which combine the FIFO and reordering behaviour (with or without loss) in a way that reduces maintenance effort as the protocol evolves, and facilitates analysis. The model also includes an arbitrary channel capacity limit, including no limit. We firstly present a simple parameterised model. Unfortunately, this model is not able to be simulated by current CPN support tools. We then modify the model to allow it to be executed in tools such as CPN Tools. The paper discusses the way in which the parameterised channel model is embedded into a protocol specification. Having a combined FIFO/reordering and possibly lossy model was useful in the case of DCCP as we tracked its development by the Internet Engineering Task Force. This paper discusses selected results of the incremental analysis of DCCP's connection management procedures, exploiting the parameterised channel model.

Keywords: Parameterized Protocol Channel Models, Coloured Petri Nets, Datagram Congestion Control Protocol, Reachability Analysis.

* Partially supported by Australian Research Council Discovery Grant DP 0559927.

K. Jensen, J. Billington, and M. Koutny (Eds.): ToPNoC III, LNCS 5800, pp. 71–97, 2009.

1   Introduction

According to Holzmann [14], protocol specifications comprise five elements: the service the protocol provides to its users; the set of messages that are exchanged between protocol entities; the format of each message; the rules governing message exchange (procedures); and the assumptions about the environment in which the protocol is intended to operate. In protocol standards documents, information related to the operating environment is usually written informally and may occur in several different places [37]. This informal specification style can lead to misunderstandings and possibly incompatible implementations. In contrast, executable formal models require precise specifications of the operating environment. Of particular significance is the communication medium or *channel* over which the protocol operates.

Channels can have different characteristics depending on the physical media (e.g. optical fibre, copper, cable or unguided media (radio)) they employ. The characteristics also depend on the level of the protocol in a computer protocol architecture. For example, the link-level operates over a single medium, whereas the network, transport and application levels may operate over a network, or network of networks such as the Internet, which could employ several different physical media. Channels (such as satellite links) can be noisy resulting in bit errors in packets. To correct bit errors in packets, many important protocols (such the Internet's Transmission Control Protocol [27]) use Cyclic Redundancy Checks (CRCs) [28] to detect errors. On detecting an error, the receiver discards the packet and relies on the sender to retransmit it for recovery, known as Automatic Repeat reQuest (ARQ) [28]. This is achieved by the receiver acknowledging the receipt of *good* packets, and by the transmitter maintaining a timer. When the timer expires before an acknowledgement has been received, the transmitter retransmits packets that have been sent but are as yet not acknowledged. It may also be possible for packets to be lost due to routers in networks discarding packets when congested. Thus at the transport level, the discarding of packets due to transmission errors (caused by noise) or due to network congestion can be lumped together and viewed as a lossy medium. When we are considering transmission errors and congestion at this level of abstraction, packet loss can be considered to occur non-deterministically.

Although packets will maintain their order of transmission on a link, this is not necessarily the case when they are routed through networks. Some routing algorithms allow packets to take different paths with different delays. This is due to some routers being lightly loaded, while others can be congested and to different speeds being used on different links in the network. Thus although packets will normally arrive in the order in which they are transmitted, this cannot be guaranteed in some networks, such as the Internet service provided at the network level (i.e. the Internet Protocol (IP)).

The time-out mechanism used for retransmitting packets in ARQ systems can lead to duplicates, as expiry of the timer cannot guarantee that a packet has been lost. It could be that the packet or its acknowledgement has been delayed, or that the acknowledgement has been lost. It may also be possible for routers to crash and, on recovery, to have lost track of which packets they have transmitted. This may also result in duplicates. For transport level protocols, such as TCP, that handle duplicates due to the ARQ mechanism, there seems little point to also consider duplication by the channel. The source of the duplication (the transmitter or a router) is not visible to the receiver, and hence only one source of duplicates is necessary to verify the protocol's operation. Finally, bit errors in addresses could cause a packet to be misdelivered at the network level. Misdelivery is handled at the network layer (e.g. by IP) by protecting the packet headers with CRC checks. Hence we do not consider that misdelivery is a property of a channel at the transport level. Thus the Internet's IP service can be considered as a reordering lossy channel.

Many protocols, including some important Internet protocols, such as TCP [27] and the Datagram Congestion Control Protocol (DCCP) [19], are designed to operate over reordering lossy channels. When analysing these protocols, we advocate [3] an incremental approach where the channel is modelled successively as First-In First-Out (FIFO), lossy FIFO, reordering and lossy reordering. This is important because loss or reordering could mask out errors such as *unspecified receptions* i.e. packets that arrive in a state where no procedure is defined for processing them. Unspecified receptions will be detected as deadlocks in the FIFO case but may go unnoticed if a lossy channel is used, because the packet that would not be received by a protocol entity (and hence would remain in the channel) will be discarded by the non-deterministic loss mechanism. It may also be received in another state that accepts the packet after undergoing reordering. Moreover, analysing the protocol over a FIFO medium first may allow detection of errors more readily (for example using state space techniques) and provide further insight into the operation of the protocol. This is because a protocol that is designed to operate over reordering channels must work over FIFO channels, and that FIFO is the usual mode of operation for most communication channels as it is only in relatively rare circumstances that reordering or loss occurs.

There are numerous recent examples of modelling and analysing protocols [26, 23, 22, 12, 13] using Coloured Petri Nets (CPNs) [15, 16]. We observe that all models assumed either a FIFO or reordering channel, but did not consider both. One factor why only one type of channel is considered is that there are significant differences between the CPN structures of reordering and FIFO channel models. At first glance, this requires two models to be built and maintained: reordering and FIFO. Maintaining two models (instead of one) doubles the amount of time required to update the model, which is a significant task when tracking the development of a new protocol. Discrepancies between the two models can also occur as changes are made. To overcome these difficulties, this paper[1] proposes a parameterised model that combines the reordering and FIFO channel models into a single net structure for ease of model maintenance and to facilitate incremental analysis. We discuss a relatively simple model that uses two parameters: one, called FIFO, that can be used to select FIFO or reordering behaviour; and another, lossy, to optionally select lossy or lossless behaviour. This is further

---

[1] This paper is a major expansion and refinement of the initial work [35].

enhanced by including another parameter (called limit) that allows the channel capacity to be unbounded, or limited to any finite size. These parameters are easily input to the model, without requiring any change to the model's structure or inscriptions. We also discuss how this simple model is not able to be executed by two tools that support CPNs, and develop a more complex model that can be executed by these tools. The suggested pattern of a combined FIFO/reordering CPN model may assist those who follow the incremental approach suggested in [3] to build and analyse a complex protocol model using CPNs. Previous work on CPN patterns is reported in [25]. It considers various queues and capacity which are well known in the literature (e.g. [1]), but it does not provide a pattern for lossy channels nor does it combine models with different ordering behaviour, using a parameterised model. Hence we believe the parameterised models considered here for combining channels with loss and capacity are new, and provide a basis for protocol specification that supports incremental analysis. We illustrate the use of the parameterised channel model by including a CPN model of DCCP's connection management procedures. Our previously published CPN model of DCCP connection management (DCCP-CM) [34] only operates over a reordering channel without loss. By extending the CPN model from [34] with the proposed combined channel pattern, we easily obtain analysis results when DCCP operates over the four different channel types. The results for the lossy channels (and FIFO without loss) are new.

The rest of this paper is organised as follows. Section 2 describes our attempts to obtain a combined FIFO/reordering channel model that includes loss and capacity parameters. The way to embed such a model within a protocol specification is then discussed in section 3. We introduce DCCP in section 4 and discuss differences between the CPN model of DCCP connection management in this paper and in [34]. We also discuss the analysis results obtained for scenarios when the protocol operates over FIFO without loss; FIFO with loss; reordering without loss; and reordering with loss. We provide some conclusions in section 5. Although presented in a tutorial style, the paper assumes some familiarity with CPNs [15, 16].

2   Parameterised Channel Models

This section firstly introduces CPN models of reordering and FIFO channels that are possibly lossy. It then discusses how to combine these models and shows how the combined model can be enhanced with a capacity limit. We then discuss implementing these models in tools such as CPN Tools [7].

2.1   CPN Model of a Reordering and Lossy Channel

A reordering channel is easily modelled by a high-level net place and is well known in the literature. Figure 1 illustrates a CPN model of a sender transmitting messages to a receiver via a reordering and possibly lossy channel. The model comprises one place, Channel, representing the channel from the sender

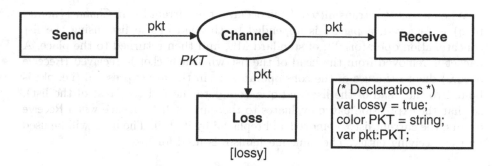

Fig. 1. CPN model of a reordering and possibly lossy channel

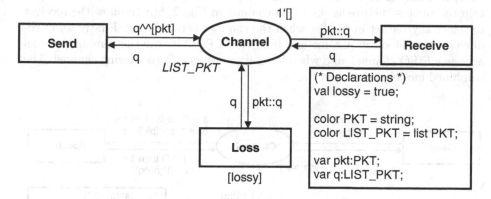

Fig. 2. A CPN model of a possibly lossy FIFO channel

to the receiver, and three transitions (Send, Loss and Receive). Channel stores in-transit packets and is typed by *PKT* which for simplicity is the set of strings (see the Declarations). Channel is initially empty. The channel is reordering because packets in the channel can be accepted by Receive in any order. Loss models the loss of any packet in the channel. Loss can only occur when the Boolean parameter, lossy, is true (see the Declarations). Hence the channel can be switched from lossy to non-lossy by changing the value of lossy.

2.2 CPN Model of a FIFO Channel

Modelling FIFO channels with CPNs is also very well known [1, 15]. It is considerably more complex than the reordering channel, requiring additional arcs with more complex arc inscriptions using the list data structure and its associated functions (concatenation and cons). Figure 2 depicts the same sender and receiver operating over a FIFO channel. The type of the channel place, Channel, is now a list of packets, *LIST_PKT*, and the place is initially marked by the empty list, []. Variable, q, of type *LIST_PKT*, is introduced to facilitate queue operations.

When a packet is transmitted (Send occurs), the current list in Channel (bound to q) is removed, a packet is appended to the end of the list, using the list concatenation operator, $^{\wedge\wedge}$, of standard ML, and then returned to the place. A packet is removed from the head of the list when a packet is received (Receive occurs) due to the use of the cons operator, ::, in the arc expression (i.e. pkt is bound to the head of the list, and q is bound to the tail (the rest of the list), so that the expression pkt::q evaluates to the current list). Hence when Receive occurs, the current list is removed and replaced by its tail. The head will be used by the receiving entity. The same mechanism is used for loss.

2.3   Modelling a Combined FIFO/Reordering Channel

Simple Combined Model. A simple way to combine these models is to maintain the same structure as the FIFO channel in Fig. 2, but to allow the receiver to select any packet in the list when the channel is reordering. Firstly we introduce another Boolean parameter, FIFO. When FIFO is true the combined model acts as a FIFO channel, and when false, it operates as a reordering channel. The combined model is shown in Fig. 3.

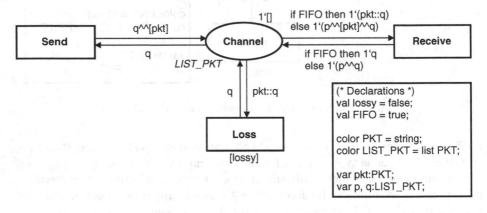

Fig. 3. Simple CPN model combining FIFO and reordering channels

The inscriptions on the arcs between Channel and Receive are if-then-else statements, which reduce to the FIFO model when FIFO is true. When FIFO is false, Receive is enabled if there is a packet in the channel. We now use two list variables, p and q. For a given packet (bound to pkt) positioned anywhere in the list, we bind p to the portion of the list from its head up to pkt, and q to the rest of the list (not including pkt). For example, if there was only one packet in the channel, then both p and q would be bound to the empty list. When Receive occurs, the packet is removed from the list, by returning $p^{\wedge\wedge}q$ to the channel.

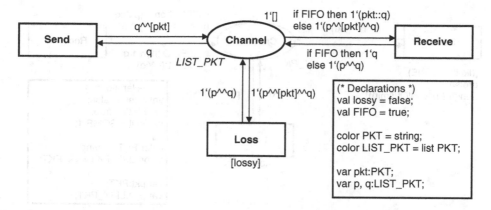

Fig. 4. Simple CPN model combining FIFO and reordering channels: loss anywhere in the queue

Simple Combined Model with Loss from Anywhere in the List. A refinement of this model (see Fig. 4) allows any packet in the channel to be lost, not just the one at the head of the list. To do this we make the following changes to Fig. 3:

1. the inscription on the arc from Channel to Loss becomes $p^{\wedge\wedge}[pkt]^{\wedge\wedge}q$, so that a packet anywhere in the channel can be selected to be lost; and
2. the inscription on the arc from Loss to Channel becomes $p^{\wedge\wedge}q$ so that pkt is lost when Loss occurs.

Simple Combined Model Including Channel Capacity. Our final refinement to this model is to introduce a capacity limit on the channel size as shown in Fig. 5. This is easily incorporated by including a parameter, limit, which limits the number of packets in the channel to the value of limit. We use the ML length of a list function, length, in a guard to prevent the number of packets in the channel from exceeding its limit. This is implemented by including a guard on the Send transition (as it is the only transition that adds items to the list).

We have declared limit as an optional integer in ML, using the keyword SOME. Optional integers are integers together with the constant, NONE. This allows there to be no limit on the size of the list. If there is no limit, then limit = NONE, and the guard is true. If there is a limit (limit<>NONE) then the length of the list in the channel (length(q)) must be less than the (value of the) limit. This allows the capacity of the channel to be unbounded or any finite size, depending on the value of limit defined in the declarations.

Implementing the Simple Combined Model in Tools. The specification, analysis and maintenance of complex practical protocols, such as TCP [5] and DCCP [6], require the use of computer aided tools. The most sophisticated tools for Coloured Petri Nets, currently, are CPN Tools [7, 16] and its predecessor,

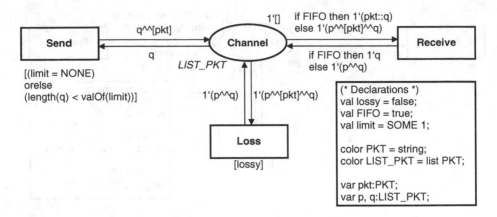

Fig. 5. Simple CPN model combining FIFO and reordering channels, with optional loss and capacity limit

Design/CPN [8]. Although the nets in Figs. 3 – 5 are syntactically correct, they produce errors when attempting to use the simulator or Occurrence Graph tool in Design/CPN or using the state space tool of CPN Tools. CPN Tools will not simulate these nets. The errors are caused by *unbounded non-determinism*. The tools have too many choices to make when trying to bind the variables pkt, q or p. Even if one limits the colour set *PKT* to small integers, such as {1,2}, and alters *LIST_PKT* to lists of length 2, neither CPN Tools nor Design/CPN will handle constructs such as $p^{\wedge\wedge}[pkt]^{\wedge\wedge}q$. The model is thus not (reliably) executable in these tools and therefore cannot be analysed by them using simulation or state spaces. This is a known limitation of these tools (see the reference manual of [8] or the help pages of [7]). In general, variables can only appear as arguments to functions in arc inscriptions if they are bound somewhere else, i.e. on another arc or in a guard.

In an attempt to overcome this problem, we can consider removing the queue from Channel, using q on the arcs to transitions Receive and Loss, and returning it after having removed an item from the queue using a function, as shown in Fig. 6. In the case of a reordering medium, both Receive and Loss will remove an arbitrarily chosen item from the queue. In the case of FIFO operation, Receive will remove the first item of the queue whereas Loss will still arbitrarily choose the packet to lose. This is similar to our work on removing the first urgent data item in the queue of TCP's Data Transfer Service (see Fig. 2 in [4]).

Note that we need to add the restriction, pkt=ExtractItem(q,i), to the guard, to extract the ith item from the queue so that it is accessible to the receiver. This restriction is not required for loss, as the value of the lost packet is not needed. Also, to allow for any sized queue, i needs to range over the positive integers.

Unfortunately this attempt has two serious problems. Due to unbounded (or too large) non-determinism, both i and pkt produce syntax errors and will not work if their types are 'large' colour sets (both the positive integers and strings are large colour sets). If we restrict both i and pkt to 'small' colour sets, the model

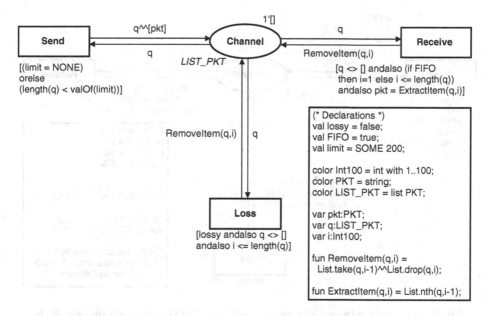

Fig. 6. Combined FIFO and Reordering Channel using a Function to remove its ith Item

runs without syntax errors, but is not correct when the queue length exceeds the maximum value of i. For example, assume the type of i is 'int with 1..100' as shown in the Declarations of Fig. 6. When the queue length grows to more than 100, we can only receive or lose items from the first 100 items in the queue. The remaining items cannot be lost or received. This means that the implementation is not correct in general. Further, with practical protocols such as TCP and DCCP, which have 32 bit or 48 bit sequence numbers, we cannot restrict pkt to a small colour set. This is because their packets contain these large sequence numbers. Hence Fig. 6 does not provide a solution for the practical systems we are trying to model. Further, we feel that having the description of the packet in a guard, instead of on the arc from the queue, is less intuitive, as one usually expects the packet being received to be on the input arc.

A Tool-Friendly Combined Model. We now present a workaround which overcomes the inability of CPN Tools (or Design/CPN) to execute the combined models presented above. This model is depicted in Fig. 7. It combines reordering and FIFO channels when loss is only from the head of the FIFO queue. The channel capacity is also included as a parameter (limit) and is allowed to be unlimited as previously (using NONE). As before, if FIFO = true then the channel will be FIFO, otherwise it operates as a reordering channel.

The net structure of this model is similar to the simple model (Fig. 5), but the inscriptions are more complex. The main idea has been to separate the FIFO and reordering data structures. For FIFO, there is just one list which stores all the

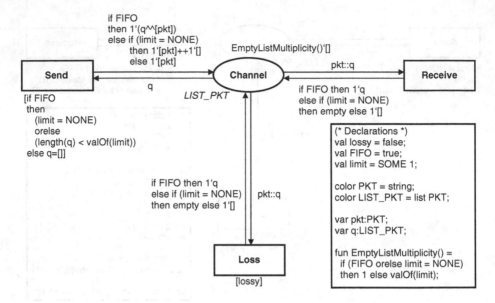

Fig. 7. A tool-executable CPN model combining FIFO and reordering channels

packets in the channel. For reordering, there is one list per packet in the channel. Capacity is implemented in the same way as that for the simple model for a FIFO channel (just limiting the length of the list). However, it is more complex for a reordering channel. In this case, the initial marking of place Channel contains as many empty lists as is required by the capacity limit, i.e. the number of empty lists equals the limit. Then each time Send occurs, an empty list is replaced by a singleton list of the packet inserted into the channel. The reordering channel now contains a singleton list (comprising a packet) for each packet inserted, instead of just the packets themselves, as in Fig. 1, or just one list of packets as in Fig. 5.

The initial marking of the channel is the empty list for the FIFO case and for the case when there is no capacity limit. However, as discussed above, when there is a capacity limit in the reordering case, the number of empty lists equals the limit. This is captured in the EmptyListMultiplicity function shown in the Declarations.

The reordering channel operates slightly differently when there is no capacity limit. This is captured in the inscription on the arc from Send to Channel. When there is no limit, the empty list is removed from *and* returned to Channel (together with the singleton list containing the packet being sent) on the occurrence of Send. This is to allow the next packet to be sent. The guard on Send guarantees that in the case of a reordering channel (FIFO is false), that q is always bound to the empty list (q = []). Returning the empty list thus allows Send to be enabled (if its input is ready). A similar situation occurs when a packet is received or lost. When there is no capacity limit, nothing is returned to the (reordering) channel, but when there is a limit, the empty list is returned, allowing another packet to be inserted into the channel.

3   Embedding the Channel Model

This section discusses two approaches to embedding the channel model into a CPN model of a protocol specification that we wish to analyse.

3.1   Embedding the Channel Model as a Module

A simple and elegant way of embedding the channel model in the protocol CPN model is to consider it as a module implemented by a substitution transition in the case of hierarchical CPNs [15]. Each protocol entity will have an output buffer that stores packets (in FIFO order) waiting to be sent to the channel (and onto the peer entity), and an input buffer (FIFO) that accepts packets from the reverse channel, and stores them waiting for the protocol entity to process them. Two channel modules (substitution transition instances) are then inserted: one between the output buffer of the local entity and the input buffer of the peer entity; and the other between the output buffer of the peer entity and input buffer of the local entity.

Although readily understood from the specification perspective, this approach suffers from significant and unnecessary state explosion when considering the utility of the model from the analysis perspective. This is because packets are stored in 3 places: the output buffer of the sender; the channel; and the input buffer of the receiver. This is compounded by the fact that it occurs for both directions of information flow, when considering a two party system. This duplication of storage is unnecessary because we can consider all storage between the sender and receiver to be in the channel.

We have done some experiments on using this modular approach with the Stop-and-Wait Protocol (see e.g. [2] for a non-modular CPN model). We consider the most conservative case where the buffers are limited to one message for the modular approach. We find that in the FIFO case, the explosion depends on the number of retransmissions. When the channel is not limited and the maximum number of retransmissions (MR) is 6, then the modular approach has over 11 times as many states (45168 versus 3920) as the integrated approach described below. For the reordering channel, the situation is worse. For example, using a channel capacity of 2, a maximum sequence number of 1 (alternating bit protocol) and MR=2, there are over 50 times as many states in the modular approach (42516 versus 844) compared with the integrated approach.

3.2   Integrating the Channel Model with the Protocol Entities

When using reachability analysis to explore properties of the model we advocate using an integrated model, rather than the modular approach above. As indicated above, we consider that all storage of packets between the protocol entities is provided in the Channel place. We then create transitions for input to the channel (and output from the channel) following the approach indicated in Fig. 7. We illustrate how this is done using the Datagram Congestion Control Protocol in the next section, where we assume the capacity of the channel is unbounded.

4   A Parameterised CPN Model of DCCP Connection Management

4.1   Introduction to the Datagram Congestion Control Protocol

Overview. The Datagram Congestion Control Protocol [17, 19] is a point-to-point transport protocol operating over the Internet between two DCCP entities, the Client and Server. It provides a bidirectional flow of data for applications, such as voice and video, that prefer timeliness to reliability. DCCP is designed to provide congestion control for these applications [9]. Its congestion control algorithms require statistics on packet loss because loss is related to the level of congestion in the network. DCCP uses sequence and acknowledgement numbers in packets to detect and report loss, and includes state variables in each protocol entity to keep track of these numbers. State variables on both sides must be synchronised, otherwise DCCP may misinterpret loss information. Thus DCCP needs mechanisms to set up, synchronise and clear state variables in both the Client and Server. We shall refer to these mechanisms in general as connection management procedures.

Connection Management Procedures. The procedures involve the exchange of packets between the Client and Server. DCCP defines 10 different packet types for this purpose: Request, Response, Data, DataAck, Ack, CloseReq, Close, Reset, Sync, and SyncAck. Figure 8 depicts a state diagram illustrating DCCP's connection establishment and release procedures for both the Client and Server. Ellipses in Fig. 8 represent states while arrows represent state transitions. There are nine states where CLOSED is both an initial and a final state. The inscription on each arrow describes the input and output actions, if any. For instance, the inscription on the arc from LISTEN to RESPOND is "rcv Request snd Response". This means that when the Server receives a DCCP-Request from the Client while in the LISTEN state, it returns a DCCP-Response and moves to the RESPOND state. The Server always passes through the LISTEN state after a "passive open" command from its application, whereas the Client is identified by an "active open" from its application and passes through the REQUEST state. Applications on both sides can issue an "active close" command but only the Server's application can issue the "server active close" command. For more details of these procedures, see [29, 19].

4.2   CPN Model of DCCP Connection Management

Previous Work. We have followed the development of DCCP since Internet Draft version 5 [20] was released in 2003. We used Design/CPN [8] to construct, maintain and analyse CPN models of DCCP's connection management procedures in order to detect errors or deficiencies in the protocol. Analysis of the connection establishment procedures of version 5 discovered a deadlock [31]. When version 6 [18] was released, we revised our CPN model to reflect the changes

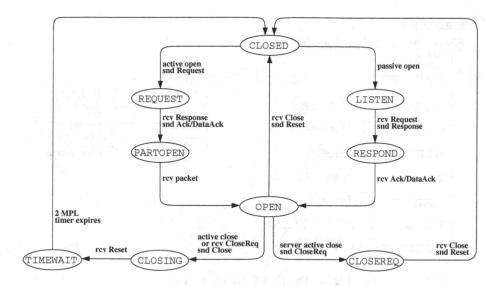

Fig. 8. DCCP state diagram (redrawn from [19])

and found similar deadlocks [21]. Version 11 of the DCCP specification was submitted to IETF for approval as a standard. We updated our model accordingly and included the synchronisation mechanism. We found that the deadlocks were removed but we discovered a serious problem known as *chatter*[2] [36, 32]. On updating the CPN model to RFC 4340, we investigated the connection establishment procedure when sequence numbers wrap [30] and found scenarios in which connection setup could fail. In [11, 10] we defined the DCCP service and confirmed that the sequence of user observable events of the protocol conformed to its service. In [33] we reported our experience with the incremental enhancement and iterative modelling of the connection management procedures as the DCCP specification was developed. Insight into the decisions behind the modelling choices can also be found in [33]. The full CPN specification of the connection management procedures can be found in Section 2 of [34]. Section 4 of [34] also explains the development of progress mappings for sweep-line state space analysis [24] of DCCP. Recently we have published an enhanced version of [33] which also discusses a procedure-based model of DCCP's connection management procedures [6].

All of our previous work was restricted to DCCP operating over a reordering channel without loss. However, following the approach advocated in [3], we have modelled and analysed the connection management procedures when operating over a FIFO channel with and without loss and also reordering with loss, but these models and their analysis results have never been published.

---

[2] The chatter scenario comprised undesired interactions between Reset and Sync packets that could involve extremely long but finite exchanges of these packets where no progress was made, until finally the system corrected itself.

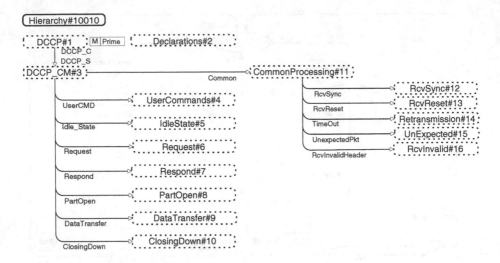

Fig. 9. The DCCP Hierarchy Page

DCCP Overview. Our CPN model from Section 3 of [34] has been extended to incorporate a combined FIFO/reordering and possibly lossy channel. It uses ideas behind the tool-friendly model described at the end of Section 2.3, but without the capacity constraint, which is not needed for the connection management procedures. This is because only a few packets are exchanged during connection management, and hence it is more interesting to determine channel bounds rather than introduce an artificial capacity constraint, as the capacity of the Internet is very large. The model comprises four hierarchical levels, shown in Fig. 9. The complete model comprises 6 places, 15 substitution transitions, 52 executable transitions and 19 ML functions. The DCCP#1 page, shown in Fig. 10, presents an abstract view of DCCP. The top layer represents users or applications issuing commands to DCCP entities. Two places, App_Client and App_Server, typed by *COMMAND*, store tokens representing user commands. In the middle layer, the two substitution transitions (rectangles with an HS tag), DCCP_C and DCCP_S, represent the Client and Server respectively. Both are linked to the same second level page, named DCCP_CM (see Fig. 9), that represents DCCP functionality. The bottom layer models the service provider, i.e. the service provided by the Internet Protocol, by two combined FIFO/reordering and possibly lossy channels, one from the Server to the Client (Ch_S_C) and the other from the Client to the Server (Ch_C_S). The channel places are typed by *PACKETS*. The transitions Loss_1 and Loss_2 model loss, and can be switched on or off depending on the value of the Boolean parameter, lossy, as described in Section 2. However, a slightly different approach was used in DCCP. When there is at least one packet in the channel (rq <> []), a packet can be lost by removing the list and returning its tail (tl(rq)). In this way we do not need to identify what the head of the list is when it is lost.

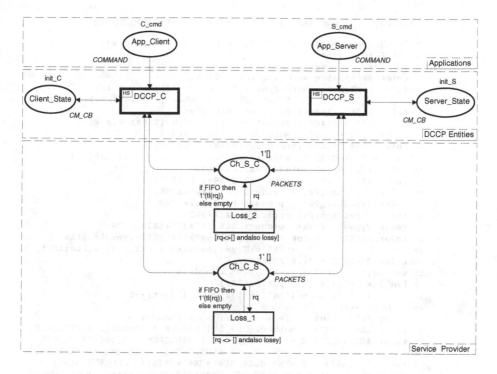

Fig. 10. The DCCP Overview Page

Declarations. There are only three minor differences (apart from declaring the Booleans FIFO and lossy) between the declarations of this *combined* model and the CPN model in [34]. All three relate to packets. Firstly, the combined model defines function insert (lines 22-23 of Listing 1), which captures the behaviour of the model in Fig. 7 regarding how an entity sends packets into the channel depending on the value of FIFO. Secondly, the packet structure of the combined model does not include the Extended Sequence Number bit, as it is not required. Thirdly, the combined model defines PACKETS (line 20) as a list of PACKET, in line with the combined FIFO/Reordering channel model in Section 2.3. Packets may have short (24 bit) or long (48 bit) sequence numbers (lines 9 to 17). Long sequence numbers (line 9) require the use of infinite integers (IntInf) in Design/CPN as normal integers are restricted to 31 bits. DCCP's control block (state information) is defined in lines 24-35. States are divided into 3 groups, depending on their data structures. Idle states (CLOSED, LISTEN and TIME-WAIT, see line 26) do not have any sequence number state variables (line 30) associated with them, whereas Active states (lines 27-28) do, as seen in line 32. The REQUEST state only includes the initial sequence number sent (ISS) and the greatest sequence number sent (GSS), see line 29. For more details of these and other declarations, see the Appendix of [34].

Listing 1. Declarations for Parameters, Packets and States

```
1  (* Channel Parameters , Packet structure , Queuing *)
2  val FIFO = true;
3  val lossy = true;
4  color PktType1 = with Request | Data;
5  color PktType2 = with Sync | SyncAck | Response | Ack
6                       | DataAck | CloseReq | Close | Rst;
7  color DATA = subset PktType1 with [Data];
8  color ACK_DATAACK = subset PktType2 with [DataAck,Ack];
9  color SN48 = IntInf with ZERO..MaxSeqNo48;
10 color SN48_AN48 = record SEQ:SN48*ACK:SN48;
11 var sn48_an48:SN48_AN48; (*Seq No. and Ack No. variable*)
12 color SN24 = int with 0..max_seq_no24;
13 color SN24_AN24 = record SEQ:SN24*ACK:SN24;
14 color Type1LongPkt = product PktType1*SN48;
15 color Type2LongPkt = product PktType2*SN48_AN48;
16 color Type1ShortPkt= product DATA*SN24;
17 color Type2ShortPkt= product ACK_DATAACK*SN24_AN24;
18 color PACKET = union PKT1:Type1LongPkt + PKT2:Type2LongPkt
19                      + PKT1s:Type1ShortPkt + PKT2s:Type2ShortPkt;
20 color PACKETS = list PACKET;
21 var sq,rq:PACKETS; (*Queuing Variables*)
22 fun insert(q:PACKETS,pkt:PACKET)
23     = if FIFO then 1'(q^^[pkt]) else 1'[pkt]++1'[];
24 (* DCCP state information *)
25 color RCNT = int;    (* Retransmission Counter *)
26 color IDLE_STATE = with CLOSED_I | LISTEN | TIMEWAIT | CLOSED_F;
27 color ACTIVE_STATE = with RESPOND | PARTOPEN | S_OPEN | C_OPEN
28                       | CLOSEREQ | C_CLOSING |S_CLOSING;
29 color RCNTxGSSxISS = product RCNT*SN48*SN48;(* REQUEST state *)
30 color GS = record GSS:SN48*GSR:SN48*GAR:SN48;(*state variables*)
31 color ISN = record ISS:SN48*ISR:SN48;(*initial sequence numbers*)
32 color ActiveStatexRCNTxGSxISN = product ACTIVE_STATE*RCNT*GS*ISN;
33 color CM_CB = union IdleState:IDLE_STATE
34                      +ReqState:RCNTxGSSxISS
35                      +ActiveState:ActiveStatexRCNTxGSxISN;
36 (* State Variables *)
37 var rcnt:RCNT;
38 var active_state:ACTIVE_STATE;
39 var gss,gsr,gar,iss,isr:SN48;
40 var g:GS;
41 var isn:ISN;
```

RcvSync Page. We illustrate the use of the combined FIFO/reordering channel constructs in Fig.11, which depicts the RcvSync page, an executable page at the 4th level of the hierarchy (see Fig. 9). This page models the receipt of a Sync or SyncAck packet in any active state (i.e. one of RESPOND, PARTOPEN, OPEN, CLOSEREQ or CLOSING). The RcvSync transition processes an incoming Sync packet. If the packet is valid, then a SyncAck packet is output to the channel using the insert function. If FIFO is true, then the SyncAck packet is concatenated to the end of the queue, otherwise both the singleton list, comprising the SyncAck packet (including its sequence and acknowledgement numbers) and an empty list, [], are added to the output place. If the packet is not valid, then the original list is returned if FIFO is true, otherwise the empty list is returned, which is equivalent to there being no output. In both cases, the Sync packet is removed from the input channel, by either returning the tail of the queue (rq), if FIFO, or nothing (empty), if reordering.

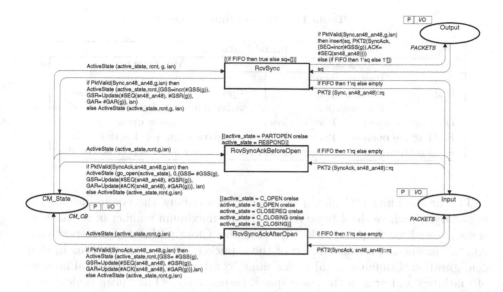

Fig. 11. The RcvSync page

4.3   Analysis Results

This section contains selected analysis results for the DCCP connection management procedures when operating over the four different kinds of channel. For further results see [29]. All results are obtained using sweep-line state space analysis [24] and the progress measures used are as discussed in Section 4 of [34]. We investigated 6 cases (A–F) as shown in Table 1. They have different (combinations of) user commands on the Client and Server sides. Case A is used to test the connection establishment procedures. The Client user issues an "active open" command while the Server user issues a "passive open". Cases B–F define scenarios in which the connection can be closed by either entity during (or after) establishment. Cases B and C allow only the Server user to attempt to close the connection using the "active close" and "server active close" commands respectively. Case D represents the situation where only the Client user may attempt to close the connection with an "active close". Cases E and F are when both sides may attempt to close the connection. The Server user issues an "active close" command in Case E but a "server active close" command in Case F. In all configurations the initial markings of the channel places, Ch_C_S and Ch_S_C, are empty, the initial state of each side is CLOSED and the initial send sequence number (ISS) is set to 5 on both sides. The sweep-line analysis results when DCCP operates over the four different channels are illustrated in Tables 2 – 5. All experiments[3] were conducted on a AMD 64x2 Dual 2.6 GHz PC with 1.87 GByte RAM.

---

[3] To alleviate state explosion, these experiments use long (48 bit) sequence numbers only and do not include DCCP-DataAck packets.

Table 1. Initial configurations

Case	Initial Markings	
	App_Client	App_Server
A	1'active open	1'passive open
B	1'active open	1'passive open ++ 1'active close
C	1'active open	1'passive open ++ 1'server active close
D	1'active open ++ 1'active close	1'passive open
E	1'active open ++ 1'active close	1'passive open ++ 1'active close
F	1'active open ++ 1'active close	1'passive open ++ 1'server active close

The first column ("Config.") in these tables defines the configurations being analysed, where the 4-tuple represents the maximum number of retransmissions allowed for Request, Ack, CloseReq and Close packet types respectively. An "x" means the retransmission of those packet types never happens in that configuration. Columns "total nodes" and "total arcs" record the total number of markings and arcs in the state space, respectively, while "peak nodes" lists the peak number of nodes stored in main memory at any one time. The time (hours:minutes:seconds) to sweep the state space using the sweep-line algorithm is given in Column "time".

The number of terminal (dead) markings is given in the next 5 columns. Terminal markings are classified into 5 types (I, II, III, IV and V). Type I terminal markings are desirable and correspond to successful connection establishment where both the Client and Server are in the OPEN state. In Type II terminal markings both sides are CLOSED. For Type III markings, the Client is CLOSED, but the Server is in the LISTEN state. This can happen when the Server is initially CLOSED (down for maintenance or busy) and rejects the connection request. The Server then recovers and moves to the LISTEN state (waiting for the next request). Type I, II and III terminal markings are expected, depending on which case is being examined. However, Type IV terminal markings are deadlocks in which the Client is CLOSED but the Server remains OPEN. Type V terminal markings are also deadlocks, in which the Client remains OPEN but the Server is CLOSED. All terminal markings have no packets left in the channels.

Table 2. Results: DCCP connection management over FIFO channels with no loss

Config.	Sweep-line				Terminal Markings					Ch. Bounds		%
	total nodes	total arcs	peak nodes	time	I	II	III	IV	V	Ch C_S	Ch S_C	space
A-(1,0,x,x)	290	532	74	00:00:00	4	1	1	0	0	3	4	25.52
B-(1,0,x,0)	1,080	2,128	229	00:00:00	0	2	1	0	0	4	5	23.23
C-(1,0,0,0)	1,270	2,516	290	00:00:00	0	2	1	0	0	4	5	22.83
D-(1,0,x,0)	743	1,402	121	00:00:00	0	2	2	0	0	4	4	16.29
E-(1,0,x,0)	2,891	5,841	374	00:00:01	0	4	2	0	0	5	5	12.94
F-(1,0,0,0)	3,457	7,031	493	00:00:03	0	4	2	0	0	5	5	14.26

Table 3. Results: DCCP connection management over FIFO channels with loss

| Config. | Sweep-line | | | | Terminal Markings | | | | | Ch. Bounds | | % |
	total nodes	total arcs	peak nodes	time	I	II	III	IV	V	Ch C_S	Ch S_C	space
A-(1,0,x,x)	715	2,076	218	00:00:00	15	1	1	12	0	3	4	30.49
B-(1,0,x,0)	3,298	11,120	960	00:00:01	0	2	1	0	14	4	5	29.11
C-(1,0,0,0)	4,104	13,825	1,218	00:00:01	0	2	1	0	14	4	5	29.68
D-(1,0,x,0)	2,221	6,951	591	00:00:01	0	2	2	26	0	4	4	26.61
E-(1,0,x,0)	10,991	39,132	2,197	00:00:04	0	4	2	0	0	5	5	19.99
F-(1,0,0,0)	14,097	50,073	2,963	00:00:05	0	4	2	0	0	5	5	21.02

Table 4. Results: DCCP connection management over reordering channels, no loss

| Config. | Sweep-line | | | | Terminal Markings | | | | | Ch. Bounds | | % |
	total nodes	total arcs	peak nodes	time	I	II	III	IV	V	Ch C_S	Ch S_C	space
A-(1,0,x,x)	2,974	8,160	1,098	00:00:01	15	1	1	0	0	3	4	36.92
B-(1,0,x,0)	33,567	120,526	12,551	00:00:18	0	2	1	0	0	4	5	37.39
C-(1,0,0,0)	59,845	217,858	23,300	00:00:39	0	2	1	0	0	4	5	38.93
D-(1,0,x,0)	23,212	74,628	7,061	00:00:10	0	2	2	0	0	4	4	34.42
E-(1,0,x,0)	459,195	1,911,473	150,257	00:10:19	0	4	2	0	0	5	5	32.72
F-(1,0,0,0)	904,036	3,870,467	303,528	00:39:40	0	4	2	0	0	5	5	33.58

Table 5. Results: DCCP connection management over reordering channels with loss

| Config. | Sweep-line | | | | Terminal Markings | | | | | Ch. Bounds | | % |
	total nodes	total arcs	peak nodes	time	I	II	III	IV	V	Ch C_S	Ch S_C	space
A-(1,0,x,x)	4,645	20,781	1,379	00:00:02	54	1	1	35	4	3	4	29.69
B-(1,0,x,0)	48,393	302,801	16,305	00:00:35	0	2	1	0	74	4	5	33.69
C-(1,0,0,0)	87,376	554,793	28,607	00:01:13	0	2	1	0	74	4	5	32.74
D-(1,0,x,0)	37,763	203,937	8,796	00:00:22	0	2	2	125	0	4	4	23.29
E-(1,0,x,0)	676,497	5,028,011	199,394	00:20:22	0	4	2	0	0	5	5	29.48
F-(1,0,0,0)	1,321,656	10,139,416	409,296	01:35:10	0	4	2	0	0	5	5	30.97

The next column, "Ch. Bounds", records the maximum number of packets that can occur in the channel places Ch_C_S and Ch_S_C. The last column (% space) shows the ratio of the number of peak states stored in main memory to the total number of states in the state space. Thus the smaller this number the more efficient the sweep-line algorithm is.

4.4 Discussion

In order to compare the results shown in Tables 2, 3, 4 and 5, we selected the retransmission configuration (1,0,0,0) corresponding to being able to retransmit just one Request packet.

State Space Size and Complexity. The state space sizes and execution times indicate how complex each scenario is. We can see that for each of the channel models that complexity increases for the cases considered in the following order: A,D,B,C,E,F. It is expected that Case A is the simplest, as it only has user commands for opening the connection. Cases B–D are for scenarios when one user closes the connection. We expect that Case C will be more complex than Cases B and D, because the "server active close" entails a 3 way handshake, rather than a two way handshake for "active close". However, it was a little surprising that the state space for Case B was more than 40% larger than that of Case D, i.e. the server closing (with an active close) is more complex than the client closing, revealing significant asymmetry in the procedures. We expect that allowing closure by both ends will be more complex than at a single end only, and that a "server active close" will be more complex again than using the "active close", and this is reflected in Cases E and F.

It is clear from the Tables that a FIFO Channel provides for the least complex behaviour while reordering is much more complex. In both cases, loss also increases the size of the state space. It is also clear that for DCCP a FIFO lossy channel is less complex than a reordering channel with no loss, which is expected. Reordering adds so much complexity that for Case F, we had to use the sweep-line library to obtain the results, and that was only for one retransmission of the Request packet. We thus used the sweep-line library for all the results for comparison of memory consumption (% space) for the different cases and channels.

Unspecified Receptions. The ability to perform incremental analysis by varying the channel type is useful during model development. For example, it is useful for discovering unspecified receptions that may be masked by a reordering and/or lossy channel, and hence go undetected. We recall that unspecified receptions occur when a protocol entity receives a packet in a particular state but has no procedures for receiving and processing it. When analysing a model with FIFO channels, unspecified receptions correspond to the packet at the head of the FIFO channel and the state of the receiving protocol entity in a terminal marking. As stated previously, for all terminal markings, including the FIFO case, there were no packets left in the Channel. This means that there are no unspecified receptions in this protocol.

Expected Terminal Markings: Types I-III. For Case A in Table 2 (FIFO) we observe that there are 4 Type I and one each of Type II and Type III terminal markings. These are all acceptable. The reason for 4 Type I markings (i.e. markings where the connection is established) is due to different values of the state variables in both the client and server. There are 3 main state variables of interest: the greatest sequence number sent (GSS); the greatest sequence number received (GSR); and the greatest acknowledgment received (GAR). The differences are caused by two effects. The first is that the server can acknowledge the receipt of an Ack packet from the Client with either a Data packet (if it has data to send) or an Ack packet if it doesn't. Because Data packets do not include an acknowledgment, GAR is not increased on its receipt by the client, whereas GAR is increased on receipt of an Ack packet. The second effect concerns the number of retransmissions. Because sequence numbers are incremented on each retransmission, the state variables are also increased. For Case A, if the connection is

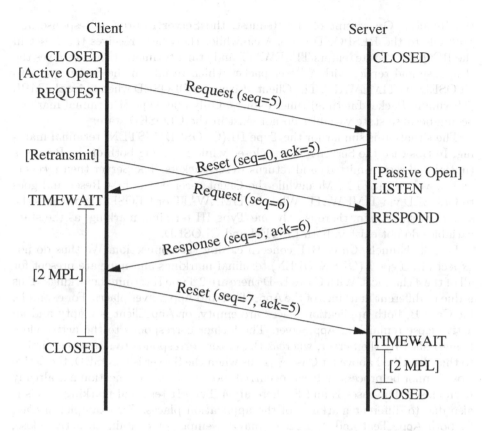

Fig. 12. Time Sequence Diagram illustrating reaching a Type II Terminal Marking

established without retransmitting the Request packet, then the vector of the values of the state variables (GSS,GSR,GAR) is (6,6,5) in both the Client and Server, when a Data packet is used, and (6,6,6) in the Client and (6,6,5) in the Server, when an Ack is used to complete the establishment of the connection. If the Request is retransmitted, then the values of the state variables are (7,8,6) in the Server, but (8,7,6) or (8,7,7) in the Client depending on whether it received a Data packet or an Ack packet. From the results we have generated, for various numbers of retransmissions of Request packets (but zero retransmissions for the other 3 packet types), the number of Type I terminal markings is given by 2(MaxRetransRequest +1) for when the Server can reply with either a Data or an Ack packet in the RESPOND state.

It is quite possible that the connection is not established, for example, due to the Server being down for maintenance. The situation is illustrated in Fig. 12. A Type II terminal marking (CLOSED-CLOSED) occurs when the Client sends its first Request when the Server is CLOSED. Being closed, the Server returns a Reset packet when it receives the Request. The Server then comes up and goes into the LISTEN state. The Reset is delayed and the Client retransmits the Request. On receipt of the Request, the Server returns a Response and proceeds to the RESPOND state. Meanwhile, the Client receives the Reset in the REQUEST state, enters TIMEWAIT and starts a timer. It then receives the Response and replies with a Reset packet which results in the Server going to CLOSED via TIMEWAIT. The Client also enters CLOSED when twice the MPL (Maximum Packet Lifetime) timer expires. Only one Type II terminal marking occurs because state variables do not exist in the CLOSED state.

The situation is similar for the Type III (CLOSED–LISTEN) terminal marking. In this case the Server remains closed while receiving both of the Requests (initial and retransmitted) and returns Reset packets. The Server then recovers and moves to LISTEN. Meanwhile the Client receives the first Reset and goes to CLOSED via TIMEWAIT. While in TIMEWAIT or CLOSED it discards the second Reset. Again, there is only one Type III terminal marking, as the state variables do not exist in both LISTEN and CLOSED.

For all channels, Cases B–F concern closing the connection. We thus do not expect any Type I (OPEN–OPEN) terminal markings and none are present for all of the 4 channels. With Cases B–D there are 2 Type II terminal markings. This is due to different markings of the App_Client and App_Server places. For example, for Case B, both application places are empty, or App_Client is empty and an active close remains in App_Server. The former corresponds to the active close being issued by the Server, whereas the latter corresponds to situations similar to that described above for Case A. Thus when the Server is CLOSED, the active close cannot be processed. It has occurred too late, as the connection has already been closed. For cases E and F, there are 4 Type II terminal markings. This is also due to different markings of the application places. For example, in Case E, both App_Client and App_Server may be empty or contain an active close, leading to four combinations. For Type III markings the Server must always be in LISTEN. In our model no close commands are processed in LISTEN (similar to CLOSED) so that App_Server always contains the close command (active or server active) in a Type III terminal marking. For cases A–C, App_Client must be empty (hence only one Type III), while for cases D–F, App_Client could be empty or contain an active close, due to it being too late. This then explains why there are two Type III terminal markings for these cases.

For Case A in Tables 3 to 5, one can see a growth in the number of Type I terminal markings. Our investigations show that this is due to different combinations of the sequence number state variables (GSS,GSR,GAR) that can occur when packets are lost or reordered.

Half-Open Connections. Tables 2 and 4 have no undesired terminal markings, whereas Tables 3 and 5 reveal the existence of Type IV and V terminal markings. Thus, when DCCP connection management operates over lossy (either FIFO or reordering) channels, the connection can fail, resulting in half-open connection deadlocks. A half-open connection is when one side is OPEN while the other is CLOSED. A simple example is when the Client is in PARTOPEN and the Server is in OPEN and sends an Ack packet to complete the establishment procedure. This Ack is lost and thus does not reach the Client. The Client runs a timer and when it expires the Client goes to CLOSED and sends a Reset packet, which is also lost. Thus the Server remains in OPEN while the Client is CLOSED (a Type IV terminal marking).

DCCP can recover from the half open connection in two ways. Firstly, the entity in the OPEN state may transmit further Data, DataAck or Ack packets. When the other side is in CLOSED and receives a non-reset packet, it replies with a Reset packet thus closing the other end. Secondly, even if all returned DCCP-Reset packets are lost or no DCCP-reset packet is sent (due to loss of the Data and Ack packets), the application on the Server or the Client can issue a close command to abort the connection [34]. The second scenario is confirmed by our analysis results in Tables 3 and 5. The entities that issue the close command eventually enter the CLOSED state.

Channel Bounds. The results in the tables (Ch. Bounds) show that the maximum number of packets that can accumulate in the channels is the same for each of the 4 different channel types. The bound depends on the number of re-transmissions and the number of user commands. In Table 2 we see that the bound on the channel from the Client to the Server is 3 for Case A. A simple example of this is where the network is congested when the Client attempts to establish the connection. The Client transmits the first Request and then a second Request before finally timing out, sending a Reset and moving to the CLOSED state. Thus the two Requests and the Reset can be in the Channel at the same time. An active close by an application will give rise to a Close packet in a channel, whereas a server active close will create a CloseReq in the channel from the Server to Client and a Close packet in the reverse channel. This gives rise to up to 5 packets in the channels at any one time. Thus we can see that the channels are bounded and that there is no need to introduce a capacity limit on the channel places to obtain results.

5   Conclusions and Future Work

For protocols that operate over lossy reordering channels it is useful to analyse the protocol when operating over four kinds of channels: FIFO without loss; FIFO with loss; reordering without loss; and reordering with loss. Using a lossless FIFO channel allows unspecified receptions to be detected, whereas lossy or reordering channels can mask out this behaviour. Incrementally analysing the protocol using the four different channel models allows further insight into the protocol's behaviour. Once undesired behaviour is discovered, the modeller can check whether it is caused by modelling errors or by incompleteness of the protocol specification. Having only one model to correct, rather than four, significantly eases the maintenance burden of the modeller.

To ease maintenance during the process of protocol development, it is beneficial to use one channel model that includes all desired channel behaviour. This paper has explored the possibilities of combining the channel properties of FIFO or reordering, the presence or absence of loss, and unbounded or finite capacity into one parameterised CPN model.

We proposed a model that includes 3 parameters: one for ordering behaviour (FIFO); one for loss (lossy); and one for the capacity limit (limit). The channel comprises one input transition, a place for storing packets using lists, and two output transitions, one for loss and the other for packet reception. The fact that there is only one input and one reception transition, avoids duplication of arcs (and their complex inscriptions), thus easing maintenance. Allowing the channel's characteristics to be switched between FIFO and reordering, and lossy or non-lossy, by changing the values of a parameters rather than the net structure, greatly facilitates incremental analysis and model maintenance.

Our channel model can be easily encapsulated by a substitution transition. This would be easy to incorporate into a protocol specification as a separate component or module. This is an elegant approach for protocol specification. However, when the aim of the model is to analyse the protocol, it is important to embed the parameterised model into the protocol specification in such a way that state space explosion is minimised. We show that a modular approach using substitution transitions exacerbates state space explosion and describe how to reduce the number of states by using an integrated channel model. We illustrate how this is done using a new Internet transport protocol, called the Datagram Congestion Control Protocol (DCCP). This model is then used to analyse the connection management procedures of DCCP. The analysis has provided some insight in to DCCP's procedures, including showing that half-open connection deadlocks can occur for lossy (FIFO or reordering) channels.

We therefore conclude that with respect to modelling a complex protocol, having a single parameterised channel model is beneficial for development, maintenance and analysis. We hope the patterns suggested in this paper will assist other modellers when considering channel models in other contexts.

Future work could include developing tool support for the simple parameterised channel model. One would first consider cases where colour sets are small and the lists used to store packets are of limited size, to overcome the unbounded non-determinism problem. Another direction would be to explore the use of transition fusion or substitution places to encapsulate the parameterised model, to allow a modular approach to be used for specification, one that did not exacerbate the state explosion problem.

Acknowledgements. The authors are grateful to Lars Kristensen whose ideas and comments assisted us in creating the tool-friendly combined CPN model. We also gratefully acknowledge the constructive comments of the reviewers of the original workshop paper and the paper submitted for this special issue. The comments have motivated many enhancements to this paper. For example, the inclusion of Fig. 6 was inspired by a suggestion from one of the reviewers.

References

1. Billington, J.: Extensions to Coloured Petri Nets and their Application to Protocols. PhD thesis, University of Cambridge, Computer Laboratory (May 1990); Published as Technical Report No. 222 (May 1991)
2. Billington, J., Gallasch, G.E.: How Stop and Wait Protocols Can Fail Over The Internet. In: König, H., Heiner, M., Wolisz, A. (eds.) FORTE 2003. LNCS, vol. 2767, pp. 209–223. Springer, Heidelberg (2003)
3. Billington, J., Gallasch, G.E., Han, B.: A Coloured Petri Net Approach to Protocol Verification. In: Desel, J., Reisig, W., Rozenberg, G. (eds.) Lectures on Concurrency and Petri Nets. LNCS, vol. 3098, pp. 210–290. Springer, Heidelberg (2004)
4. Billington, J., Han, B.: Formalising TCP's Data Transfer Service Language: A Symbolic Automaton and its Properties. Fundamenta Informaticae 80(1-3), 49–74 (2007)
5. Billington, J., Han, B.: Modelling and Analysing the Functional Behaviour of TCP's Connection Management Procedures. International Journal on Software Tools for Technology Transfer 9(3-4), 269–304 (2007), http://dx.doi.org/10.1007/s10009-007-0034-1
6. Billington, J., Vanit-Anunchai, S.: Coloured Petri Net Modelling of an Evolving Internet Standard: the Datagram Congestion Control Protocol. Fundamenta Informaticae 88(3), 357–385 (2008)
7. CPN Tools home page, http://wiki.daimi.au.dk/cpntools/cpntools.wiki
8. Design/CPN Online, http://www.daimi.au.dk/designCPN/
9. Floyd, S., Handley, M., Kohler, E.: Problem Statement for the Datagram Congestion Control Protocol (DCCP), RFC 4336 (March 2006), http://www.rfc-editor.org/rfc/rfc4336.txt
10. Gallasch, G.E., Billington, J., Vanit-Anunchai, S., Kristensen, L.M.: Checking Safety Properties On-The-Fly with the Sweep-line Method. International Journal on Software Tools for Technology Transfer 9(3-4), 374–391 (2007), http://dx.doi.org/10.1007/s10009-007-0031-4
11. Gallasch, G.E., Vanit-Anunchai, S., Billington, J., Kristensen, L.M.: Checking Language Inclusion On-The-Fly with the Sweep-line Method. In: The Sixth Workshop and Tutorial on Practical Use of Coloured Petri Nets and the CPN Tools. DAIMI PB 576, October 24-26, pp. 1–20. Department of Computer Science, University of Aarhus (2005), http://www.daimi.au.dk/CPnets/workshop05/cpn/papers/
12. Gordon, S.: Verification of the WAP Transaction Layer uisng Coloured Petri Nets. PhD thesis, Institute for Telecommunications Research and Computer Systems Engineering Centre, School of Electrical and Information Engineering, University of South Australia, Adelaide, Australia (November 2001)
13. Han, B.: Formal Specification of the TCP Service and Verification of TCP Connection Management. PhD thesis, Computer Systems Engineering Centre, School of Electrical and Information Engineering, University of South Australia, Adelaide, Australia (December 2004)
14. Holzmann, G.J.: Design and Validation of Computer Protocols. Prentice-Hall International Editions, Englewood Cliffs (1991)
15. Jensen, K.: Coloured Petri Nets: Basic Concepts, Analysis Methods and Practical Use, 2nd edn. Basic Concepts. Monographs in Theoretical Computer Science, vol. 1. Springer, Heidelberg (1997)
16. Jensen, K., Kristensen, L.M., Wells, L.: Coloured Petri Nets and CPN Tools for Modelling and Validation of Concurrent Systems. International Journal on Software Tools for Technology Transfer 9(3-4), 213–254 (2007)

17. Kohler, E., Handley, M., Floyd, S.: Designing DCCP: Congestion Control Without Reliability. In: Proceedings of the 2006 ACM Conference on Applications, Technologies, Architectures, and Protocols for Computer Communications (SIGCOMM 2006), Pisa, Italy, September 11-15, pp. 27–38 (2006)
18. Kohler, E., Handley, M., Floyd, S.: Datagram Congestion Control Protocol, draft-ietf-dccp-spec-6 (February 2004),
    http://www.read.cs.ucla.edu/dccp/draft-ietf-dccp-spec-06.txt
19. Kohler, E., Handley, M., Floyd, S.: Datagram Congestion Control Protocol, RFC 4340 (March 2006), http://www.rfc-editor.org/rfc/rfc4340.txt
20. Kohler, E., Handley, M., Floyd, S., Padhye, J.: Datagram Congestion Control Protocol, draft-ietf-dccp-spec-5 (October 2003),
    http://www.read.cs.ucla.edu/dccp/draft-ietf-dccp-spec-05.txt
21. Kongprakaiwoot, T.: Verification of the Datagram Congestion Control Protocol using Coloured Petri Nets. Master's thesis, Computer Systems Engineering Centre, School of Electrical and Information Engineering, University of South Australia, Adelaide, Australia (November 2004)
22. Kristensen, L.M., Jensen, K.: Specification and Validation of an Edge Router Discovery Protocol for Mobile Ad Hoc Networks. In: Ehrig, H., Damm, W., Desel, J., Große-Rhode, M., Reif, W., Schnieder, E., Westkämper, E. (eds.) INT 2004. LNCS, vol. 3147, pp. 248–269. Springer, Heidelberg (2004)
23. Liu, L.: Towards Parametric Verification of the Capability Exchange Signalling Protocol. PhD thesis, Computer Systems Engineering Centre, School of Electrical and Information Engineering, University of South Australia, Adelaide, Australia (May 2006)
24. Mailund, T.: Sweeping the State Space - A Sweep-Line State Space Exploration Method. PhD thesis, Department of Computer Science, University of Aarhus (February 2003)
25. Mulyar, N.A., van der Aalst, W.M.P.: Patterns in Coloured Petri Nets. Technical report, Department of Technology Management, Eindhoven University, P.O. Box 513, NL-5800 MB, Eindhoven, The Netherlands (April 2005)
26. Ouyang, C.: Formal Specification and Verification of the Internet Open Trading Protocol using Coloured Petri Nets. PhD thesis, Computer Systems Engineering Centre, School of Electrical and Information Engineering, University of South Australia, Adelaide, Australia (June 2004)
27. Postel, J.: Transmission Control Protocol (TCP), RFC793 (September 1981), http://www.rfc-editor.org/rfc/rfc793.txt
28. Tanenbaum, A.: Computer Networks, 4th edn. Prentice Hall, Englewood Cliffs (2003)
29. Vanit-Anunchai, S.: An Investigation of the Datagram Congestion Control Protocol's Connection Management and Synchronisation Procedures. PhD thesis, Computer Systems Engineering Centre, School of Electrical and Information Engineering, University of South Australia, Adelaide, Australia (November 2007)
30. Vanit-Anunchai, S., Billington, J.: Effect of Sequence Number Wrap on DCCP Connection Establishment. In: Proceedings of the 14th IEEE International Symposium on Modeling, Analysis, and Simulation of Computer and Telecommunication Systems (MASCOTS), Monterey, California, USA, September 11-13, pp. 345–354. IEEE Computer Society Press, Los Alamitos (2006)
31. Vanit-Anunchai, S., Billington, J.: Initial Result of a Formal Analysis of DCCP Connection Management. In: Proceedings of the Fourth International Network Conference (INC 2004), July 6-9, pp. 63–70. University of Plymouth, Plymouth (2004)

32. Vanit-Anunchai, S., Billington, J.: Chattering Behaviour in the Datagram Conges-
    tion Control Protocol. IEE Electronics Letters 41(21), 1198–1199 (2005)
33. Vanit-Anunchai, S., Billington, J.: Modelling the Datagram Congestion Control
    Protocol's Connection Management and Synchronisation Procedures. In: Kleijn,
    J., Yakovlev, A. (eds.) ICATPN 2007. LNCS, vol. 4546, pp. 423–444. Springer,
    Heidelberg (2007)
34. Vanit-Anunchai, S., Billington, J., Gallasch, G.E.: Analysis of the Datagram Con-
    gestion Control Protocol's Connection Management Procedures using the Sweep-
    line Method. International Journal on Software Tools for Technology Trans-
    fer 10(1), 29–56 (2008), http://dx.doi.org/10.1007/s10009-007-0050-1
35. Vanit-Anunchai, S., Billington, J., Gallasch, G.E.: A Combined Protocol Channel
    Model and its Application to the Datagram Congestion Control Protocol. In: Pro-
    ceedings of the International Workshop on Petri Nets and Distributed Systems,
    Xi'an, China, June 23-24, pp. 32–46 (2008)
36. Vanit-Anunchai, S., Billington, J., Kongprakaiwoot, T.: Discovering Chatter and
    Incompleteness in the Datagram Congestion Control Protocol. In: Wang, F. (ed.)
    FORTE 2005. LNCS, vol. 3731, pp. 143–158. Springer, Heidelberg (2005)
37. Wheeler, G.: The Modelling and Analysis of IEEE 802.6's Configuration Control
    Protocol with Coloured Petri Nets. In: Billington, J., Díaz, M., Rozenberg, G.
    (eds.) APN 1999. LNCS, vol. 1605, pp. 69–92. Springer, Heidelberg (1999)

On Modelling and Analysing the Dynamic MANET On-Demand (DYMO) Routing Protocol

Jonathan Billington and Cong Yuan

Computer Systems Engineering Centre
University of South Australia
Mawson Lakes Campus, SA 5095, Australia
Jonathan.Billington@unisa.edu.au, Cong.Yuan@postgrads.unisa.edu.au

Abstract. The Dynamic MANET On-demand (DYMO) routing protocol, being developed by the Internet Engineering Task Force, is a reactive routing protocol for mobile ad-hoc networks (MANETs). The basic operations of DYMO are route discovery and route maintenance. Constructing an analysable model of the DYMO protocol specification is a challenge because the routing operations are complex and the network topology changes dynamically. This paper presents a formal model of DYMO using Coloured Petri Nets. The model has a compact net structure, with functions in the arc inscriptions representing DYMO's routing algorithms. The paper shows how careful crafting of the model results in smaller state spaces, compared with models using intuitively appealing hierarchical constructs. Initial results of state space analysis of the model are presented.

Keywords: MANETs, On-Demand Routing, DYMO, Coloured Petri Nets, Reachability Analysis.

1   Introduction

A Mobile ad hoc network (MANET) [1] consists of mobile devices that communicate with each other without any pre-existing infrastructure or centralized administration. This is achieved by using radio transmission between mobile devices (nodes) which implement routing procedures, so that mobile nodes double as network routers. Conventional routing protocols cannot be directly used in MANETs, because the topology changes unpredictably due to the arbitrary movement of mobile nodes [21]. The Internet Engineering Task Force (IETF) MANET Working Group (WG) [13] has developed two approaches for routing protocols known as: Proactive MANET Protocols (PMP) and Reactive MANET Protocols (RMP). In the PMP approach, routes are attempted to be established between each mobile node pair irrespective of user demand and maintained if possible as nodes move. This results in a large amount of routing traffic. To overcome this problem the RMP approach was developed. It only finds routes on demand, i.e. when packets require the route. So far, no routing protocol

K. Jensen, J. Billington, and M. Koutny (Eds.): ToPNoC III, LNCS 5800, pp. 98–126, 2009.
© Springer-Verlag Berlin Heidelberg 2009

for MANETs has been issued as an Internet Standard, however, the Dynamic MANET On-demand (DYMO) routing protocol [4] is a hot topic of RMP, being developed by the IETF MANET WG currently.

Formal verification of a routing protocol is highly desirable, since correct routing is vital for the success of MANETs. There have been some previous attempts to formally model and analyse MANET routing protocols. Bhargavan et al [2] and Obradovic [20] verified the Ad-hoc On-demand Distance Vector (AODV) routing protocol [22] using SPIN [12] and the HOL Theorem Proving System [11]. Cavalli et al [3] presented a validation model for Dynamic Source Routing (DSR) [17] using the Specification and Description Language [14]. Xiong et al [26, 25] presented a timed Coloured Petri Net (CPN) [15, 18, 16] model for AODV. A topology approximation mechanism is proposed to address mobility of nodes, rather than accurately modelling dynamic topologies. S. Chiyangwa et al [5] have applied UPPAAL [19] to consider the effect of the protocol parameters on the timing behaviour of AODV. Wibling et al [24] consider an automatic verification strategy, and use two model checking tools, SPIN and UPPAAL, to verify both the data and control aspects of the Lightweight Underlay Network Ad hoc Routing (LUNAR) protocol. Renesse et al [23] model and verify the Wireless Adaptive Routing Protocol (WARP) using SPIN. However, until recently [27, 8, 9, 28], there have been no attempts to formally model and analyse DYMO.

In previous work [27], we presented the first CPN model of DYMO according to Internet Draft version 5. Both route discovery and maintenance procedures were modelled, including message loss and retransmission. In the paper, we developed an abstract modelling approach where a generic net structure (of 4 places and 9 transitions) is employed for modelling routing messages and basic operations (such as transmitting and receiving routing messages), while ML functions are used in arc expressions and guards to implement the detailed routing procedures. A limitation of our model in [27] is that routing messages were not properly broadcast, with a view to reducing state space explosion, while still providing the ability to test the basic DYMO routing procedures of route discovery and maintenance. This was done using simulation. Several ambiguous statements in the Internet draft were discovered that could lead to incorrect routes.

Recently, Espensen et al [8, 9] presented a hierarchical CPN model of DYMO [4]. Their approach is motivated by an interest in implementing the protocol. In their model much greater use is made of net structure (places, transitions and arcs). The model is organised into 4 hierarchical levels and comprises 14 modules. However, the model is not complete, as the route maintenance procedures are ignored, and only the core parts of the route discovery procedure are considered. (It does not include additional procedures introduced in DYMO from version 7.) The network topology and topology changes are directly modelled, whereas this is not part of the DYMO specification. Although the additional net structure allows easier visualisation of the separate parts of the model and possibly easier validation against the Internet Draft, the intermediate processing and storage of information has negative consequences for state space analysis.

To create a model of DYMO more suited to state space analysis, our work [27, 28] has focussed on modelling the protocol using a compact but powerful net structure, which may be considered as a general template to model RMPs. This is because RMP actions are aggregated into the fundamental operations of transmitting and receiving routing messages and their consequences for updating routing tables. Each operation is modelled by a CPN transition. The complicated routing algorithms corresponding to a particular RMP are represented by ML functions and included in the model as arc expressions or guards. This provides a form of data hiding. Hence, when modelling different RMPs, the different routing algorithms are realised by just modifying the functions without the need to change the structure of the CPN model. Compared with [27], the model in [28] adds several new procedures developed by the MANET working group: a route request (RREQ) flood; route replies (RREP) created by intermediate nodes; and route error (RERR) messages created during the route discovery procedure. It also models the broadcast in a real MANET: a message broadcast by a node is either received by all, none or any number of nodes. In [28] we also showed that the state spaces of our compact model for a set of scenarios devised by Espensen et al [9] are significantly smaller than those published in [9]. In the process of developing and analysing the CPN model, we sent several suggestions regarding DYMO drafts to the IETF MANET working group. Some were accepted in version 11, where our work is acknowledged by the protocol developers.

This paper is a substantially expanded version of our workshop paper [28]. A simple example of a route discovery and maintenance scenario is now included in the description of DYMO to aid understanding of its operation. If the route discovery procedure does not find a route within a certain period of time, then the source of the data packets (requiring the route) is informed that the destination cannot be reached. Our CPN model has been enhanced to include this notification to more closely model DYMO and to provide a better comparison of state space performance with [9]. The description of our CPN model has been expanded to include details of some of the functions that are used to process routing messages, as these functions are important for DYMO's operation. Our state space results now include values of retransmissions that were not included in [9] for each of the scenarios, demonstrating that our compact model can provide additional results. We also include a substantial discussion of the analysis results.

The rest of the paper is organised as follows. Section 2 reviews the basic operations of DYMO as a prelude to presenting our CPN model in Section 3. In Section 4, we discuss our modelling approach and compare it in some detail with that of Espensen et al [9]. Our state space results are analysed in Section 5 to provide further insight into DYMO's route discovery procedures. Finally, conclusions and future work are presented in Section 6.

2 Dynamic MANET On-demand (DYMO) Routing Protocol

In [27], the DYMO routing protocol was introduced in detail with an example, whereas only a brief introduction to DYMO was given in [28]. In this section, we use a simple example (similar to that in [27]) to illustrate DYMO's basic route discovery and route maintenance procedures [4].

Consider a MANET comprising 4 nodes each running DYMO. Fig. 1 shows a series of MANET topologies as time increases. Each node is represented by a circle with its identity (address) inside the circle. Nodes that can communicate with each other are connected by dashed lines. In DYMO, each node has a routing table, comprising a list of route entries, where each route entry has the following fields:

- *Address*: the IP address of the destination node.
- *SeqNum*: the sequence number of the destination known by the node.
- *NextHop*: the IP address of the next hop on the path towards the destination.
- *Broken*: a flag indicating whether the route entry is valid.
- *Dist*: a metric indicating the distance traversed from the node to reach the destination. We use *HopCnt*, the number of intermediate hops, as the distance metric in this paper.
- *Prefix*: indicates that the associated address is a network address, rather than a host address. We omit this field in our simple example, as we only consider situations where all addresses are host addresses.

In Fig. 1, a routing table is associated with each node where each route entry is given by a 5-tuple: (Address, SeqNum, NextHop, Broken, HopCnt). The Broken flag is set to *broken* if the route entry is no longer valid (the route is considered to be broken) and to *valid* otherwise. The initial state of the MANET is shown in Fig. 1(1). We assume that each node has one route entry to itself, in which the destination and the next hop are itself, the hop count is 0, and its initial sequence number is assumed to be 1 (since sequence number 0 is reserved and is used to represent an unknown sequence number). Of course each entry to itself will always be *valid*.

DYMO defines 3 types of routing messages: a route request (RREQ); a route reply (RREP); and a route error (RERR) message. In Fig. 1, a DYMO message sent by a node is indicated by a solid arrow with its type next to the arrow.

Assume that node 1 wishes to initiate the sending of a data packet to node 2. Node 1, the origin node, does not have a (valid) route entry for its destination (node 2), called the target node. Hence, node 1 invokes DYMO's route discovery procedure by generating a route request (RREQ) message, represented by RREQ1 in Fig. 1(2). The RREQ includes the following attributes: (RREQ, hoplimit, Numadds, (TargetAdd, TargetSeqNum, TargetDist), (OrigAdd, OrigSeqNum, OrigDist)). RREQ is the type of the message and hoplimit indicates the remaining number of hops the message is allowed to traverse. Here it is set to 3 initially. Numadds records the number of addresses contained in the message, which in

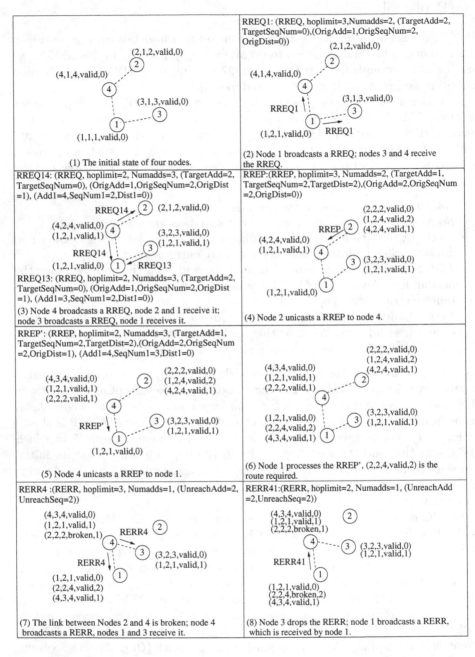

Fig. 1. A Simple Example of DYMO's Route Discovery and Maintenance Procedures

this case is two. The target node's address TargetAdd and sequence number TargetSeqNum are included in the message. Here, TargetSeqNum is set to 0 as it is unknown. Because the distance to the target (number of hops) is unknown, TargetDist is not included in the RREQ. Node 1 increments its sequence number by one, and sets the originator node's routing information, OrigAdd, OrigSeqNum and OrigDist, to values associated with the origin node (e.g. the distance to the origin is zero hops). Node 1 broadcasts the created RREQ and then waits for a reply. At this time, only nodes 3 and 4 are in range of node 1 and receive RREQ1.

When a node receives a RREQ, the hop limit of the message is decremented by one. If the node is the RREQ originator node, the message is dropped. Otherwise, for each address (except that of the target node) that includes distance information, its hop count is incremented by one. The node then determines if its routing table has a route entry to the RREQ originator. If this entry does not exist, a new route entry to the originator node is created. This is an important assumption in DYMO: links are assumed to be bidirectional [4]. (If a node receives a RREQ, it assumes that its sender will be able to receive a message from itself.) If the entry does exist, the node compares the routing information of the originator with its route entry. The node updates the route entry only if the incoming information is *superior*. Superior routing information has a higher sequence number, or a smaller hop count in the case of equal sequence numbers.

In our example, Nodes 3 and 4 each receive and process RREQ1 to form a new RREQ as shown in Fig. 1(3). The hoplimit is decremented by one and OrigDist is incremented. Then each node creates a route entry to node 1. To facilitate future route discovery, the node appends its address Add1, incremented sequence number SeqNum1 and initial distance information Dist1 to the new RREQ. Hence, NumAdds is incremented. The new RREQ formed at node 3 is denoted by RREQ13 and that at node 4 by RREQ14. Then, nodes 3 and 4 broadcast these RREQs. Their updated routing tables are shown in Fig. 1(3). Nodes 1 and 2 both receive RREQ14 and node 1 also receives RREQ13. Node 1 discards these RREQs, since it is the originator. In contrast, node 2 processes the received RREQ.

As shown in Fig. 1(4), node 2 creates route entries to nodes 1 and 4. Because it is the RREQ target node, it generates a route reply (RREP) and discards RREQ14. Node 2 increases its own sequence number before creating the RREP, because TargetSeqNum is unknown (zero) in the received RREQ. The created RREP, which has a similar format to the RREQ, is given in Fig. 1(4). The RREQ originator, node 1, now becomes the target node of the RREP. Node 2 unicasts the RREP to node 4, because it is the next hop of the route entry for node 1.

As shown in Fig. 1(5), node 4 receives this RREP and processes it. The hoplimit is decremented by one and OrigDist is incremented by one. It also inserts a route entry to node 2 in its routing table, and appends its routing information, Add1, SeqNum1 and Dist1, to the RREP after incrementing its own sequence number. Then the processed message, RREP' is unicast to node 1.

Node 1 receives RREP' and updates its routing table as shown in Fig. 1(6). This completes the route discovery procedure because node 2 now has the required route entry, (2,2,4,valid,2), which is used to deliver the data packet to node 2 (not shown).

There are two other mechanisms included in DYMO's route discovery procedures. Firstly, instead of using the RREP, the target node may reply with a RREQ flood on receipt of a RREQ. DYMO just defines the RREQ flood to be the broadcasting of a RREQ which is addressed to itself (as the target node). It gives no further information. Secondly, if the node is not the target node but has a valid route entry to the target node, it may respond with an intermediate node RREP.

Now consider that when receiving data packets from node 1, node 4 cannot deliver them to node 2 because of changes to the topology of the MANET (see Fig. 1(7)). In this situation, node 4 flags the route entry for node 2 as broken, and then creates a route error message (RERR), denoted by RERR4. The RERR includes: (RERR, hoplimit, Numadds, (UnreachAdd, UnreachSeq)), where (UnreachAdd, UnreachSeq) represents the address and sequence number of the unreachable node (node 2). As shown in Fig. 1(7), node 4 broadcasts the created RERR, and nodes 1 and 3 receive this message as node 3 has moved into the transmission range of node 4. Node 1 processes the unreachable node address contained in the RERR by marking the affected route entry as broken. In contrast, node 3 drops the received RERR, because no route entry in its routing table is affected by this unreachable node address. As shown in Fig. 1(8), node 1 broadcasts the processed RERR after decrementing the hoplimit of the message. Node 4 receives the RERR and then drops it, since the unreachable destination has been marked in its routing table. This completes the route maintenance procedure.

3   CPN Model of DYMO

3.1   Design Rationale and Assumptions

As mentioned in the introduction, our CPN model has been developed as a general template to model reactive MANET protocols, having been used previously (in unpublished work) to model the Ad hoc On-Demand Distance Vector routing protocol. The model has been developed to be relatively complete by including all of DYMO's major routing mechanisms, including those that are optional, such as the route request flood, route replies from intermediate nodes, and the addition of routing information in routing messages as the message traverses the network. In this paper we have also included notification to the sender of the data packets that a route cannot be established to the destination, if the originator of the route discovery procedure does not receive a reply from the destination within a specified period.

The purpose of the CPN model is to facilitate analysis of the protocol using state space analysis. This is very important for a protocol as complex as DYMO. We have therefore been careful to avoid any intermediate storage of messages within the model that would lead to state space explosion. We also represent DYMO messages in a simpler way. A DYMO message uses a very flexible and extensible but complex structure [6], including address block and address tlv-block, where a field may be included or not. In our model the elements related to size or length and the message semantics are omitted. To cover the most general case, we keep all the routing information (*Address, SeqNum, Dist, Prefix*) for each node in the message. This scheme has no influence on the protocol's behaviour but simplifies modelling.

There is no initial state given in DYMO Internet draft 10 [4]. We assume that each node only has a route entry to itself in its routing table, where its own sequence number is set to one and the hop count to itself is set to zero. This corresponds to all nodes being isolated initially. Furthermore, we consider the communication environment to be lossy, corresponding to packets being discarded if bit errors occur due to noise and interference.

3.2   CPN Model

Our CPN model of DYMO is created using Design/CPN [7]. It is a hierarchical model organised into 3 modules, as shown in Fig. 2.

DYMO Protocol represents the top-level module of the CPN model, as shown in Fig. 3. The two basic routing operations of DYMO are modelled in the submodules, Route_Discovery and Route_Maintenance, which are represented by substitution transitions indicated by rectangles with HS tags. The model comprises

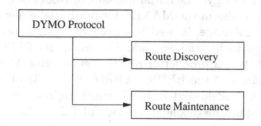

Fig. 2. Module Hierarchy for the DYMO CPN Model

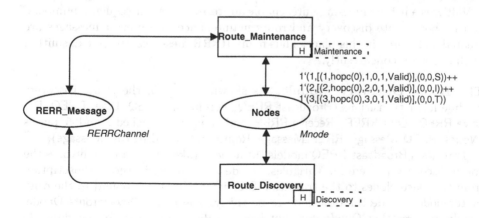

Fig. 3. The DYMOProtocol Module

5 places, 2 substitution transitions, 10 executable transitions and more than 70 ML functions to realise DYMO's routing algorithms.

The place, Nodes, stores all mobile nodes in the MANET. Nodes is typed by Mnode, as shown in line 16 of Fig. 4, which lists the data types of the model. Mnode is a product of NodeId, RT and RREQTries. NodeId (see line 2 of Fig. 4) is the set of node addresses and is represented by an integer from zero to MaxNodes, where zero is a reserved value and MaxNodes indicates the maximum number of nodes in the network. The *declare ms* means that the name NodeId is declared as a multiset. This is used in the function that models the broadcast, which we discuss later. RT (see line 12 in Fig. 4) is the node's routing table, comprising a list of route entries, one for each destination. The route entry, RTEntry, is a 6-tuple comprising DestAddress, HopCnt, NextHopAddress, Prefix, SeqNum and REstate. The route entry state (REstate line 9) is a flag indicating whether the route entry is broken or not. RREQTries (see line 15 in Fig. 4) is a triple containing the target IP address (DestAddress), the number of transmissions (Tries) and the role of a node (Role) during a particular route discovery. The role of the originator or source node is represented by S, an intermediate node by I and the target by T. Role allows us to only consider a single source-target pair for analysis. If the role is ignored, several route discovery procedures may occur simultaneously. The marking of Nodes indicates the number of nodes in the network and their view of current network topology. The initial marking of Nodes (see Fig. 3), considers a configuration of 3 nodes in the MANET for our initial experiments.

The place, RERR_Message, is used to model the communication channel for route error messages within the network. The colour set RERRChannel is given in line 39 of Fig. 4. It is a product of RERRMess and NodeId, which represents the destination address of the RERR. RERRMess (line 38 of Fig. 4) comprises the sender's address NodeId and RERRM, which represents a RERR message comprising errormessage, msghoplimit and AddinfoRERR. As shown in line 36 of Fig. 4, errormessage is a singleton set comprising the message's type, RERR. AddinfoRERR (see line 35 of Fig. 4) consists of Numadds and UnreAddtlvs, where Numadds indicates the number of addresses contained in the message, and UnreAddtlvs is a list, comprising unreachable addresses and their sequence numbers. During both route discovery and route maintenance, route error messages are created and stored in this place. Thus many RERR messages can be transmitted in the network concurrently.

The Route Discovery Module. As shown in Fig. 5, the Route Discovery module has 7 transitions (Broadcast_RREQ, Retransmit_RREQ, Lose_ RREQ, Receive_RREQ, Lose_RREP, Receive_RREP and MaxTries_Reached), and 5 places (Nodes, RREQ_Message, RREP_Message, Route_Status and RERR_ Message).

Transition Broadcast_RREQ models the actions taken when a node invokes the route discovery procedure. Variables, Onode and Tnode, in the arc inscription from the place Nodes to the transition Broadcast_RREQ, are bound to the originator node and destination node respectively. The guard Novalidroute(Onode, Tnode) ensures that Onode does not have a valid route entry for Tnode in its routing table. It also ensures that only a node with role S can be bound to

```
------------------------------------------------------------------------
 1  (* each node: IP address, the routing table, additional information *)
 2  color NodeId = int with 0..MaxNodes declare ms;
 3  color DestAddress = NodeId;
 4  color Hopcount = int with 0..MaxNodes;
 5  color HopCnt = union hopc:Hopcount + unknown declare of_hopc;
 6  color NextHopAddress = NodeId;
 7  color Prefix = int;
 8  color SeqNum = int with 0..MaxSeqNr;
 9  color REstate = with Valid | Broken;
10  color RTEntry = product DestAddress * HopCnt * NextHopAddress*
11                           Prefix * SeqNum * REstate;
12  color RT = list RTEntry;
13  color Tries = int with 0..MaxTries;
14  color Role = with S | T | I;
15  color RREQTries = product DestAddress * Tries * Role;
16  color Mnode = product NodeId * RT * RREQTries;
17  (* msg-header *)
18  color msgtype = with RREQ | RREP;
19  color msghoplimit = int with 0..Netdiameter;
20  color msgheader= product msgtype * msghoplimit;
21  (* address-block and address-tlv-block *)
22  color Numadds = int;
23  color Addtlv = product NodeId * SeqNum * HopCnt * Prefix;
24  color Addtlvs= list Addtlv;
25  color TarAddtlv = Addtlv;
26  color Addinfo = product Numadds * TarAddtlv * Addtlvs;
27  (* General message format: RREQ/RREP message *)
28  color RMessage = product msgheader * Addinfo;
29  color RRMess = product NodeId * RMessage;
30  color RMChannel = product RRMess * NodeId declare mult;
31  color BOOL = bool;
32  (* General message format: RERR message *)
33  color UnreAddtlv = product NodeId * SeqNum;
34  color UnreAddtlvs = list UnreAddtlv;
35  color AddinfoRERR = product Numadds * UnreAddtlvs;
36  color errormessage = with RERR;
37  color RERRM = product errormessage * msghoplimit * AddinfoRERR;
38  color RERRMess = product NodeId * RERRM;
39  color RERRChannel = product RERRMess * NodeId declare mult;
40  (* Route status *)
41  color Status = with RREQsent | ICMPsent;
42  color RouteStatus = product NodeId * NodeId * Status;
43  var n, d: NodeId;
44  var Onode, Tnode, node: Mnode;
45  var rreq, rrep: RRMess;
46  var rerr: RERRMess;
47  var RREQflood: BOOL;
48  var status: Status;
------------------------------------------------------------------------
```

Fig. 4. Data Types of the CPN Model

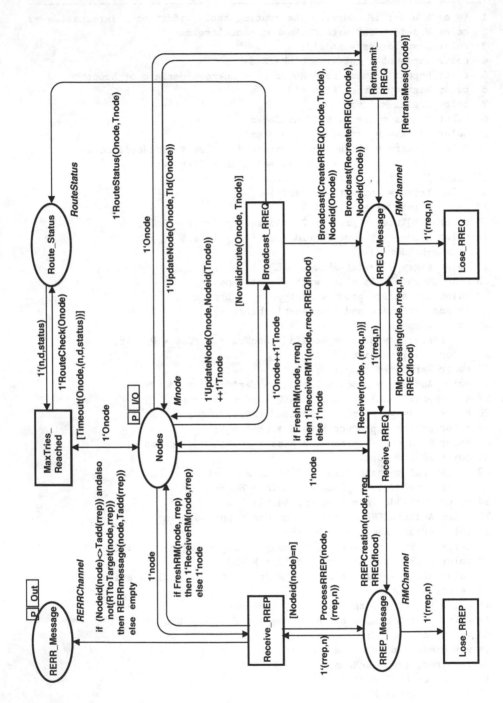

Fig. 5. The Route Discovery Module

Onode and one with role T can be bound to Tnode when we fix the originator/target nodes for analysis. Function, UpdateNode, on the return arc to place Nodes returns the node bound to Onode with its sequence number and number of transmissions incremented. Broadcast is a function which occurs in the expression on the arc from Broadcast_RREQ to the place RREQ_Message. This function makes (MaxNodes-1) copies of an RREQ created by the function CreateRREQ, where MaxNodes is the number of nodes in the MANET. Each copy is for a node in the system except for the sender. Due to loss, the number of nodes receiving the broadcast can vary from zero to (n-1) in a n-node network. The created RREQ messages are stored in the place, RREQ_Message, which is typed by the colour set RMChannel (see line 30 in Fig. 4), similar to RERRChannel.

According to sections 5.4 and 5.2.2 of DYMO [4], after issuing a RREQ, the originator node waits for a reply (from the destination) that includes the requested route. If the originator does not receive the reply within the retransmission timeout period, it may try again to discover a route by sending another RREQ. This continues up to the maximum transmission limit MaxTries. If the node receives a reply within the total waiting time, it forwards the data packets to the next hop using the new route entry. If the route reply is not received in time, the node informs the source of the data packets that a route was not found using a 'destination unreachable' Internet Control Message Protocol (ICMP) message. The expression on the arc from Broadcast_RREQ to the place Route_Status is the function RouteStatus. This function returns a triple comprising the originator's address, the target's address and RREQsent, to Route_Status. RREQsent indicates that the originator node has sent a RREQ and is waiting for a route reply. Route_Status thus records the status (see line 42 of Fig. 4) of each route discovery attempt. If the route reply message is not received after MaxTries attempts, MaxTries_Reached occurs and records the status as ICMPsent to indicate that the source of the data packets has been notified of the failure.

Retransmission of RREQs is implemented by transition Retransmit_ RREQ. Its guard, RetransMess(Onode), ensures that Onode is the originator node and MaxTries has not been reached. When the RREQ is retransmitted, the originator node increments its sequence number and number of transmissions using function UpdateNode.

Receive_RREQ models the actions taken by a node when it receives a RREQ. The inscription on the arc from the place RREQ_Message to this transition is a pair, (rreq,n), representing the transmitted RREQ and its destination's address. The guard Receiver(node,(rreq,n)) ensures that node is this destination node. The inscription on the arc from this transition to Nodes indicates that if the received RREQ is superior (determined by FreshRM), the node updates its routing table accordingly, using function ReceiveRM1. Otherwise, the node is returned without any change. RMprocessing (arc to RREQ_Message) rebroadcasts the received RREQ. If the node is the target then RMprocessing implements a RREQ flood when the RREQflood argument is true. Otherwise, no message is returned by RMprocessing. If the node is an intermediate node that does not have a route to the target, the function appends its own routing information and broadcasts the updated RREQ. RREPCreation ceates a route reply destined for the origin and sent via RREP_Message. It returns a RREP message if the node is the target when RREQflood is false, or an intermediate node RREP message if the node has a valid route to the target node. Lose_RREQ deletes an RREQ, which corresponds to node n being currently out of range of the sender.

So far, we have only given brief descriptions of the functions that are used in the model, without giving the ML code. We now provide some insight into the complexity of these functions by considering the functions Broadcast and RMprocessing, which are important for route discovery.

```
1   fun Broadcast(message,originNodeID) =
2     mult'RMChannel(1'message, NodeId -- 1'0 -- 1'originNodeID);
```

Fig. 6. The Broadcast Function

To broadcast a message, we need to pair the message with each of its destination addresses (node ids). In CPN ML this is done most directly by multiset operations. The Broadcast function defined in Fig. 6 is for broadcasting a RREQ message. It uses the function mult'RMChannel which takes a multiset over RRMess and a multiset over NodeId (see line 30 of Fig. 4) and produces a multiset over RMChannel by pairing each element of the first argument with each element of the second argument, where the multiplicities of the pairs are the products of the multiplicities of each component of the pair in the arguments. In our case, we are really dealing with sets, so each of the multiplicities is one. Hence this amounts to creating a product set from the two arguments and representing it as a multiset. This is further simplified by the first argument just containing one element, the message to be broadcast. The second argument is the set of node identifiers (addresses), except for the reserved value zero, and the address of the node originating the broadcast, which will not receive it. NodeId is declared a multiset (line 2 of Fig. 4) and multiset subtraction is used to remove 0 and the address of the originating node (represented as multisets). Thus the function produces the appropriate pairing of addresses and the message to broadcast the RREQ.

As shown in Fig. 7, RMprocessing models that the receiving node (node) processes the received RREQ ((rreq,n)), and (if appropriate) broadcasts the processed message. The function firstly checks if the received message is superior (function FreshRM in line 2). If so, it checks whether node is the target node. If node is not the target, the function checks whether it can respond with an intermediate node RREP (determined by function Intermediatereply, see line 4). If there is no route entry for the target, the node broadcasts the processed RREQ if the hop limit is positive (function checkhoplimit in line 5). The RREQ is processed by function UpdateRM (see line 6) as discussed below. If the hop limit is reached, the RREQ is dropped. On the other hand, if the node is an intermediate node that can respond with an intermediate node RREP, the function returns empty (see line 8), i.e. the node does not rebroadcast the RREQ. If node is the target and RREQflood is true, node broadcasts an RREQ assembled by function

```
1  fun RMprocessing(node,rreq,n,RREQflood)=
2  if  FreshRM(node,rreq)
3  then if not(Tadd(rreq) = Nodeid(node))
4        then if not(Intermediatereply(node,rreq))
5              then if checkhoplimit(rreq)
6                    then Broadcast(UpdateRM(node,rreq),Nodeid(node))
7                    else empty
8              else empty
9        else if RREQflood
10            then Broadcast(CreateRREQ(node,node),Nodeid(node))
11            else empty
12 else empty;
```

Fig. 7. The RMprocessing Function

CreateRREQ, as shown in line 10. Otherwise, if RREQflood is false, the target node responds with a RREP, so this function returns empty (see line 11). This is also the case when the received message is not superior (see line 12).

As shown in Fig. 8, UpdateRM returns a routing message (RREQ or RREP) (line 8) processed by a node. When a routing message is received, the hop limit of the message hoplimit is decremented by one (line 8). Then for each address (except the target node), its hop count is incremented by one by function UpdateHop (see line 4).

The node removes any additional address from the message if the routing information of this address is not superior to that in its routing table. This is done by function RemoveAdd (Fig. 9), which returns the remaining addresses (see line 3). Then the node increments its own sequence number and appends its own routing information to the address block (a list of addresses and their routing information, see lines 23 and 24 of Fig. 4), using function AddOwnRouteInfo. The updated address block is stored in *NewAdd* (line 5). In line 6, the number of addresses contained in the new address block plus one (the target address), is stored in NUM, the number of addresses in the new message. As shown in line 8, UpdateRM returns the processed message, where the node's address n is the new source address of the message.

```
1  fun UpdateRM((n,rt,tries),(s,((type,hoplimit),(num, tadd, adds))))=
2  let
3  val RemainingAdd = RemoveAdd((n,rt,tries),(s,((type,hoplimit),
4                                    (num,tadd,UpdateHop(adds)))))
5  val NewAdd = AddOwnRouteInfo((n,rt,tries),RemainingAdd)
6  val NUM = length(NewAdd)+1
7  in
8  (n,((type, (hoplimit-1)),(NUM,tadd,NewAdd)))
9  end;
```

Fig. 8. The UpdateRM Function

```
1 fun RemoveAdd(node,(s,(mheader,(num,tadd,[]))))= []
2 | RemoveAdd(node,(s,(mheader,(num,tadd,add1::adds))))=
3 if not(SuperiorRM(node,(s,(mheader,(num,tadd,add1::adds)))))
4 then RemoveAdd(node,(s,(mheader,(num,tadd,adds))))
5 else add1::RemoveAdd(node,(s,(mheader,(num,tadd,adds))));
```

Fig. 9. The RemoveAdd Function

When a node processes a routing message, if the routing information for the originator node is not superior to that stored in its routing table, the message is discarded as already discussed (see line 2 of Fig. 7). Further, if the routing information for the originating node is superior, but that of an additional address is not, then that address is removed from the message to ensure that old routing information is not propagated. Function RemoveAdd does this recursively, as shown in Fig. 9. The function checks each address in the address block in turn (function SuperiorRM in line 3) and returns a new address block in which the routing information of each address is superior.

When a node receives a RREP (transition Receive_RREP) that is destined for it (implemented by its guard), the following actions result. If the node is not the target, but it can forward the RREP, it updates the destination address to the next hop towards the RREP target, and sends it on (implemented in function ProcessRREP). If the node cannot forward the RREP, determined by RTtoTarget, (and it is not the target, determined by Tadd), the node broadcasts a RERR message (function RERRmessage). The RERR contains the RREP target node as an unreachable destination. If the RREP is considered superior (determined by function FreshRM), node also updates its routing table based on the message using function ReceiveRM. If the node is the target, it updates its routing table with the desired entry. Transition Lose_RREP models that the transmitted RREP is lost via an unreliable channel.

The Route Maintenance Module. As shown in Fig. 10, the Route Maintenance module comprises 3 transitions (Route Broken or Route Error, Receive_RERR and Lose_RERR) and 2 places (Nodes and RERR_Message).

Transition Route Broken or Route Error models RERR creation. This occurs when a link to a neighbour (i.e. a next hop in the routing table) is reported as broken. (DYMO actively monitors links to its next hops to determine if they are broken [4].) In this case, valid route entries to all destinations that use that next hop are declared Broken. The event of reporting that a next hop link is broken is modelled non-deterministically. The guard, ARoute, ensures that the route entry for a destination node, n, in the routing table of node is marked as Valid (if a route entry is valid, then it may become broken). When Route Broken or Route Error occurs, all the route entries for the destinations that have the same next hop as destination n are set to Broken, using function RouteError, and all other nodes are informed of this condition by broadcasting a RERR (implemented by function RERRmessage). The RERR includes all the destination addresses (and their sequence numbers) that use the same next hop which has the broken link.

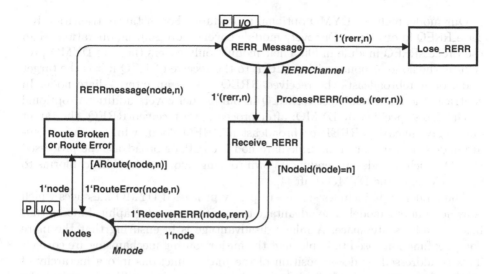

Fig. 10. The Route Maintenance Module

On receiving a RERR (transition Receive_RERR), a node marks the route entries for all destinations affected by the RERR as Broken in its routing table. This is implemented by function ReceiveRERR. The function ProcessRERR returns (MaxNodes-1) copies of the processed RERR. In the processed RERR, the message hop limit is decremented and any unreachable destination that does not result in a change is removed. Lose_RERR models that a node (n) does not receive the RERR. We note that before a node receives a RERR, the wrong routing information is kept in its routing table but it is still flagged as Valid. Hence, the node's view of the environment can be out of date, because it does not have a central view of the whole network.

4 Discussion

In this section we compare and discuss our modelling approach with that of Espensen et al [9]. We concentrate on several important modelling issues, including model structure and MANET topology and demonstrate their effect on the size of the state space.

In [9], hierarchical CPNs with a complex net structure, comprising 4 levels of hierarchy and 14 modules, are used to model operations of the protocol. Their approach may be considered a natural way of modelling a protocol using hierarchical CPNs. The route discovery procedure is divided into several routing operations and each is realised by a substitution transition. Usually, a complicated operation is also divided into several actions, where each action is modelled by a transition or another substitution transition if necessary. In contrast, our model only has 2 hierarchical levels including 3 modules.

Our model includes DYMO optional procedures. For instance, transition Receive_RREQ in our Route Discovery module covers more routing operations than the ProcessRREQ module in [9]. Their module only covers the core DYMO procedure: the node responds with a RREP to the received RREQ if it is the target node, or it rebroadcasts the received RREQ if it is an intermediate node. In contrast, the transition Receive_RREQ in our model covers additional optional mechanisms specified by DYMO: after processing a received RREQ, the target node may unicast a RREP or broadcast a RREQ flood, while an intermediate node can transmit an intermediate node RREP, or broadcast the processed RREQ which includes appending its own routing information. This conforms to section 5.3.4 of the DYMO draft [4].

The model in [9] includes storage of partly processed DYMO messages which is avoided in our model. The advantage is a more compact graphical model that has a smaller state space. A minor disadvantage is in ensuring that the more complex functions used to implement the major routing mechanisms are correct. This is addressed by decomposition of the major functions into a hierarchy of smaller functions, each of which is tested separately and validated against the specification. These smaller functions will basically correspond to the functions used in a model with more net structure.

The topology of a MANET changes due to the movement of nodes. In [9], the marking of a place called Topology indicates the current connection of nodes, and that of the place Topology Changes specifies the possible link changes that can occur. We consider the network topology in a different way. Our model captures the dynamically changing topology by allowing broadcast messages to be received by an arbitrary subset of nodes. This implicitly captures all possible MANET topologies. A partial view of the topology can be derived from a node's routing table entries. This seems to be an accurate representation for the purpose of specification, but will lead to extensive state space explosion. We see the explicit introduction of topology to be useful for state space analysis of DYMO for various fixed topologies, allowing state space explosion to be controlled. However, we do not see it as a necessary part of the specification of the environment of the protocol. Further, the perceived topology of the network by each mobile node at any instant in time may be partial and different from that of other nodes, as it depends on the information stored in each node's routing table. Including a place that explicitly stores topology and allows arbitrary topology changes will include a large number of states that the nodes of the MANET are not aware of (they do not correspond to changes in routing tables). These states are not represented in our model. We conjecture that this will significantly increase the state space in comparison with our model.

We also consider that the extra intermediate events introduced in [9] will lead to an increase in the state space of the model. To test this hypothesis under the same conditions, we modify the model in Fig. 5. In the analysis of [9], only scenarios with a static topology are considered for one route request operating over a reliable network. Our model covers more DYMO routing operations than [9]. Hence, we just consider the route discovery procedure part of the model and fix the originator/target nodes by setting the element Role. The transitions Lose_RREQ and Lose_RREP are omitted as are the procedures for the RREQ flood, intermediate node RREP and broadcasting the RERR. Our analysis model is presented in Fig. 11, in which a place Topology is added in the same way as the topology place is used in [9]. It is typed by the colour set TOP (see line 2 of Fig. 12).

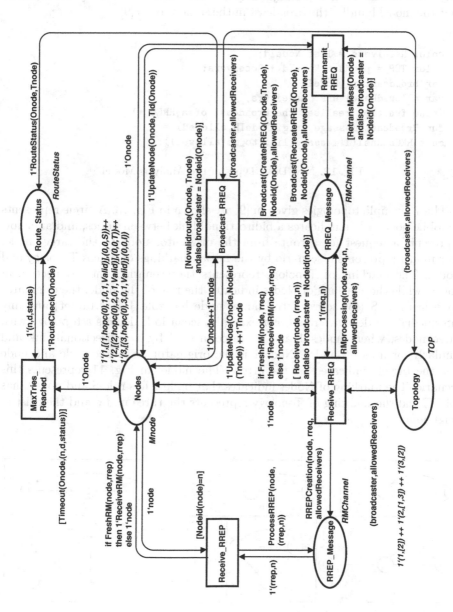

Fig. 11. The Analysis Model of DYMO's Route Discovery Procedure

TOP is a product of a node's address NodeId and its adjacency list Receivers. The expressions on the arcs from Topology to the transitions Broadcast_RREQ, Receive_RREQ and Retransmit_RREQ comprise a pair, (broadcaster, allowedReceivers). Guards of these transitions ensure that Onode or node is the node denoted by broadcaster. The updated function Broadcast (see lines 6-7 of Fig. 12) ensures that only the nodes in the sender's adjacency list can receive the RREQ broadcast, so that the model handles the broadcast in the same way as [9].

```
1   color Receivers = list NodeId;
2   color TOP = product NodeId * Receivers;
3   var broadcaster: NodeId;
4   var allowedReceivers: Receivers;
5   (*the function realises the broadcast of a RREQ *)
6   fun Broadcast(message,originNodeID,allowed) =
7   mult'RMChannel(1'message,list_to_ms(allowed));
```

Fig. 12. New Data Types of the Analysis Model

The 7 example topologies given in [9] are shown in Fig. 13. A circle represents a mobile node, a line indicates a bidirectional link between nodes, and an arrow represents a request for a route from the originator node to the target node. Our model represents a scenario by the initial markings of places Topology and Nodes. The marking of Topology represents the scenario's interconnection of nodes, while the marking of Nodes indicates the role of the node: the originator node has role S (for source), the target node has role T (for target) and any intermediate nodes have role I. For example, scenario B depicts a topology with three nodes, where nodes 1 and 2 are connected by a bidirectional link and similarly for nodes 2 and 3. Node 1 is the originator or source node and node 2 is the target, indicated by the arrow. The model in Fig. 11 represents this scenario. The marking of Nodes indicates that node 1 has role S and node 2 has role T, and the marking of Topology represents the three nodes and their list of neighbours.

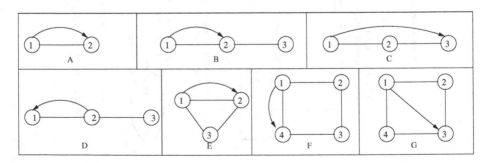

Fig. 13. Example Topologies with One Route Request, from [9]

We investigate the behaviour of DYMO's route discovery procedures for each of the scenarios in Fig. 13. Table 1 records the state space statistics for small values of the transmission attempts parameter, MaxTries. The first two columns give the identification of the scenario as shown in Fig. 13 and the number of Max-Tries (as Tries). The next three columns list the number of Nodes (markings) of the occurrence graph (OG) of our model, that given in [9] (Nodes [9]) and the improvement factor (F1 = Nodes [9]/Nodes). The following three columns represent the arcs of the OGs for the two models and the corresponding improvement factor (F2). The last column lists the dead markings of our model.

Table 1. Comparison of OG Results for the Route Discovery Procedure

Scenario	Tries	Nodes	Nodes [9]	F1	Arcs	Arcs [9]	F2	DeadMarkings
A	1	7	18	2.8	7	24	3.4	[5,7]
	2	26	145	5.6	36	307	8.5	[5,19,22,25,26]
	3	97			179			[5,23..] (9)
B	1	7	18	2.8	7	24	3.4	[5,7]
	2	26	145	5.6	36	307	8.5	[5,19,22,25,26]
	3	97			179			[5,23..] (9)
C	1	19	50	2.6	30	90	3	[16,19]
	2	292	1,260	4.3	722	3,875	5.4	[38,150..] (11)
	3	5413			18,659			[993,5400..] (34)
D	1	19	74	3.9	31	156	5.0	[16,19]
	2	252	2,785	11.1	646	10,203	15.8	[32,217..] (9)
	3	3493			12,578			[35,846..] (23)
E	1	67	446	6.7	129	1,172	9.1	[46,59..] (6)
	2	4,660	166,411	35.7	14,944	804,394	53.8	[471,472..] (67)
	3	373,421			1,749,458			[992,993..] (395)
F	1	163	1,098	6.7	407	3,444	8.5	[91,118..] (6)
	2	36,892			158,064			[494,5157..] (119)
G	1	139	558	4.0	358	1,606	4.9	[128,133,138,139]
	2	21,536			92,261			[21536,21535..] (66)

The results indicate that state spaces of our CPN model are smaller than that in [9], especially when retransmission occurs. For the simplest model (scenario A(1)), the number of nodes for our model is reduced by a factor of 2.8, whereas for the most complex scenario (E(2)), the factor is 35.7. This confirms our hypothesis that the additional net structure (corresponding to storage of partly processed DYMO messages) will cause state space explosion. Also the trend is for the situation to get significantly worse as the number of nodes and retries increases. Espensen et al [9] do not provide any results for Tries > 2, for scenarios, A to E, and for Tries > 1, for scenarios with 4 nodes (F and G), probably because of state space explosion. This is indicated by blanks in Table 1.

The size of the state spaces presented in [28] are smaller than those presented here. This is because we enhanced our model to include the transition MaxTries_Reached and the place Route_Status to model that the source of the data packets is informed when route discovery fails (via an ICMP message). This now provides a better comparison with those of [9], who also model this behaviour. Our results include the appending of address blocks by intermediate nodes, which are not included in [9]. Hence our results are for a more complex route discovery protocol than that modelled in [9].

5   Analysis

This section provides some insight into DYMO's route discovery procedure by examining some of the reachability analysis results. In particular, we consider scenarios B and E in some detail.

5.1   Scenario B

The occurrence graph (OG) generated by Design/CPN for Scenario B with Max-Tries = 1 is shown in Fig. 14 with the details of its markings and binding elements (arcs) in Fig. 15.

This graph comprises 7 markings (global states of the system) represented by rounded boxes and 7 arcs labelled by their associated transition names (sometimes abbreviated). The rounded boxes are labelled by a marking number (to identify the marking) followed by two integers separated by a colon. The first integer represents the number of incoming arcs, and the second integer, the number of outgoing arcs. Hence the initial marking has no incoming arcs (see Marking 1) and terminal or dead markings have no outgoing arcs (see Markings 5 and 7). In Fig. 15, markings are defined in dashed boxes, headed by their marking number. For example, the first dashed box, headed by 1, defines the initial marking. The box lists all the places and the marking of each place. We shall use Mn to denote marking n.

Each arc represents the occurrence of a binding element of the CPN. Each binding element is defined in a dashed box in Fig. 15, headed by its identifier and positioned after its source marking. For example, in the initial marking, M1, the only enabled binding element is transition Broadcast_RREQ (abbreviated BroadcastRQ in Fig. 14) with the binding listed in the dashed box, headed by 1:1→2.

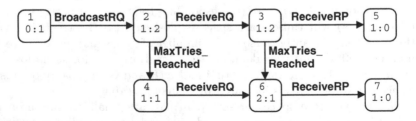

Fig. 14. Occurrence Graph of the Model for Scenario B with MaxTries=1

```
1
RouteDiscovery'Nodes 1: 1'(1,[(1,hopc(0),1,0,1,Valid)],(0,0,S))++ 1'(2,[(2,hopc(0),2,0,1,Valid)],(0,0,T))++
1'(3,[(3,hopc(0),3,0,1,Valid)],(0,0,I))
RouteDiscovery'Route_Status 1: empty
RouteDiscovery'RREP_Message 1: empty
RouteDiscovery'Topology 1: 1'(1,[2])++ 1'(2,[1,3])++ 1'(3,[2])
RouteDiscovery'RREQ_Message 1: empty
```

```
1:1->2
RouteDiscovery'Broadcast_RREQ 1: {broadcaster=1,allowedReceivers=[2],
Tnode=(2,[(2,hopc(0),2,0,1,Valid)],(0,0,T)),Onode=(1,[(1,hopc(0),1,0,1,Valid)],(0,0,S))}
```

```
2
RouteDiscovery'Nodes 1: 1'(1,[(1,hopc(0),1,0,2,Valid)],(2,1,S))++ 1'(2,[(2,hopc(0),2,0,1,Valid)],(0,0,T))++
1'(3,[(3,hopc(0),3,0,1,Valid)],(0,0,I))
RouteDiscovery'Route_Status 1: 1'(1,2,RREQsent)
RouteDiscovery'RREP_Message 1: empty
RouteDiscovery'Topology 1: 1'(1,[2])++ 1'(2,[1,3])++ 1'(3,[2])
RouteDiscovery'RREQ_Message 1: 1'((1,((RREQ,3),(2,(2,0,unknown,0),[(1,2,hopc(0),0)]))),2)
```

```
2:2->3
RouteDiscovery'Receive_RREQ 1: {rreq=(1,((RREQ,3),(2,(2,0,unknown,0),[(1,2,hopc(0),0)]))),
node=(2,[(2,hopc(0),2,0,1,Valid)],(0,0,T)),n=2,broadcaster=2,allowedReceivers=[1,3]}
```

```
3:3->4
RouteDiscovery'MaxTries_Reached 1: {status=RREQsent,n=1,d=2,Onode=(1,[(1,hopc(0),1,0,2,Valid)],(2,1,S))}
```

```
3
RouteDiscovery'Nodes 1: 1'(1,[(1,hopc(0),1,0,2,Valid)],(2,1,S))++
1'(2,[(1,hopc(1),1,0,2,Valid),(2,hopc(0),2,0,2,Valid)],(0,0,T))++ 1'(3,[(3,hopc(0),3,0,1,Valid)],(0,0,I))
RouteDiscovery'Route_Status 1: 1'(1,2,RREQsent)
RouteDiscovery'RREP_Message 1: 1'((2,((RREP,3),(2,(1,2,hopc(1),0),[(2,2,hopc(0),0)]))),1)
RouteDiscovery'Topology 1: 1'(1,[2])++ 1'(2,[1,3])++ 1'(3,[2])
RouteDiscovery'RREQ_Message 1: empty
```

```
4:3->5
RouteDiscovery'Receive_RREP 1: {rrep=(2,((RREP,3),(2,(1,2,hopc(1),0),[(2,2,hopc(0),0)]))),
node=(1,[(1,hopc(0),1,0,2,Valid)],(2,1,S)),n=1}
```

```
5:3->6
RouteDiscovery'MaxTries_Reached 1: {status=RREQsent,n=1,d=2,Onode=(1,[(1,hopc(0),1,0,2,Valid)],(2,1,S))}
```

```
5
RouteDiscovery'Nodes 1: 1'(1,[(1,hopc(0),1,0,2,Valid),(2,hopc(1),2,0,2,Valid)],(0,0,S))++
1'(2,[(1,hopc(1),1,0,2,Valid),(2,hopc(0),2,0,2,Valid)],(0,0,T))++ 1'(3,[(3,hopc(0),3,0,1,Valid)],(0,0,I))
RouteDiscovery'Route_Status 1: 1'(1,2,RREQsent)
RouteDiscovery'RREP_Message 1: empty
RouteDiscovery'Topology 1: 1'(1,[2])++ 1'(2,[1,3])++ 1'(3,[2])
RouteDiscovery'RREQ_Message 1: empty
```

```
4
RouteDiscovery'Nodes 1: 1'(1,[(1,hopc(0),1,0,2,Valid)],(2,1,S))++ 1'(2,[(2,hopc(0),2,0,1,Valid)],(0,0,T))++
1'(3,[(3,hopc(0),3,0,1,Valid)],(0,0,I))
RouteDiscovery'Route_Status 1: 1'(1,2,ICMPsent)
RouteDiscovery'RREP_Message 1: empty
RouteDiscovery'Topology 1: 1'(1,[2])++ 1'(2,[1,3])++ 1'(3,[2])
RouteDiscovery'RREQ_Message 1: 1'((1,((RREQ,3),(2,(2,0,unknown,0),[(1,2,hopc(0),0)]))),2)
```

```
6:4->6
RouteDiscovery'Receive_RREQ 1: {rreq=(1,((RREQ,3),(2,(2,0,unknown,0),[(1,2,hopc(0),0)]))),
node=(2,[(2,hopc(0),2,0,1,Valid)],(0,0,T)),n=2,broadcaster=2,allowedReceivers=[1,3]}
```

```
6
RouteDiscovery'Nodes 1: 1'(1,[(1,hopc(0),1,0,2,Valid)],(2,1,S))++
1'(2,[(1,hopc(1),1,0,2,Valid),(2,hopc(0),2,0,2,Valid)],(0,0,T))++ 1'(3,[(3,hopc(0),3,0,1,Valid)],(0,0,I))
RouteDiscovery'Route_Status 1: 1'(1,2,ICMPsent)
RouteDiscovery'RREP_Message 1: 1'((2,((RREP,3),(2,(1,2,hopc(1),0),[(2,2,hopc(0),0)]))),1)
RouteDiscovery'Topology 1: 1'(1,[2])++ 1'(2,[1,3])++ 1'(3,[2])
RouteDiscovery'RREQ_Message 1: empty
```

```
7:6->7
RouteDiscovery'Receive_RREP 1: {rrep=(2,((RREP,3),(2,(1,2,hopc(1),0),[(2,2,hopc(0),0)]))),
node=(1,[(1,hopc(0),1,0,2,Valid)],(2,1,S)),n=1}
```

```
7
RouteDiscovery'Nodes 1: 1'(1,[(1,hopc(0),1,0,2,Valid),(2,hopc(1),2,0,2,Valid)],(0,0,S))++
1'(2,[(1,hopc(1),1,0,2,Valid),(2,hopc(0),2,0,2,Valid)],(0,0,T))++ 1'(3,[(3,hopc(0),3,0,1,Valid)],(0,0,I))
RouteDiscovery'Route_Status 1: 1'(1,2,ICMPsent)
RouteDiscovery'RREP_Message 1: empty
RouteDiscovery'Topology 1: 1'(1,[2])++ 1'(2,[1,3])++ 1'(3,[2])
RouteDiscovery'RREQ_Message 1: empty
```

Fig. 15. Markings and Binding Elements for Fig. 14

Inspecting the place Topology in M1, shows that the topology corresponds to scenario B. The marking of place Nodes shows that there are 3 nodes, each with only one route entry to itself in its routing table. Node 1 is the originator or source node (indicated by S), node 2 is the target node (indicated by T) and node 3 is an intermediate node (indicated by I). The rest of the places are empty.

The OG (Fig. 14) shows that after initiating route discovery in node 1, the procedure ends in either M5 or M7. The path M1, M2, M3, M5 results in successful route establishment (and the data packets being forwarded to node 2). The marking of Nodes in M5 (see Fig. 15) shows that node 1 now has a valid route entry for node 2 (and vice versa). This corresponds to the sequential occurrence of transitions Broadcast_RREQ (by node 1), Receive_RREQ (by node 2) and Receive_RREP (by node 1). In contrast, M7 corresponds to the situation where the RREP was not received before a timeout occurred in node 1 resulting in the source of the data packets being informed (by an ICMP message) that no route had been found in time. This results in the data packets being discarded. However (because no loss has been considered in this model) the RREP does arrive after the ICMP message has been sent, and the route is established. This is useful for when the source tries again. This corresponds to the occurrence of transition MaxTries_Reached after Broadcast_RREQ (which sets MaxTries = 1) but before Receive_RREP.

5.2   Scenarios A,C and D

The two dead markings for scenarios A, C and D with MaxTries = 1, listed in Table 1, are similar to markings M5 and M7 for Scenario B. This corresponds to the results in [9] except that in their work (due to a modelling error) the ICMP message can be sent after the RREP message has arrived, as the routing table is not checked to ensure that a route has not been discovered.

The situation when there are retransmissions (MaxTries>1) is more complex. More terminal markings are involved as sequence numbers are incremented when a retransmission occurs. Thus a set of terminal markings, similar to M5 and M7 for Scenario B, are obtained with different values of sequence numbers.

5.3   Scenario E

We now examine Scenario E when MaxTries = 1. A partial OG is given in Fig. 16. It shows the initial marking (M1) and traces to all 6 dead markings (M46, M59, M60, M61, M66, M67). M1 in Fig. 17 shows that all nodes only have route entries to themselves and that the originator is node 1 which requires a route to the target, node 2.

Terminal markings M46 and M59 (see Fig. 17) are states where nodes 1 and 2 have route entries to each other, and node 3 has a route entry for node 1. M46 corresponds to successful route discovery whereas M59 shows that the RREP arrived from node 2 after MaxTries was reached, and hence the data packets were discarded and an ICMP message was sent (Route_Status marked by (1, 2, ICMPsent)). A possible scenario is given by the left branch of Fig. 16. Node 1

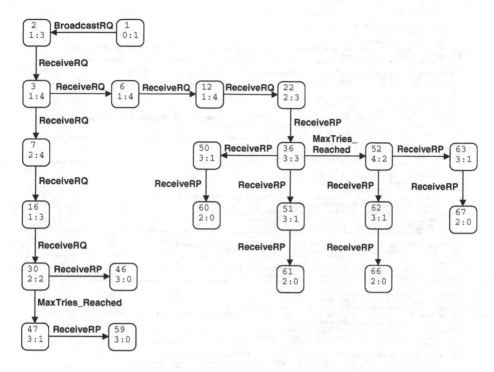

Fig. 16. Partial Occurrence Graph for Scenario E with MaxTries=1

broadcasts a RREQ which is received by nodes 2 and 3. Node 3 rebroadcasts the RREQ to nodes 1 and 2, and node 2 sends a RREP to node 1. Node 2 drops the RREQ sent by node 3, since the message is not superior, and node 1 discards the RREQ since it is the originator. Then node 1 receives the RREP and updates its routing table (marking M46). If transition MaxTries_Reached occurs before the RREP is received, then marking M59 results.

As shown in Fig. 18, M60 and M66 are terminal states where each node has an entry to all other nodes in the network. Scenarios in which this can occur are shown by the right branch of the OG in Fig. 16, which includes M6. The scenario requires that the RREQ first gets to node 2 via node 3, and an RREP is sent back to node 1 via node 2. Meanwhile node 2 receives the RREQ from node 1 and sends back a second RREP directly to node 1. If the RREP sent via node 3 is received before the direct RREP by node 1 before the timeout occurs, then the route discovery is successful, and all route entries are included in the routing tables. This corresponds to terminal marking M60. If the timeout occurs before the RREPs are received (transition from M36 to M52) then the data packets are dropped and an ICMP message is sent. The route entries are still updated as seen in marking M66.

M67 and M61 are similar except that node 1 does not have a route entry for node 3, as shown in Fig. 18. This is because the RREP sent by node 2 directly to node 1 is received before the RREP sent via node 3. Because the hop count for

```
1
RouteDiscovery'Topology 1: 1'(1,[2,3])++ 1'(2,[1,3])++ 1'(3,[1,2])
RouteDiscovery'Route_Status 1: empty
RouteDiscovery'RREP_Message 1: empty
RouteDiscovery'Nodes 1: 1'(1,[(1,hopc(0),1,0,1,Valid)],(0,0,S))++ 1'(2,[(2,hopc(0),2,0,1,Valid)],(0,0,T))++ 1'(3,[(3,hopc(0),3,0,1,Valid)],(0,0,I))
RouteDiscovery'RREQ_Message 1: empty
```
```
1:1->2
RouteDiscovery'Broadcast_RREQ 1: {broadcaster=1,allowedReceivers=[2,3],Tnode=(2,[(2,hopc(0),2,0,1,Valid)],(0,0,T)),
Onode=(1,[(1,hopc(0),1,0,1,Valid)],(0,0,S))}
```
```
2
RouteDiscovery'Topology 1: 1'(1,[2,3])++ 1'(2,[1,3])++ 1'(3,[1,2])
RouteDiscovery'Route_Status 1: 1'(1,2,RREQsent)
RouteDiscovery'RREP_Message 1: empty
RouteDiscovery'Nodes 1: 1'(1,[(1,hopc(0),1,0,2,Valid)],(2,1,S))++ 1'(2,[(2,hopc(0),2,0,1,Valid)],(0,0,T))++ 1'(3,[(3,hopc(0),3,0,1,Valid)],(0,0,I))
RouteDiscovery'RREQ_Message 1: 1'((1,((RREQ,3),(2,(2,0,unknown,0),[(1,2,hopc(0),0)]))),2)++
1'((1,((RREQ,3),(2,(2,0,unknown,0),[(1,2,hopc(0),0)]))),3)
```
```
2:2->3
RouteDiscovery'Receive_RREQ 1: {rreq=(1,((RREQ,3),(2,(2,0,unknown,0),[(1,2,hopc(0),0)]))),
node=(3,[(3,hopc(0),3,0,1,Valid)],(0,0,I)),n=3,broadcaster=3,allowedReceivers=[1,2]}
```
```
3
RouteDiscovery'Topology 1: 1'(1,[2,3])++ 1'(2,[1,3])++ 1'(3,[1,2])
RouteDiscovery'Route_Status 1: 1'(1,2,RREQsent)
RouteDiscovery'RREP_Message 1: empty
RouteDiscovery'Nodes 1: 1'(1,[(1,hopc(0),1,0,2,Valid)],(2,1,S))++ 1'(2,[(2,hopc(0),2,0,1,Valid)],(0,0,T))++
1'(3,[(1,hopc(1),1,0,2,Valid),(3,hopc(0),3,0,2,Valid)],(0,0,I))
RouteDiscovery'RREQ_Message 1: 1'((1,((RREQ,3),(2,(2,0,unknown,0),[(1,2,hopc(0),0)]))),2)++
1'((3,((RREQ,2),(3,(2,0,unknown,0),[(1,2,hopc(1),0),(3,2,hopc(0),0)]))),1)++
1'((3,((RREQ,2),(3,(2,0,unknown,0),[(1,2,hopc(1),0),(3,2,hopc(0),0)]))),2)
```
```
5:3->6
RouteDiscovery'Receive_RREQ 1: {rreq=(3,((RREQ,2),(3,(2,0,unknown,0),[(1,2,hopc(1),0),(3,2,hopc(0),0)]))),
node=(2,[(2,hopc(0),2,0,1,Valid)],(0,0,T)),n=2,broadcaster=2,allowedReceivers=[1,3]}
```
```
6:3->7
RouteDiscovery'Receive_RREQ 1: {rreq=(1,((RREQ,3),(2,(2,0,unknown,0),[(1,2,hopc(0),0)]))),
node=(2,[(2,hopc(0),2,0,1,Valid)],(0,0,T)),n=2,broadcaster=2,allowedReceivers=[1,3]}
```
```
7
RouteDiscovery'Topology 1: 1'(1,[2,3])++ 1'(2,[1,3])++ 1'(3,[1,2])
RouteDiscovery'Route_Status 1: 1'(1,2,RREQsent)
RouteDiscovery'RREP_Message 1: 1'((2,((RREP,3),(2,(1,2,hopc(1),0),[(2,2,hopc(0),0)]))),1)
RouteDiscovery'Nodes 1: 1'(1,[(1,hopc(0),1,0,2,Valid)],(2,1,S))++ 1'(2,[(1,hopc(1),1,0,2,Valid),(2,hopc(0),2,0,2,Valid)],(0,0,T))++
1'(3,[(1,hopc(1),1,0,2,Valid),(3,hopc(0),3,0,2,Valid)],(0,0,I))
RouteDiscovery'RREQ_Message 1: 1'((3,((RREQ,2),(3,(2,0,unknown,0),[(1,2,hopc(1),0),(3,2,hopc(0),0)]))),1)++
1'((3,((RREQ,2),(3,(2,0,unknown,0),[(1,2,hopc(1),0),(3,2,hopc(0),0)]))),2)
```
```
18:7->16
RouteDiscovery'Receive_RREQ 1: {rreq=(3,((RREQ,2),(3,(2,0,unknown,0),[(1,2,hopc(1),0),(3,2,hopc(0),0)]))),
node=(2,[(1,hopc(1),1,0,2,Valid),(2,hopc(0),2,0,2,Valid)],(0,0,T)),n=2,broadcaster=2,allowedReceivers=[1,3]}
```
```
16
RouteDiscovery'Topology 1: 1'(1,[2,3])++ 1'(2,[1,3])++ 1'(3,[1,2])
RouteDiscovery'Route_Status 1: 1'(1,2,RREQsent)
RouteDiscovery'RREP_Message 1: 1'((2,((RREP,3),(2,(1,2,hopc(1),0),[(2,2,hopc(0),0)]))),1)
RouteDiscovery'Nodes 1: 1'(1,[(1,hopc(0),1,0,2,Valid)],(2,1,S))++ 1'(2,[(1,hopc(1),1,0,2,Valid),(2,hopc(0),2,0,2,Valid)],(0,0,T))++
1'(3,[(1,hopc(1),1,0,2,Valid),(3,hopc(0),3,0,2,Valid)],(0,0,I))
RouteDiscovery'RREQ_Message 1: 1'((3,((RREQ,2),(3,(2,0,unknown,0),[(1,2,hopc(1),0),(3,2,hopc(0),0)]))),1)
```
```
45:16->30
RouteDiscovery'Receive_RREQ 1: {rreq=(3,((RREQ,2),(3,(2,0,unknown,0),[(1,2,hopc(1),0),(3,2,hopc(0),0)]))),
node=(1,[(1,hopc(0),1,0,2,Valid)],(2,1,S)),n=1,broadcaster=1,allowedReceivers=[2,3]}
```
```
30
RouteDiscovery'Topology 1: 1'(1,[2,3])++ 1'(2,[1,3])++ 1'(3,[1,2])
RouteDiscovery'Route_Status 1: 1'(1,2,RREQsent)
RouteDiscovery'RREP_Message 1: 1'((2,((RREP,3),(2,(1,2,hopc(1),0),[(2,2,hopc(0),0)]))),1)
RouteDiscovery'Nodes 1: 1'(1,[(1,hopc(0),1,0,2,Valid)],(2,1,S))++ 1'(2,[(1,hopc(1),1,0,2,Valid),(2,hopc(0),2,0,2,Valid)],(0,0,T))++
1'(3,[(1,hopc(1),1,0,2,Valid),(3,hopc(0),3,0,2,Valid)],(0,0,I))
RouteDiscovery'RREQ_Message 1: empty
```
```
81:30->46
RouteDiscovery'Receive_RREP 1: {rrep=(2,((RREP,3),(2,(1,2,hopc(1),0),[(2,2,hopc(0),0)]))),node=(1,[(1,hopc(0),1,0,2,Valid)],(2,1,S)),n=1}
```
```
82:30->47
RouteDiscovery'MaxTries_Reached 1: {status=RREQsent,n=1,d=2,Onode=(1,[(1,hopc(0),1,0,2,Valid)],(2,1,S))}
```
```
46
RouteDiscovery'Topology 1: 1'(1,[2,3])++ 1'(2,[1,3])++ 1'(3,[1,2])
RouteDiscovery'Route_Status 1: 1'(1,2,RREQsent)
RouteDiscovery'RREP_Message 1: empty
RouteDiscovery'Nodes 1: 1'(1,[(1,hopc(0),1,0,2,Valid),(2,hopc(1),2,0,2,Valid)],(0,0,S))++
1'(2,[(1,hopc(1),1,0,2,Valid),(2,hopc(0),2,0,2,Valid)],(0,0,T))++ 1'(3,[(1,hopc(1),1,0,2,Valid),(3,hopc(0),3,0,2,Valid)],(0,0,I))
RouteDiscovery'RREQ_Message 1: empty
```
```
47
RouteDiscovery'Topology 1: 1'(1,[2,3])++ 1'(2,[1,3])++ 1'(3,[1,2])
RouteDiscovery'Route_Status 1: 1'(1,2,ICMPsent)
RouteDiscovery'RREP_Message 1: 1'((2,((RREP,3),(2,(1,2,hopc(1),0),[(2,2,hopc(0),0)]))),1)
RouteDiscovery'Nodes 1: 1'(1,[(1,hopc(0),1,0,2,Valid)],(2,1,S))++ 1'(2,[(1,hopc(1),1,0,2,Valid),(2,hopc(0),2,0,2,Valid)],(0,0,T))++
1'(3,[(1,hopc(1),1,0,2,Valid),(3,hopc(0),3,0,2,Valid)],(0,0,I))
RouteDiscovery'RREQ_Message 1: empty
```
```
111:47->59
RouteDiscovery'Receive_RREP 1: {rrep=(2,((RREP,3),(2,(1,2,hopc(1),0),[(2,2,hopc(0),0)]))),node=(1,[(1,hopc(0),1,0,2,Valid)],(2,1,S)),n=1}
```
```
59
RouteDiscovery'Topology 1: 1'(1,[2,3])++ 1'(2,[1,3])++ 1'(3,[1,2])
RouteDiscovery'Route_Status 1: 1'(1,2,ICMPsent)
RouteDiscovery'RREP_Message 1: empty
RouteDiscovery'Nodes 1: 1'(1,[(1,hopc(0),1,0,2,Valid),(2,hopc(1),2,0,2,Valid)],(0,0,S))++
1'(2,[(1,hopc(1),1,0,2,Valid),(2,hopc(0),2,0,2,Valid)],(0,0,T))++ 1'(3,[(1,hopc(1),1,0,2,Valid),(3,hopc(0),3,0,2,Valid)],(0,0,I))
RouteDiscovery'RREQ_Message 1: empty
```

Fig. 17. Trace of the Left Branch of the OG for Scenario E with MaxTries=1

```
┌60 ─────────────────────────────────────────────────────────┐
│ RouteDiscovery'Topology 1: 1'(1,[2,3])++ 1'(2,[1,3])++ 1'(3,[1,2])                        │
│ RouteDiscovery'Route_Status 1: 1'(1,2,RREQsent)                                           │
│ RouteDiscovery'RREP_Message 1: empty                                                       │
│ RouteDiscovery'Nodes 1: 1'(1,[((1,hopc(0),1,0,2,Valid),(2,hopc(1),2,0,3,Valid),(3,hopc(1),3,0,3,Valid)],(0,0,S))++ │
│ 1'(2,[((1,hopc(1),1,0,2,Valid),(2,hopc(0),2,0,3,Valid),(3,hopc(1),3,0,2,Valid)],(0,0,T))++ │
│ 1'(3,[((1,hopc(1),1,0,2,Valid),(2,hopc(1),2,0,2,Valid),(3,hopc(0),3,0,3,Valid)],(0,0,I))   │
│ RouteDiscovery'RREQ_Message 1: empty                                                       │
├61 ─────────────────────────────────────────────────────────┤
│ RouteDiscovery'Topology 1: 1'(1,[2,3])++ 1'(2,[1,3])++ 1'(3,[1,2])                        │
│ RouteDiscovery'Route_Status 1: 1'(1,2,RREQsent)                                           │
│ RouteDiscovery'RREP_Message 1: empty                                                       │
│ RouteDiscovery'Nodes 1: 1'(1,[((1,hopc(0),1,0,2,Valid),(2,hopc(1),2,0,3,Valid)],(0,0,S))++ │
│ 1'(2,[((1,hopc(1),1,0,2,Valid),(2,hopc(0),2,0,3,Valid),(3,hopc(1),3,0,2,Valid)],(0,0,T))++ │
│ 1'(3,[((1,hopc(1),1,0,2,Valid),(2,hopc(1),2,0,2,Valid),(3,hopc(0),3,0,3,Valid)],(0,0,I))   │
│ RouteDiscovery'RREQ_Message 1: empty                                                       │
├66 ─────────────────────────────────────────────────────────┤
│ RouteDiscovery'Topology 1: 1'(1,[2,3])++ 1'(2,[1,3])++ 1'(3,[1,2])                        │
│ RouteDiscovery'Route_Status 1: 1'(1,2,ICMPsent)                                           │
│ RouteDiscovery'RREP_Message 1: empty                                                       │
│ RouteDiscovery'Nodes 1: 1'(1,[((1,hopc(0),1,0,2,Valid),(2,hopc(1),2,0,3,Valid),(3,hopc(1),3,0,3,Valid)],(0,0,S))++ │
│ 1'(2,[((1,hopc(1),1,0,2,Valid),(2,hopc(0),2,0,3,Valid),(3,hopc(1),3,0,2,Valid)],(0,0,T))++ │
│ 1'(3,[((1,hopc(1),1,0,2,Valid),(2,hopc(1),2,0,2,Valid),(3,hopc(0),3,0,3,Valid)],(0,0,I))   │
│ RouteDiscovery'RREQ_Message 1: empty                                                       │
├67 ─────────────────────────────────────────────────────────┤
│ RouteDiscovery'Topology 1: 1'(1,[2,3])++ 1'(2,[1,3])++ 1'(3,[1,2])                        │
│ RouteDiscovery'Route_Status 1: 1'(1,2,ICMPsent)                                           │
│ RouteDiscovery'RREP_Message 1: empty                                                       │
│ RouteDiscovery'Nodes 1: 1'(1,[((1,hopc(0),1,0,2,Valid),(2,hopc(1),2,0,3,Valid)],(0,0,S))++ │
│ 1'(2,[((1,hopc(1),1,0,2,Valid),(2,hopc(0),2,0,3,Valid),(3,hopc(1),3,0,2,Valid)],(0,0,T))++ │
│ 1'(3,[((1,hopc(1),1,0,2,Valid),(2,hopc(1),2,0,2,Valid),(3,hopc(0),3,0,3,Valid)],(0,0,I))   │
│ RouteDiscovery'RREQ_Message 1: empty                                                       │
└─────────────────────────────────────────────────────────────┘
```

Fig. 18. Dead Markings: 60, 61, 66 and 67

the direct route from node 1 to node 2 (which has been established by the receipt of the first RREP) is less than that via node 3, the second RREP is dropped, and hence the routing table in node 1 is not updated with an entry to node 3.

In an environment without packet loss, it is highly unlikely that terminal markings M60, M61, M66 and M67 will occur as the direct path should be faster than the indirect path via node 3. However, in the case of loss (and retransmission), these terminal states are possible, as multipath fading interference could cause loss on the direct path and not on the indirect path.

We observe that DYMO's procedure of dropping packets, such as the returned broadcast from node 3 to node 1, means that the routing table will not be updated with the route entry to the node that returned the broadcast. This seems inefficient, as the routing table may as well be updated, so that data packets can be forwarded to the new node, without having to action the route discovery procedure again. Again the situation is very complex if retransmissions are allowed, as incrementing sequence numbers exacerbates the state space explosion problem.

5.4 Strongly Connected Components

For all scenarios given in Table 1 the strongly connected component graphs were generated and shown to have the same number of nodes and arcs as the occurrence graphs (OGs). This means that there are no cycles in the OGs and thus the route discovery procedure always terminates in the dead markings indicated which all have the desired route entries for the requested route.

6   Conclusions and Future Work

In this paper, we have illustrated how a reactive MANET routing protocol, DYMO, can be formalised using Coloured Petri Nets, in a way that reduces state space explosion, so that the model is more amenable to analysis. Although the model looks deceptively simple, it models most of the DYMO protocol specification, including options and route maintenance procedures. We provide a simple way of modelling the dynamically changing network topology, by allowing messages that are broadcast to be received by an arbitrary subset of the rest of the nodes.

We compare our approach with that of Espensen et al [8, 9]. Their model includes much more net structure (14 modules instead of 3) and therefore allows easier visualisation of some of the procedures and how different parts of the model interact. This is eminently suited to implementation, but has the drawback of increased state space explosion. Their model covers less routing operations developed for the DYMO specification than ours. A further difference is that they model the topology of the MANET explicitly in a place. This is useful for guided state space analysis of the model for fixed topologies, but is not part of the DYMO specification. To demonstrate the difference in state space performance due to the different structure of the models, we present some initial results for the route discovery procedure for a fixed set of network topologies and designated originator and target nodes as documented in [9]. We find that the compact net structure leads to significantly smaller state spaces. We can thus provide more results than those presented in [9]. We also believe we have found a minor error in the CPN model of [9], where an ICMP message can be sent after the RREP has been received by the origin node.

Finally, we investigate the operation of DYMO's route discovery procedures using state spaces for two of the fixed topology scenarios, when there is no loss and no retransmission. This provides insight into DYMO's core routing procedures and demonstrates that under ideal conditions, the protocol will establish the desired routes. This occurs whether or not the source of the data packets is informed of failure. Although this appears strange at first sight, this is reasonable behaviour. The procedure can fail due the RREP not being received in time, but later the RREP is successfully received, and the routing tables are updated, ready for the next set of data packets for that destination. Scenario E shows that the route discovery procedure can provide more route entries in the routing tables of the nodes than are necessary for obtaining a route to the target, depending on different delays through the network nodes. This can provide an advantage when these routes are required for new data packet streams to other destinations.

In the future we plan to extend these results to consider an unreliable network (allowing for loss), optional DYMO procedures and to verify the loop freedom property. This may also include topology guided analysis as well as allowing for arbitrary topology changes. The use of advanced methods of state space analysis, such as symmetries or parametric verification ideas [10], may also be employed.

Acknowledgments. The authors would like to express their deep appreciation for Guy Gallasch's generous assistance with many aspects of this paper, including: fruitful discussions on technical content; advice on modelling, particularly concerning the use of ML; and his invaluable help with the LaTeX word processsing system. We are grateful for the constructive comments of the anonymous reviewers which have helped to improve this paper. Cong Yuan has been supported by an Australian Government International Postgraduate Research Scholarship.

References

1. Basagni, S., Conti, M., Giordano, S., Stojmenovic, I.: Mobile Ad Hoc Networking. IEEE Press, New York (2004)
2. Bhargavan, K., Obradovic, D., Gunter, C.A.: Formal Verification of Standards for Distance Vector Routing Protocols. JACM 49(4), 538–576 (2002)
3. Cavalli, A., Grepet, C., Maag, S., Tortajada, V.: A Validation Model for the DSR Protocol. In: 24th International Conference on Distributed Computing Systems Workshops, pp. 768–773 (2004)
4. Charkeres, I.D., Perkins, C.E.: Dynamic MANET On-demand (DYMO) Routing Protocol. IETF Internet Draft, draft-ietf-manet-dymo-10.txt (2007)
5. Chiyangwa, S., Kwiatkowska, M.: A Timing Analysis of AODV. In: Steffen, M., Zavattaro, G. (eds.) FMOODS 2005. LNCS, vol. 3535, pp. 306–321. Springer, Heidelberg (2005)
6. Clausen, T., Dearlove, C., Dean, J., Adjih, C.: Generalized MANET Packet/Message Format (2007),
   http://www.ietf.org/internet-drafts/draft-ietf-manet-packetbb-07.txt
7. Design/CPN, http://www.daimi.au.dk/designCPN
8. Espensen, K.L., Kjeldsen, M.K., Kristensen, L.M.: Towards Modelling and Verification of the DYMO Routing Protocol for Mobile Ad-hoc Networks. In: 8th Workshop and Tutorial on Practical Use of Coloured Petri Nets and the CPN Tools, Aarhus, Denmark, pp. 243–262 (2007)
9. Espensen, K.L., Kjeldsen, M.K., Kristensen, L.M.: Modelling and Initial Validation of the DYMO Routing Protocol for Mobile Ad-Hoc Networks. In: van Hee, K.M., Valk, R. (eds.) PETRI NETS 2008. LNCS, vol. 5062, pp. 152–170. Springer, Heidelberg (2008)
10. Gallasch, G.E., Billington, J.: A Parametric State Space for the Analysis of the Infinite Class of Stop-and-Wait Protocols. In: Valmari, A. (ed.) SPIN 2006. LNCS, vol. 3925, pp. 201–218. Springer, Heidelberg (2006)
11. Gordon, M.J.C., Melham, T.F.: Introduction to HOL: A Theorem Proving Environment for Higher Order Logic. Cambridge University Press, Cambridge (1993)
12. Holzmann, G.J.: The Spin Model Checker, Primer and Reference Manual. Addison-Wesley, Reading (2003)
13. IETF. Mobile Ad-hoc Networks (manet),
    http://www.ietf.org/html.charters/manet-charter.html

14. ITU-T. Recommendation Z.100: Specification and Description Language. International Telecomunication Union, Geneva (2007)
15. Jensen, K.: Coloured Petri Nets: Basic Concepts, Analysis Methods and Practical Use, vol. 1–3. Springer, Heidelberg (1997)
16. Jensen, K., Kristensen, L.M., Wells, L.: Coloured Petri Nets and CPN Tools for Modelling and Validation of Concurrent Systems. International Journal on Software Tools for Technology Transfer 9(3-4), 213–254 (2007)
17. Johnson, D.B., Maltz, D.A., Hu, Y.C.: The Dynamic Source Routing Protocol for Mobile Ad hoc Networks (DSR). IETF MANET Working Group, draft-ietf-manet-dsr-09.txt (2003)
18. Kristensen, L.M., Christensen, S., Jensen, K.: The Practitioner's Guide to Coloured Petri Nets. International Journal on Software Tools for Technology Transfer 2(2), 98–132 (1998)
19. Larsen, K.G., Pettersson, P., Wang, Y.: UPPAAL in a Nutshell. International Journal on Software Tools for Technology Transfer 1(1-2), 134–152 (1997)
20. Obradovic, D.: Formal Analysis of Routing Protocols. PhD thesis, University of Pennsylvania (2002)
21. Perkins, C.E.: Ad Hoc Networking. Addison-Wesley, Reading (2001)
22. Perkins, C.E., Royer, E.M.: Ad-hoc On-demand Distance Vector Routing. In: 2nd IEEE Workshop on Mobile Computing Systems and Applications, New Orleans, Louisiana, USA, pp. 90–100 (1999)
23. de Renesse, R., Aghvami, A.H.: Formal Verification of Ad-hoc Routing Protocols Using SPIN Model Checker. In: 12th IEEE Mediterranean Electrotechnical Conference, Dubrovnik, Croatia, Piscataway, N.J, vol. 3, pp. 1177–1182. IEEE, Los Alamitos (2004)
24. Wibling, O., Parrow, J., Pears, A.: Automatized Verification of Ad Hoc Routing Protocols. In: de Frutos-Escrig, D., Núñez, M. (eds.) FORTE 2004. LNCS, vol. 3235, pp. 343–358. Springer, Heidelberg (2004)
25. Xiong, C., Murata, T., Leigh, J.: An Approach for Verifying Routing Protocols in Mobile Ad Hoc Networks Using Petri Nets. In: Proceedings of IEEE 6th CAS Symposium on Emerging Technologies: Frontiers of Mobile and Wireless Communication, Shanghai, China, pp. 537–540 (2004)
26. Xiong, C., Murata, T., Tsai, J.: Modeling and Simulation of Routing Protocol for Mobile Ad Hoc Wireless Networks Using Colored Petri Nets. In: Proceedings of Workshop on Formal Methods Applied to Defence Systems in Formal Methods in Software Engineering and Defence Systems, Conferences in Research and Practice in Information Technology, vol. 12, pp. 145–153 (2002)
27. Yuan, C., Billington, J.: A Coloured Petri Net Model of the Dynamic MANET On-demand Routing Protocol. In: 7th Workshop and Tutorial on Practical Use of Coloured Petri Nets and the CPN Tools, Department of Computer Science Technical Report, DAIMI PB 579, Aarhus, Denmark, pp. 37–56 (2006)
28. Yuan, C., Billington, J.: On Modelling the Dynamic MANET On-demand (DYMO) Routing Protocol. In: International Workshop on Petri Nets and Distributed Systems (PNDS 2008), Xi'an, China, pp. 47–66 (2008)

Modelling Mobile IP with Mobile Petri Nets

Charles Lakos

School of Computer Science, University of Adelaide
Adelaide, SA 5005, Australia

Abstract. Mobile systems explore the interplay between *locality* and *connectivity*. A subsystem may have a connection to a remote subsystem and use this for communication. It may be necessary or desirable to move the subsystem *close to* the other in order to communicate. Alternatively, the method of communication may vary depending on proximity. This paper reviews a Petri Net formalisation for mobile systems which is intended to harness the intuitive graphical representation of Petri Nets and the long history of associated analysis techniques.

The main contribution of the current paper is to assess the above formalism by using it to model and simulate Mobile IP, an Internet standard which caters for mobile nodes using IP version 4 addresses. These addresses indicate a fixed point of attachment to the Internet and the protocol caters for nodes being *away from home*. By defining the model as a Mobile Petri Net, the graphical notation helps to convey the flow of information, and the executable nature of the model opens the way to simulation, state space exploration and model checking.

1 Introduction

This paper presents an experiment in modelling and simulation of a realistic mobile system in a Petri Net formalism, called *Mobile Petri Nets*. This formalism was proposed in [14] and was intended to build on the intuitive graphical representation of Petri Nets, the long history of associated analysis techniques [23,24], and to be able to reason about causality and concurrency [4].

Mobile Petri Nets attempt to capture the interplay between locality and connectivity, which is the essence of mobile systems [20]. Connectivity allows one subsystem to interact with another by virtue of having a connection or reference. In a distributed environment, the access from one subsystem to another via a reference is then constrained by locality or proximity.

An earlier paper which presented a model of the Mobile IP protocol in Mobile Unity [18] has provided the motivation for this paper — both the title of the paper and the overall structure of the model, but not the formalism. The main contribution of the current paper is to present a model in the Mobile Petri Net formalism, with a view to assessing the suitability of that formalism — firstly, whether the graphical Petri Net notation helps to display the flow of information in the model in contrast to Mobile Unity; secondly, whether it is possible to harness existing Petri Net analysis techniques and tools for exploring the properties of the system; thirdly, whether the formalism is effective in capturing the interplay between locality and connectivity in this realistic mobile system.

K. Jensen, J. Billington, and M. Koutny (Eds.): ToPNoC III, LNCS 5800, pp. 127–158, 2009.

It will probably not be possible to give an unequivocal answer to the first question, since the value of a graphical notation will be a matter of personal taste and will depend on the facilities provided by a particular graphical editor. The answer to the second question will determine the ongoing applicability of this kind of Petri Net formalism.

The third question is, in some ways, fundamental. The case study is considered to be a genuine example of mobility — mobile nodes communicate with each other via a reference, in this case an Internet Protocol or IP address. The Internet is partitioned into a number of networks, with full connectivity between the networks, though a more constrained configuration would also be possible. Mobile nodes are associated with a home network (as indicated by the IP address) and any messages sent to them are sent via that home network. If the mobile node moves to another network, then it is the responsibility of the home network to forward the messages (after the move has been suitably registered). In other words, the accessibility to a mobile node with a given IP address depends on the location of the node.

Note that the term *subnet* is applicable both to the Internet and to Petri Net formalisms. In this paper, we will speak of the Internet consisting of networks, and we will reserve the term *subnet* for the Petri Net context. Also note that we will use the terms *datagram* and *message* interchangeably.

The paper is organised as follows: section 2 presents a motivating example of a simple mail agent system. Section 3 presents the formal definitions for a colourless version of Mobile Petri Nets. (Only the colourless version is presented here — the reader is referred to the earlier paper for the coloured version [14].) In section 4 we give an informal introduction to Mobile IP (from [18] and [22]), and in section 5 we present a Mobile Petri Net model of Mobile IP. In section 6 we consider some aspects of mapping the model into a form acceptable to the *Maria* tool [15], together with some state space exploration results. We consider related work in section 7 and finish with conclusions in section 8. The reader is assumed to have a basic understanding of Petri Nets and Internet protocols.

2 Example

We informally introduce the concepts of Mobile Petri Nets by considering a small example with elements drawn from the Mobile IP model presented in detail in Section 5. The excerpt identifies three networks and one mobile node which travels between the networks. The mobile node acts as a simple mail agent which travels to the different networks in order to deliver mail.

The initial form of the mail agent is shown in Fig. 1. It is drawn with a rounded rectangle border because we use the boundary to indicate externally accessible nodes and because Mobile Petri Nets retain the nested structure of subnets (unlike Modular Petri Nets [3]). There is one place *idle* which is marked with a token, and thus allows transitions to fire. There are six transitions — transition *recv* can fire to indicate the reception of a mail message (by the mail agent), while transitions *sendA*, *sendB* and *sendC* can fire to deliver a message to one of three destinations. Transitions *occupy* and *vacate* add or remove a token from place *idle*, which has the effect of activating or deactivating the mail agent. They are drawn in grey because they may be implicit, as we shall see later.

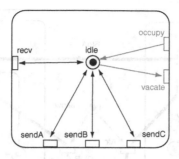

Fig. 1. Simple mail agent

One of the key notions of Mobile Petri Nets is the classification of transitions according to how they interact with local places. A transition which takes a subnet with a non-empty marking (of local places) and replaces it with an empty marking (for all enabled firings) is called a *vacate* transition, while a transition which takes a subnet with an empty marking and installs a non-empty marking (for all enabled transitions) is called an *occupy* transition. (Hence the naming of these transitions in Fig. 1.) Finally, a transition which takes a non-empty marking and replaces it with another non-empty marking (for all enabled transitions) is called a *regular* transition. In Fig. 1, transitions *recv*, *sendA*, *sendB* and *sendC* are all regular.

All the transitions are drawn adjacent to the subnet boundary to indicate that they will be fused with transitions in the environment. In other words, the firing of these transitions will be synchronised with the firing of transitions external to the subnet.

The composite mail system is shown in Fig. 2. Each rounded rectangle is a *location*, which is an instance of a subnet together with a fusion environment. The main or root location is labelled *System*, and it contains three locations labelled *NetA*, *NetB* and *NetC*. Within each of these locations is a nested location for the mail agent, labelled *LocA*, *LocB* and *LocC*, respectively.

A location is occupied if at least one of the local places is marked. In our example, the initial marking indicates that locations *System*, *NetA*, *NetB*, *NetC* and *LocA* are occupied, while locations *LocB* and *LocC* are not. Regular transitions should only be enabled for occupied locations and hence such transitions must have at least one input arc and one output arc incident on a local place. On the other hand, vacate transitions must have at least one input arc incident on local places, but no output arc, and occupy transitions must have at least one output arc incident on local places, but no input arc. It is for this reason that some transitions, like *AtoC*, *CtoB*, *BtoA*, have trivial side conditions. (A side condition is a place with both an input and an output arc incident on an adjacent transition.)

This composite mail system uses transition fusion but not place fusion. When two transitions are fused, the transitions are synchronised and must fire simultaneously. (When two places are fused, they are treated as the one place with a common marking, so that all transitions which access the fused place compete with each other.) Fusion can be indicated by annotation or by graphical convention. We assume that a suitable annotation can be used to equate names in two (or more) locations, e.g. $System.NetA.rA = System.NetA.LocA.sendA$.

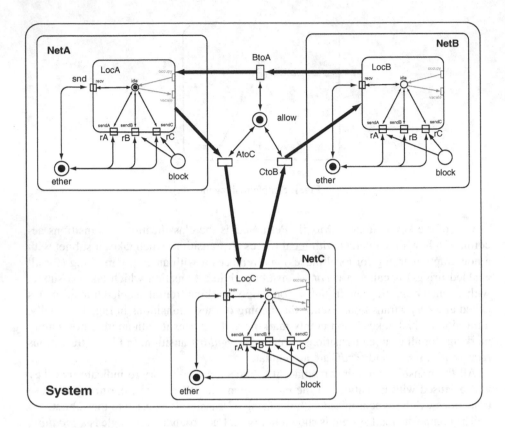

Fig. 2. Composite mail system

We also assume the availability of graphical conventions to indicate (transition) fusion. Firstly, transitions drawn adjacent to and on either side of a location boundary are assumed to be fused. Thus, in the composite mail system, location *LocA* is an instance of the mail agent where its transitions *recv*, *sendA*, *sendB*, *sendC* are fused to transitions *snd*, *rA*, *rB*, *rC*, respectively, of location *NetA*. Thus, transition *rA* of location *NetA* would be interpreted as receiving a mail message from the mail agent. Transitions *rB* and *rC* cannot fire in location *NetA* because place *block* is unmarked, and there is no way to make it marked. By contrast, transition *rB* in location *NetB* can fire (but not *rA* and *rC*), and transition *rC* in location *NetC* can fire (but not *rA* and *rB*).

Now, in order to support mobility, we assume that a vacate transition of one location will normally be fused with an occupy transition of another, matching location. In the composite mail system, this could be done by fusing the transition *vacate* of location *LocA* with the transition *occupy* of location *LocC*. Since the transition *vacate* clears the marking of location *LocA*, while the transition *occupy* installs a non-empty marking in location *LocC*, the overall effect is to move the marking (or state) of the mobile node from one location to another.

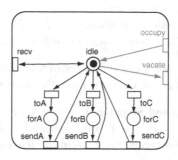

Fig. 3. Enhanced simple mail agent

However, the example of the mail agent is rather trivial, since it has only one local place. We might, for example, wish to extend the mail agent to the form of Fig. 3, so that we can capture the logic of first committing to send a message to a chosen destination, and only subsequently delivering the message. In this case, there are four possible places that can be marked, which would normally require four different vacate transitions and four matching occupy transitions. We anticipate that an implementation may elaborate all these alternatives implicitly, or may allow the user to specify explicitly a (restricted) set of alternatives. For example, we could adopt a naming convention which would deduce that *vacate0* matches *occupy0*, i.e. the prefix of the name would be one of *vacate* or *occupy* and the suffix would be used to match the two. In any case, we adopt the graphical convention of having a broad directed arc incident on a location as indicating a fusion with vacate and occupy transitions. Thus, in the composite mail system, the transition *AtoC* has the effect of moving the mail agent from location *LocA* to location *LocC*, i.e. place *idle* in location *LocA* will become unmarked, and place *idle* in location *LocC* will become marked.

The above example has been deliberately kept simple so that it can be presented as a (Colourless) Petri Net. However, even here, it quickly becomes apparent that the conciseness of Coloured Petri Nets is required for anything beyond a trivial example. Thus, it would be desirable to have one transition to send a mail message, with a colour determining the choice of destination. Similarly, the various vacate and occupy transitions could be differentiated by a colour. In the detailed model of Mobile IP presented in Section 5, we will use the capabilities of Coloured Petri Nets.

3   Mobile Petri Nets

In this section we present a definition of Mobile Petri Nets. The reader is referred to the earlier paper [14] for a definition in terms of Coloured Mobile Petri Nets.

Definition 1 (Multiset). *A multiset over set S is a mapping $m : S \rightarrow \mathbb{N}$, where $m(s)$ gives the number of occurrences of s. The set of all multisets over set S is denoted $\mu(S)$.*

Definition 2 (Petri Net). *A Petri Net (PN) is a tuple $PN = (P, T, W)$ where:*

1. *P is a finite set of nodes called places.*
2. *T is a finite set of nodes called transitions with $P \cap T = \emptyset$.*
3. *$W : (P \times T) \cup (T \times P) \to \mathbb{N}$ is an arc weight function.*

The arc weight function W indicates the number of tokens consumed or generated by the firing of the transition. If an arc weight is zero, the arc is not drawn.

Definition 3 (PN Markings and Steps). *For a Petri Net PN, a marking is a mapping $M : P \to \mathbb{N}$, i.e. $M \in \mu(P)$, and a step is a mapping $Y : T \to \mathbb{N}$, i.e. $Y \in \mu(T)$. A Petri Net System is a Petri Net together with an initial marking.*

The example of Fig. 3 is a Petri Net System where the arcs shown have a weight of *1* and the arcs which have not been drawn have a weight of *0*. Further, the initial marking of place *idle* is one token, while the other places hold no tokens.

Definition 4 (PN Behaviour). *For a Petri Net PN, a step Y is enabled in marking M, denoted $M[Y\rangle$, if $\forall p \in P : \Sigma_{t \in Y} W(p,t) \leq M(p)$. If step Y is enabled in marking M, then it may* occur, *leading to marking M', denoted $M[Y\rangle M'$, where $\forall p \in P : M'(p) = M(p) - \Sigma_{t \in Y} W(p,t) + \Sigma_{t \in Y} W(t,p)$. We write $[M\rangle$ for the set of markings reachable from M by the occurrence of zero or more steps.*

The above definitions are quite conventional. They capture the requirement that a place must have sufficient tokens to satisfy all consume demands of the step, and that when the step occurs, a place receives tokens from all generate actions. We now depart from convention by defining *locations* and *mobile systems*. These have been motivated by *Modular Petri Nets* [3], but here we retain the nested structure of the modules, which we call *locations*, and thereby capture the notion of locality.

Petri Net locations (or simply *locations*) are properly nested nets. Locations are unique, as are places and transitions (which are differentiated by the location in which they directly reside). We use *loc(X)* for the set of locations including *X* and all its nested locations. We use *plc(X)* and *trn(X)* for the places and transitions, respectively, in location *X* and all its nested locations. Formally, we have:

Definition 5 (PN Location). *A Petri Net Location is a tuple $L = (S_L, P_L, T_L, W_L)$ where:*

1. *S_L is a finite set of locations. We define $loc(L) = \bigcup_{s \in S_L} loc(s) \cup \{L\}$. We require $\forall s \in S_L : loc(s) \cap \{L\} = \emptyset$.*
2. *(P_L, T_L, W_L) is a Petri Net. We define $plc(L) = \bigcup_{s \in S_L} plc(s) \cup P_L$ and $trn(L) = \bigcup_{s \in S_L} trn(s) \cup T_L$.*

We can define markings, steps and behaviour for individual locations just as for Petri Nets, but we defer such definitions to mobile systems.

Definition 6 (Mobile System). *A Mobile System is a tuple $MS = (L_0, PF, TF, M_0)$ where:*

1. *L_0 is a Petri Net location, called the* root *location. We define $P = plc(L_0)$ and $T = trn(L_0)$.*

2. PF is a set of place fusion sets where $\bigcup_{pf \in PF} pf = P$ and $\forall pf_1, pf_2 \in PF :$ $pf_1 \cap pf_2 \neq \emptyset \Rightarrow pf_1 = pf_2$.
3. TF is a set of transition fusion sets where $\bigcup_{tf \in TF} tf = T$ and $\forall tf_1, tf_2 \in TF :$ $tf_1 \cap tf_2 \neq \emptyset \Rightarrow |tf_1| = |tf_2|$.
4. M_0 is the initial marking of the location L_0.

The set of place fusion sets covers all places and the fusion sets cannot partially overlap. This is in contrast to the definition of Modular Petri Nets [3], where the transitive closure of the fusion sets is used to determine the equivalence classes over places, which are then called place instance groups. Our approach means that each place fusion set corresponds to one place instance group. Similarly, the set of transition fusion sets is required to cover all transitions and if one transition occurs in more than one transition fusion set then these sets must have the same size. This constraint is imposed to simplify the definitions for Coloured Mobile Petri Nets. Again, this is more restricted than that of Modular Petri Nets, but it is not a theoretical restriction given that every transition can be duplicated so there is one duplicate per occurrence in a fusion set.

In a mobile system, as in modular nets, the markings of fused places are identical, and the multiplicity of fused transitions in a step are identical. This justifies extending the definition of markings and steps of such a system to include place fusion sets and transition fusion sets.

Definition 7 (MS Markings and Steps). *For mobile system MS, a marking is a mapping* $M : P \cup PF \to \mathbb{N}$, *where* $\forall pf \in PF : \forall p \in pf : M(p) = M(pf)$, *and a step is a mapping* $Y : T \cup TF \to \mathbb{N}$, *where* $\forall tf \in TF : \forall t \in tf : Y(t) = Y(tf)$.

It is then also appropriate to extend the definition of the arc weight function W to apply to place fusion sets and transition fusion sets, i.e.

$$\forall (f_1, f_2) \in (PF \times TF) \cup (TF \times PF) : W(f_1, f_2) = \Sigma_{x \in f_1, y \in f_2} W(x, y)$$

Definition 8 (MS Behaviour). *For a mobile system MS, a step Y is* enabled *in marking* M *if* $\forall pf \in PF : \Sigma_{tf \in Y} W(pf, tf) \leq M(pf)$. *If step Y is enabled in marking M, then it may* occur, *leading to marking M' where* $\forall pf \in PF : M'(pf) = M(pf) - \Sigma_{tf \in Y} W(pf, tf) + \Sigma_{tf \in Y} W(tf, pf)$.

The above definition is almost identical to that for modular nets [3].

Definition 9. *For a Mobile System MS we classify places and transitions as follows:*

1. $LP = \{p \in P \mid \exists pf \in PF : pf = \{p\}\}$ *is the set of* local places.
2. $EP = P - LP$ *is the set of* exported places.
3. $LT = \{t \in T \mid \exists tf \in TF : tf = \{t\}\}$ *is the set of* local transitions.
4. $ET = T - LT$ *is the set of* exported transitions.
5. $VT = \{t \in T \mid \exists p \in LP : W(p, t) > 0 \land \forall p \in LP : W(t, p) = 0\}$ *is the set of* vacate transitions.
6. $OT = \{t \in T \mid \exists p \in LP : W(t, p) > 0 \land \forall p \in LP : W(p, t) = 0\}$ *is the set of* occupy transitions.
7. $RT = \{t \in T \mid \exists p_1, p_2 \in LP : W(t, p_1) > 0 \land W(p_2, t) > 0\}$ *is the set of* regular transitions.

We distinguish local as opposed to exported places and transitions — exported entities are fused to at least one other. With the notion of mobility, we are interested in whether a location is occupied and this is determined by the marking of local places. Accordingly, we classify transitions by their interaction with local places.

Definition 10 (Well-behaved). *A Mobile System MS is* well-behaved *if:*

1. *All transitions are vacate, occupy or regular transitions, i.e.* $T = VT \cup OT \cup RT$.
2. *Vacate transitions empty a location for all reachable markings, i.e.* $\forall L \in loc(L_0)$:
   $\forall t \in VT \cap T_L : \forall M \in [M_0\rangle : M[t\rangle M' \Rightarrow \forall p \in LP \cap plc(L) : M'(p) = \emptyset.$
3. *Occupy transitions fill a location for all reachable markings, i.e.* $\forall L \in loc(L_0)$:
   $\forall t \in OT \cap T_L : \forall M \in [M_0\rangle : M[t\rangle M' \Rightarrow \forall p \in LP \cap plc(L) : M(p) = \emptyset.$

The above definition of a *well-behaved mobile system* is the key definition that supports mobility. We identify a location as being *occupied* if a local place is marked. A *vacate* transition has the effect of transforming an occupied location (and its nested locations) to unoccupied, while an *occupy* transition has the effect of transforming an unoccupied location (and its nested locations) to occupied. A *regular* transition has the effect of ensuring that an occupied location stays occupied.

The requirement that all transitions fall into one of these three categories ensures that occupy transitions are the only ones that can become enabled if the location to which they belong is unoccupied. The requirements that vacate transitions make a location unoccupied and that occupy transitions make a location occupied apply to all reachable markings. It is therefore debatable whether these should be classified as requirements for being *well-formed* or *well-behaved*. The choice of terminology here avoids the confusion with *Well-Formed Nets* [1]. However, our intention is that these conditions should be determined from the structure of the net, without the need for reachability analysis. Essentially, the problem is one of incorporating *clear* and *set* arcs [13] — the nature of these arcs determines whether reachability and boundedness are decidable for this form of Petri Net [5]. If all places in all locations are bounded, then it will be possible to incorporate complementary places and there will be a finite number of possibilities for the clear and set arcs, and hence for clearing and setting the marking of the location. If there are unbounded places, then we will need the generalised form of clear and set arcs [13] which make reachability and boundedness undecidable [5]. Note that if we introduce complementary places, then the complementary places should be exported or else the classification of occupied locations should be modified so as not to depend on these complementary places. In other words, we would need to identify a subset of the local places, called *base* places, and the marking of these will determine the marking of the other local places and whether a location is occupied or not.

We refer to a location together with a non-empty marking as a *subsystem*. By incorporating vacate and occupy transitions, the above formalism is sufficient for studying mobility — fusing a vacate transition in one location with an occupy transition in another corresponds to changing the location of a subsystem. For notational convenience, we introduce the broad arcs incident on locations, as in Section 2. These are a shorthand for sets of vacate and/or occupy transitions — one for each possible reachable marking of the location. They summarise the possibility of shifting the location of a subsystem

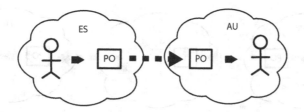

Fig. 4. Normal mail delivery

whatever its current state. Again, with bounded places, the possible alternatives can be enumerated. With unbounded places, the generalised clear and set arcs are required.

As is common with Petri Nets, the above formulation of Mobile Petri Nets appears to be rather static — the number of locations and the communication partners at each location are determined in advance. However, a more dynamic version can be achieved by suitable use of colours or types, as in [14]. Firstly, we can fold similar locations on top of each other and introduce a colour to distinguish the various subsystems resident at that location. Thus, in the example of Fig. 2, we could allow multiple mail agents to be resident at each of the locations *LocA*, *LocB* and *LocC*, using colour to distinguish which tokens belong to each mail agent. Secondly, we can use colours not just to distinguish multiple subsystems resident in the one location but also to fold multiple locations (with similar fusion contexts) onto one, with colours identifying the different locations. Thus, we could fold locations *SiteA*, *SiteB* and *SiteC* of Fig. 2 on top of each other, thus opening the way to an arbitrary number of sites. Thirdly, we can use synchronous channels [2] to determine communication partners dynamically. Note that in order to be able to classify transitions as *vacate*, *occupy* and *regular* in the coloured version, it will be necessary to relate each token and firing mode to the subsystem(s) to which they belong. This is a property we have called *consistency* [14].

4   Mobile IP

4.1   Introduction

Mobile IP can be informally introduced by analogy with the traditional mail system. The normal situation is shown in Fig. 4, with each country possessing a post office, shown as *PO*. When a sender in Spain wishes to transmit a letter (drawn as a pointed rectangle) to a person in Australia, they use the receiver's advertised mail address. The post offices take care of the mail delivery.

In Fig. 5, the Australian receiver has travelled overseas temporarily but has left forwarding instructions (in advance) with the local post office. Now, if a letter is sent from Spain, it is delivered to Australia as before, but now the Australian post office *forwards* the letter, possibly by enclosing it in an envelope with the forwarding address.

The above forwarding capability obviates the need for a receiver to notify all potential senders of the temporary change of address, but at the cost of a slower delivery. Also, the receiver normally has to register well in advance if no mail is to be lost.

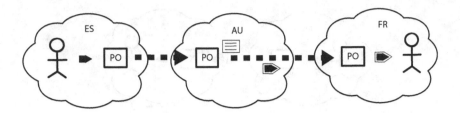

Fig. 5. Forwarded mail delivery

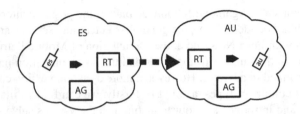

Fig. 6. Normal message delivery

Mobile IP addresses similar concerns but solves them in a slightly different way, as shown in Fig. 6. The delivery functions of the post office are now assumed by routers (depicted as boxes annotated *RT*). For scalability of the Internet, routers do not (generally) manage routes for individual nodes. Route selection is determined by the network part of the IP address. Thus, if one wishes to maintain a fixed IP address then this normally implies a fixed position in the Internet. Consequently, any registration and forwarding functions are built into separate devices or programs known as *agents* (depicted as boxes annotated with *AG*). In normal operation, one node (depicted as a mobile with annotation *es*) will send a data message to another node (depicted as a mobile with annotation *au*). The router in domain *ES* passes the message (directly or indirectly) to the router in domain *AU* which then passes it to the receiver. The agents are not involved.

When a mobile node moves away from its home network, the agents come into play. A *home agent* manages the registration for absent nodes and performs the mail forwarding. The electronic equivalent of putting mail in an envelope for transmission is known as *tunnelling*. A *foreign agent* acts as a point for initiating the registrations and providing a *care-of* address. The functionality of home agent and foreign agent can be combined, in which case we simply refer to a *mobility agent*. However, we will still differentiate the functionality of the home and foreign agent by name.

Another significant difference with Mobile IP is that registration is not done in advance, but after the event. This is because a mobile node may visit several foreign networks before returning home and it is difficult, if not impossible, to predict which in advance. Thus, when a mobile node moves to a new network, it waits for an agent advertisement (or solicits such an advertisement), as shown in Fig. 7. It then responds with a request for a registration which is forwarded to the home agent, as shown in

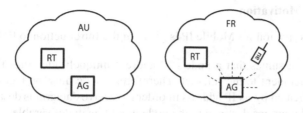

Fig. 7. Foreign agent advertisement

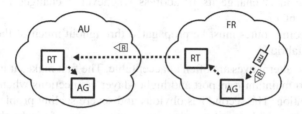

Fig. 8. Mobile registration

Fig. 8. The home agent registers the care-of address and sends an acknowledgement back to the foreign agent and thence the mobile node.

Now, if a data message is sent to the mobile node's original IP address, it is delivered to the home network, where the home agent traps the message. It then forwards the message (by tunnelling) to the care-of address. If the care-of address is that of a foreign agent, the agent will extract the body of the message and pass it to the mobile node. This is shown in Fig. 9.

Thus, Mobile IP provides a genuine example of mobility because the IP address of a mobile node acts as a reference. The delivery of datagrams to the mobile node is determined by that IP address. However, normal delivery is only possible if the mobile node is attached to its home network. Otherwise, if the mobile node has moved to a foreign network, then delivery cannot be achieved without employing some other mechanism, in this case a form of tunnelling. Mobile IP provides such a mechanism which is scalable. Mobile IP thus embodies an interplay between connectivity and locality.

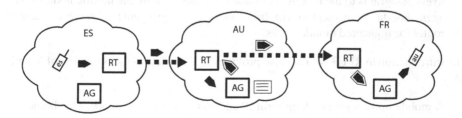

Fig. 9. Forwarded message delivery

4.2   Technical Motivation

The technical motivation for Mobile IP is given in the Introduction to RFC 3344 [22]:

> IP version 4 assumes that a node's IP address uniquely identifies the node's point of attachment to the Internet. Therefore, a node must be located on the network indicated by its IP address in order to receive datagrams destined to it; otherwise, datagrams destined to the node would be undeliverable. For a node to change its point of attachment without losing its ability to communicate, currently one of the two following mechanisms must typically be employed:
> a) the node must change its IP address whenever it changes its point of attachment, or
> b) host-specific routes must be propagated throughout much of the Internet routing fabric.
>
> Both of these alternatives are often unacceptable. The first makes it impossible for a node to maintain transport and higher-layer connections when the node changes location. The second has obvious and severe scaling problems, especially relevant considering the explosive growth in sales of notebook (mobile) computers.
>
> A new, scalable, mechanism is required for accommodating node mobility within the Internet.

4.3   Overview

Mobile IP introduces the following functional entitities [22]:

Mobile Node: A host or router that changes its point of attachment from one network or subnetwork to another. A mobile node may change its location without changing its IP address; it may continue to communicate with other Internet nodes at any location using its (constant) IP address, assuming link-layer connectivity to a point of attachment is available.

Home Agent: A router on a mobile node's home network which tunnels datagrams for delivery to the mobile node when it is away from home, and maintains current location information for the mobile node.

Foreign Agent: A router on a mobile node's visited network which provides routing services to the mobile node while registered. The foreign agent detunnels and delivers datagrams to the mobile node that were tunneled by the mobile node's home agent. For datagrams sent by a mobile node, the foreign agent may serve as a default router for registered mobile nodes.

The Introduction to RFC3344 [22] also provides an overview of the operation of Mobile IP:

> A mobile node is given a long-term IP address on a home network. This home address is administered in the same way as a "permanent" IP address is provided to a stationary host. When away from its home network, a "care-of address" is associated with the mobile node and reflects the mobile node's current

point of attachment. The mobile node uses its home address as the source address of all IP datagrams that it sends, except where otherwise described in this document for datagrams sent for certain mobility management functions.

The following steps provide a rough outline of operation of the Mobile IP protocol:

- Mobility agents (i.e., foreign agents and home agents) advertise their presence via Agent Advertisement messages. A mobile node may optionally solicit an Agent Advertisement message from any locally attached mobility agents through an Agent Solicitation message.
- A mobile node receives these Agent Advertisements and determines whether it is on its home network or a foreign network.
- When the mobile node detects that it is located on its home network, it operates without mobility services. If returning to its home network from being registered elsewhere, the mobile node deregisters with its home agent, through exchange of a Registration Request and Registration Reply message with it.
- When a mobile node detects that it has moved to a foreign network, it obtains a care-of address on the foreign network. The care-of address can either be determined from a foreign agent's advertisements (a foreign agent care-of address), or by some external assignment mechanism such as DHCP (a co-located care-of address).
- The mobile node operating away from home then registers its new care-of address with its home agent through exchange of a Registration Request and Registration Reply message with it, possibly via a foreign agent.
- Datagrams sent to the mobile node's home address are intercepted by its home agent, tunneled by the home agent to the mobile node's care-of address, received at the tunnel endpoint (either at a foreign agent or at the mobile node itself), and finally delivered to the mobile node.
- In the reverse direction, datagrams sent by the mobile node are generally delivered to their destination using standard IP routing mechanisms, not necessarily passing through the home agent.

When away from home, Mobile IP uses protocol tunneling to hide a mobile node's home address from intervening routers between its home network and its current location. The tunnel terminates at the mobile node's care-of address. The care-of address must be an address to which datagrams can be delivered via conventional IP routing. At the care-of address, the original datagram is removed from the tunnel and delivered to the mobile node.

4.4 Changes

The version of the protocol considered in the Mobile UNITY paper was associated with RFC 2002 [21]. This was superceded by RFC 3344 [22]. To a large extent the changes concern security, timing and network load, issues which are peripheral to this study.

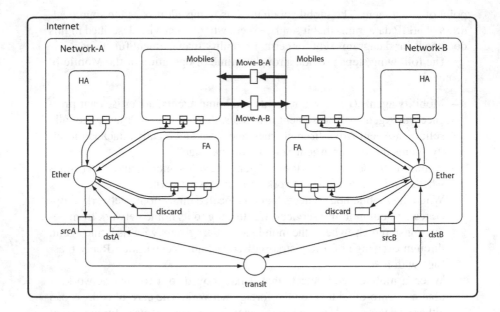

Fig. 10. Overview of the Internet

5 Mobile IP Model

We now present our Mobile Petri Net (MPN) model of Mobile IP. The model will necessarily employ the coloured extension to MPNs [14]. Also, since the complete model
cannot be displayed in one diagram, we examine one module at a time. We anticipate
that a graphical editor will facilitate this kind of manipulation.

We remind the reader that we use the term *network* when referring to a segment of
the Internet and the terms *net* and *subnet* when referring to segments of the Petri Net
model.

We also note that the Petri Nets provided in this section indicate the flow of information, but are not annotated to indicate the precise values. This has been done to avoid
cluttering the diagrams.

5.1 Internet

The model of the overall network is shown in Fig. 10. We display two local networks
called *Network-A* and *Network-B*. Each local network includes a place called *Ether*
which holds the datagrams in transit, a location called *HA* for the home agent, and a
location called *FA* for the foreign agent. These agents are stationary — there are no
capabilities for moving them. There is also a location called *Mobiles* which holds the
mobile nodes currently resident at this local network.

The networks communicate via the place *transit* which holds the datagrams in transit
from one network to another. Within each local network, we have assumed an Ethernet-
style network, with place *Ether* holding the messages in transit at the local level.

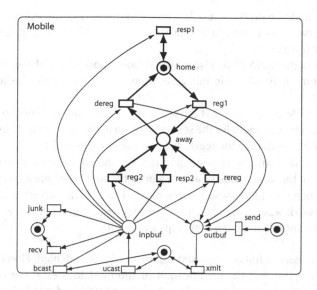

Fig. 11. Structure of a Mobile node

The transitions *move-B-A* and *move-A-B* are used to move mobile nodes from one network to another.

The Mobile Unity model ensures proper handling of the local Ethernet via its *responds-to* mechanism. This allows the action of removing a (non-null) datagram from the Ethernet to be taken in preference to any other action. One possibility for achieving a similar effect in the Petri Net model is to use transition priorities. This is considered further in subsection 6.2.

5.2   Mobile Node

The structure of each mobile node is given by the subnet of Fig. 11. It interacts with its environment by exchanging messages or datagrams — transitions *bcast* and *ucast* accept broadcast and unicast messages, respectively, while transition *xmit* sends a message. These transitions interact with places *inpbuf* and *outbuf* which then hold the incoming and outgoing messages, respectively. The use of these places is slightly different to the Mobile Unity model which allowed a mobile node to deposit a datagram directly on the Ethernet.

In Fig. 11, the net components drawn in bold highlight the lifecycle of a mobile node. There are two basic states of the mobile node, identified by the places *home* (to indicate that the mobile is assumed to be connected to its home network) and *away* (to indicate that the mobile node is assumed to be connected to a foreign network). RFC 3344 [22] gives a number of possibilities for a node to know when it has moved from (or returned to) its home network. As in the Mobile Unity model, we only consider the option of receiving an agent advertisement. There are five possible scenarios on receiving a foreign agent advertisement:

– the mobile node could have just left home and received the advertisement for a different network — in this case, transition *reg1* would fire to register the node as being away from home;
– the mobile node could have just returned home and received the advertisement from the home network — in this case, transition *dereg* would fire to cancel any registration;
– the mobile node could have just moved from one foreign network to another and received the advertisement for the second foreign network — in this case, transition *reg2* would fire to change the registration to the new network;
– the mobile node could be at home and receive the advertisement from the home network — in this case, transition *junk* would fire to discard the advertisement;
– the mobile node could be away from home and receive the advertisement from the foreign network where it is resident — in this case, transition *rereg* would fire to reregister the mobile node (and preempt a timeout).

Registration (and deregistration requests) can be accepted or denied. However, the only effect of these responses is the modification of timing delays — if the request is accepted, then the time to reregistration is set; if the request is denied, then the time to register is set. Given that we are abstracting away from timing matters, we have modelled both responses in the same way — transition *resp1* handles responses when the mobile node is at home, while *resp2* handles responses when the mobile node is away from home.

The remaining transitions have the following significance:

– transition *recv* receives a data message. The content is not of interest.
– transition *send* originates a data message destined for another mobile node. Again, its content is not of interest.
– transition *junk* discards any message which is not handled by other transitions.

Finally, we note that there are three simple side conditions so that transitions *junk*, *recv*, *bcast*, *ucast*, *xmit* and *send* qualify as *regular* transitions (as in Definition 10).

5.3 Home Agent

The structure of a home agent is shown in the subnet of Fig. 12. Again, interaction with the environment is via message exchange, firing transitions *get* and *put* which then use the places *inpbuf* and *outbuf* to hold the incoming and outgoing messages respectively.

The home agent uses two places to record the status of mobile nodes associated with this network — *home* holds one token for each mobile node which is known to be at home, while *careof* holds the care-of address for each mobile node which is known to be away from home. The home agent accepts messages for nodes which have a token in place *careof*. For this reason, the place is initialised with a pseudo entry for the agent itself. Datagrams destined for mobiles known to be away from home are forwarded to the relevant foreign agent by firing the transition *fwd*.

Again, the net components drawn in bold highlight the lifecycle of the home agent. The home agent deals with registration requests as follows:

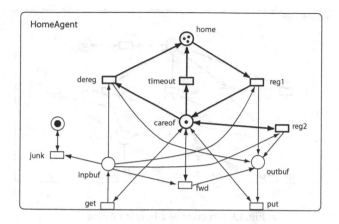

Fig. 12. Structure of a Home Agent

- transition *reg1* handles registration requests for when the mobile node has moved from its home network to a foreign network;
- transition *reg2* handles registration requests for when the mobile node has moved from one foreign network to another;
- transition *dereg* handles registration requests for when a mobile node has returned home.

The remaining home agent transitions have the following significance:

- transition *timeout* will automatically deregister a mobile node after a period of time.
- transition *junk* discards irrelevant messages.

A simple side condition has been added so that transition *junk* qualifies as a *regular* transition.

5.4   Foreign Agent

Finally, the structure of a foreign agent is shown in the subnet of Fig. 13. It interacts with the environment by firing transitions *get* to receive a message, *advert* to multi-cast an advertisement, and *put* to unicast a message. The foreign agent maintains the status of visiting mobile nodes in the place *visitor*, which is initially empty. The first registration request for a mobile node is recorded by firing the transition *reg*, while a reregistration request is recorded by transition *rereg*. A positive response from the home agent to a registration request is recorded by firing transition *ok* (which updates the visitor status), while a negative response is recorded by the transition *deny* (which discards the registration). The transition *timeout* is fired to discard the registrations after some period of time. The transition *fwd* is fired to extract an encapsulated message and forward it to the relevant mobile node. Side conditions have been added so that transitions *get, advert, put, deny* and *timeout* qualify as *regular* transitions.

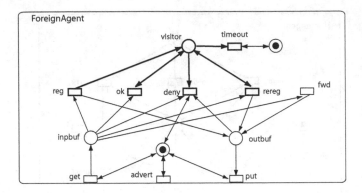

Fig. 13. Structure of a Foreign Agent

6   Exploring the Model in Maria

In order to experiment with the above model, it has been mapped into a form acceptable to the *Maria* tool. In this section, we introduce the key elements of Maria, and present the results for reachability analysis of the model.

6.1   Introduction to Maria

The tool *Maria* is a reachability analyser for High-Level Algebraic System Nets [15]. It can generate reachability graphs for these nets, detect deadlocks and check properties expressed using temporal logic formulae. Its data type system and expression syntax are heavily inspired by the C programming language.

Maria supports a rich selection of data types, including structures, unions, arrays, queues and stacks. It constrains the types so that reachability analysis can be performed.

In Maria, Petri Nets are specified in a textual notation rather than graphically. This tends to be quite verbose even though it is quite general. Maria supports a limited form of conditional compilation, which can be used to parameterise a model, according to size, features, etc.

Since 2002, Maria has supported the specification of modules and arbitrary nesting of those modules [17]. Module synchronisation is achieved by transition fusion. A transition definition is introduced by the keyword *trans* (or *transition*) followed by the transition name which is an identifier. If a transition name is preceded by a colon, e.g. *trans :callee*, then it is not an actual transition but designates a kind of a macro or a function body that can be added to other transitions, as in: *trans caller :trans callee*. The Maria parser merges the definitions of *trans :callee* to the definition of *trans caller* [16].

6.2   Mapping the Model to Maria

The Maria tool provides sufficient capabilities to perform preliminary state space explorations of the Mobile IP model developed in Section 5. The support for arbitrary nesting of modules is exactly what we require to represent locations. The transition fusion provided by Maria is sufficient for our needs, even if it is a bit clumsy to use. The tool has no support for identifying *vacate* and *occupy* transitions which are characteristic of our definition of Mobile Petri Nets. We anticipate that such a capability will, in due course, be supplied by a suitable front end to *Maria*.

As noted in subsection 5.1, the *responds-to* mechanism of Mobile Unity was used in [18] to control the access to the Internet and the local Ethernets. We have achieved the same effect in our model by using transition priorities in *Maria*. The allocated priorities are:

- priority 9 is used for mobiles reading from the Ethernet;
- priority 8 is used for agents reading from the Ethernet;
- priority 7 is used for exporting from a local Ethernet to the Internet;
- priority 6 is used for importing from the Internet to a local Ethernet;
- priority 5 is used for processing a datagram from the Ethernet at higher priority;
- priority 4 is used for processing a datagram from the Ethernet at lower priority;
- priority 2 is used for putting a datagram onto the local Ethernet;
- priority 0 is used for spontaneously generating a datagram or moving a mobile node.

Broadly, these priorities ensure proper Ethernet behaviour — datagrams are removed from the Ethernet at the highest priorities and deposited at the lowest priorities, thus ensuring that no more than one datagram is on the Ethernet at any time. While the basic logic of the subnets from the preceding subsections has been retained, some modifications have been made but have not been shown in the diagrams for the sake of simplicity. Firstly, the places *inpbuf* and *outbuf* which are used in the mobile nodes, the home agents and the foreign agents, do not hold a multiset of message tokens but rather a single queue of datagrams. This ensures that incoming messages are processed in the same order that they are received and that outgoing messages are sent in the same order that they are generated. This modification does *not* apply to the place *transit* in the Internet, because datagrams in that environment can normally be reordered.

Secondly, we do not generate an agent advertisement (in a foreign agent) or a data message (in a mobile node), unless the outgoing buffer is empty. This seems to be a reasonable constraint given that there is little point in generating a new datagram before the previous one has been dispatched. Further, to reduce the state space explosion, we insist that only one agent or mobile node has a message to transmit at a time. This does not constrain the possible order of sending messages, but simply ensures that a message is sent as soon as it is generated. This is consistent with the Mobile Unity model which did not have output buffers, but deposited a message on the Ethernet as soon as it was generated.

Thirdly, we have attempted to reduce the state space explosion by only allowing a mobile node to move between networks when its buffers are empty. It is more debatable whether this is an appropriate restriction. It is not difficult to relax this constraint in future analysis.

In the following subsection, we highlight the main features of the mapping of the model from Section 5 into Maria.

6.3   Excerpts of Maria Code for the Model

internet.pn. The module *internet.pn* is the main program and corresponds to Fig. 10. An excerpt is included below. Conditional compilation is used extensively to determine the number of networks and the number of mobiles to be incorporated into a particular configuration. Thus, the program begins by documenting the various compilation flags that can be used to constrain or modify the model. Here, defining the flag *MOVE* will determine if the mobiles are allowed to move, while defining the flag *QLEN2* will determine whether a buffer size of two (instead of one) is used for the channels.

The program then defines the parameters and message formats. It defines the place *transit* together with the callee transitions that access it from network *A* (at priorities 7 and 6). It also defines the callee transitions for moving mobile 3 from one network to another. It finishes by instantiating the networks. (The transitions and instances for other networks and mobiles are omitted.)

```
// Options for running the Mobile IP model:
// -DDATA - to generate, send and receive data messages
// -DMOVE - to allow mobile nodes to move between networks
// -DHAS3 - allow for 3 (rather than just 2) mobile phones
// -DHAS4 - allow for 4 (rather than just 2) mobile phones
// -DQLEN2 - allow for 2 (rather than just 1) element queues

// Constant definitions
unsigned MaxNetNo = 2;
unsigned MaxHostNo = 6;
#ifndef QLEN2
unsigned MaxQueue = 1;
#else
unsigned MaxQueue = 2;
#endif


// Type definitions

typedef unsigned (0..MaxNetNo) Net;
typedef unsigned (0..MaxHostNo) Host;

typedef enum { m_data, m_advert, m_reg, m_rereg, m_dereg, m_ok, m_deny } MsgKind;
typedef struct { Net net, Host host } IPaddr;

typedef struct { IPaddr mob, IPaddr fwd } CareAddr;

typedef struct { IPaddr src, IPaddr dst, MsgKind kind } SimpleMessage;
typedef struct { IPaddr src, IPaddr dst, SimpleMessage msg } Message;

typedef Message [queue MaxQueue] Buffer;


// Global interface

// Place to hold messages in transit

place transit Message;


// Callee transitions to access messages in transit

trans :srcA 7
    out {
        place transit: msg;
    }
;
```

```
...
trans :dstA 6
    in {
        place transit: msg;
    }
;
...

// Callee transitions to move a mobile from one net to another

#ifdef MOVE
trans :move3toA
;
trans :move3toB
;
...
#endif

// Network A

subnet NetworkA {
#define netA
#include "network.pn"
#undef netA
};
...
```

network.pn. The module *network.pn* corresponds to the locations labelled *Network-A* and *Network-B* in Fig. 10. (We only show the conditional compilation elements for *Network-A*.) The module defines the transitions to import and export messages — these are fused with the relevant callee transitions like *srcA* and *dstA* of *internet.pn*, which access the place *transit*. The module defines a place *ether* which holds a buffer of messages. As already noted, the priorities of transitions accessing this place ensure that the buffer is either empty or holds exactly one message.

Maria provides special operations for accessing and modifying queues: $*ebuf$ accesses the front element of the queue, $-ebuf$ removes the front element, $ebuf + msg$ adds an element to the back of the queue, $/ebuf$ and $\%ebuf$ determine the number of full and empty slots in the queue, respectively.

The module defines callee transitions *getAll* for broadcast messages and *getM3*, etc. for unicast messages for specific mobiles. It also defines callee transitions *getFA* and *getHA* for interaction with the mobility agents.

For moving the mobile nodes into and out of the network, the module defines callee transitions *arrive* and *leave*. Finally, the module instantiates the agents and (the locations for) the mobiles.

Note that the subnets for mobiles are duplicated rather than using colours to differentiate different mobiles folded onto the one subnet. These alternatives are essentially equivalent, but the unfolded version turns out to be a bit simpler to express in Maria.

```
/* Network with mobile nodes */

// Identity of network

#ifdef netA
Net net = 1;
#endif
...
```

```
// Local Ethernet - a bit strange because mobile nodes retain IP address
// - use priority to ensure they consume message before forwarded

place ether Buffer: {};

// Transfer messages into and out of network

trans import 6          // 4th highest priority to xfer from network
#ifdef netA
: trans dstA
#endif
...
    in {
        place ether: ebuf;
    }
    out {
        place ether: ebuf+msg;
    }
    gate   %ebuf > 0
    gate   msg.dst.net == net
;

trans export 7          // 3rd highest priority to consume by network
#ifdef netA
: trans srcA
#endif
...
    in {
        place ether: ebuf;
    }
    { Message msg = *ebuf; }
    out {
        place ether: -ebuf;
    }
    gate   /ebuf > 0
    gate   msg.dst.net != net
;

trans discard 7         // 3rd highest priority to discard a message
    in {
        place ether: ebuf;
    }
    { Message msg = *ebuf; }
    out {
        place ether: -ebuf;
    }
    gate   /ebuf > 0
    gate   msg.dst.net == net
; 

// Transitions to access messages within the network
// - highest priority for mobile access
// - lesser priority for agent access (cf. home agent forwarding)

trans :getAll 9         // for broadcast message
    in {
        place ether: ebuf;
    }
    { Message msg = *ebuf; }    // access front of queue
    out {
        place ether: -ebuf;     // pop the queue
    }
    gate   /ebuf > 0
;

trans :getM3 9              // for mobile to access ethernet
    in {
```

```
        place ether: ebuf;
    }
    { Message msg = *ebuf; }
    out {
        place ether: -ebuf;
    }
    gate  /ebuf > 0
;
...

trans :getHA 8              // for home agent to access ethernet
    in {
        place ether: ebuf;
    }
    { Message msg = *ebuf; }
    out {
        place ether: -ebuf;
    }
    gate  /ebuf > 0
;
...

// Transitions for moving a mobile node into or out of the network

trans :leave3
#ifdef netA
: trans move3toB
#endif
...
;

trans :arrive3
#ifdef netA
: trans move3toA
#endif
...
;

subnet homeAgent {
#include "homeAgent.pn"
};

subnet foreignAgent {
#include "foreignAgent.pn"
};

subnet mobile3 {
#define mob3
#include "mobile.pn"
#undef mob3
};
...
```

mobile.pn. The module *mobile.pn* corresponds to Fig. 11. Instead of having two places *home* and *away*, it is more convenient to define a single place *atHome* with a boolean token. The module has transitions to register, deregister, reregister the mobile and to ignore the advertisements.

Given the absence of inhibitor arcs, all mobiles receive all broadcast messages (whether or not they are in the right network to receive them). Only those which are resident in the appropriate network actually respond to the broadcast advertisements.

Transitions *vacate* and *occupy* achieve movement of the mobiles. The *vacate* transition has the effect of emptying the local marking while the *occupy* transition has the

150   C. Lakos

dual effect. A *vacate* transition is always paired with an *occupy* transition. For example, in network A, the transition *vacate* for mobile location 3 is fused with the callee transition *leave3* of network A, which is fused with the callee transition *move3toB* of the *internet* module. Similarly, in network B, the transition *occupy* for mobile location 3 is fused with the callee transition *arrive3* of network A, which is fused with the same callee transition *move3toB* of the *internet* module. Note that movement of the mobile requires that the input and output buffers are empty.

```
/* Network for a mobile node */

// Identity of network

#ifdef netA
Net net = 1;
#endif
...

// Identity of mobile

#ifdef mob3
Host host = 3;
#endif
...

// Home network of the mobile

Net home = (host-1)/2;

// Identities of home and foreign agent

Host agentH = 1;
Host agentF = 2;

place mobileNode IPaddr: {home,host};            // static identification

// Possible communication partners

place partner IPaddr: (1#{1,3}, 1#{2,5}          // always >= 2 mobiles
                       ... );

// Message buffers

place inpbuf Buffer : {};
place outbuf Buffer : {};

// Residency status of mobile - changed when mobile moves

place resident bool :
#ifdef netA
#ifdef mob3
    true;
#endif
...

// Record of registration

place atHome bool      // Includes both 'home' and 'away' places
#ifdef netA
#ifdef mob3
    : true;
#endif
...
#endif

// Transfer messages to and from the network
```

```
trans getBcast 9          // Highest priority for local access
: trans getAll
#ifdef mob3
    in {
        place resident: m3;
        place inpbuf: buf3;
    }
    out {
        place resident: m3;
        place inpbuf: (m3 && msg.dst.net==net ? buf3+msg : buf3);
    }
    gate  %buf3 > 0
#endif
...
    gate  msg.dst.host == 0
;

trans get 9          // Highest priority for local access
#ifdef mob3
: trans getM3
#endif
...
    in {
        place inpbuf: buf;
        place atHome: ah;
    }
    out {
        place inpbuf: buf+msg;
        place atHome: ah;
    }
    gate  %buf > 0
    gate  msg.dst.net == home && msg.dst.host == host
;

trans put 1          // Lower priority for local generation
:trans putAll
#ifdef mob3
    in {
        place outbuf: bufM3;
    }
    { Message msgM3 = (/bufM3 > 0 ? *bufM3 : {{0,0},{0,0},{{0,0},{0,0},m_data}}); }
    out {
        place outbuf: (/bufM3 > 0 ? -bufM3 : bufM3);
    }
    gate  /bufM3 == 0 || msg.dst.net != net || msg.dst.host != host
#endif
...
;

// Logic of mobile node

trans regM 4
    in {
        place inpbuf: buf1;
        place outbuf: buf2;
        place atHome: true;
    }
    { Message msg = *buf1; }
    { Message req = {{home,host}, {net,agentF},
                        {{home,host}, {net,agentF}, m_reg}}; }
    out {
        place inpbuf: -buf1;
        place outbuf: buf2+req;
        place atHome: false;
    }
    gate  /buf1 > 0
    gate  %buf2 > 0
```

```
        gate  msg.msg.kind == (is MsgKind m_advert)
        gate  msg.src.net != home && msg.src.host == agentF
;

trans reregM 4
    in {
        place inpbuf: buf1;
        place outbuf: buf2;
        place atHome: false;
    }
    { Message msg = *buf1; }
    { Message req = {{home,host}, {net,agentF},
                        {{home,host}, {net,agentF}, m_rereg}}; }
    out {
        place inpbuf: -buf1;
        place outbuf: buf2+req;
        place atHome: false;
    }
    gate  /buf1 > 0
    gate  %buf2 > 0
    gate  msg.msg.kind == (is MsgKind m_advert)
    gate  msg.src.net != home && msg.src.host == agentF
;

trans deregM 4
    in {
        place inpbuf: buf1;
        place outbuf: buf2;
        place atHome: false;
    }
    { Message msg = *buf1; }
    { Message req = {{home,host}, {net,agentH},
                        {{home,host}, {net,agentH}, m_dereg}}; }
    out {
        place inpbuf: -buf1;
        place outbuf: buf2+req;
        place atHome: true;
    }
    gate  /buf1 > 0
    gate  %buf2 > 0
    gate  msg.msg.kind == (is MsgKind m_advert)
    gate  msg.src.net == home && msg.src.host == agentF
;

trans ignoreFM1 4        // Ignore foreign advert - we are home and know
    in {
        place inpbuf: buf;
        place atHome: true;
    }
    { Message msg = *buf; }
    out {
        place inpbuf: -buf;
        place atHome: true;
    }
    gate  /buf > 0
    gate  msg.msg.kind == (is MsgKind m_advert)
    gate  net == home
;
...

// Message sending and receiving
#ifdef DATA

trans recvM 4
    in {
        place atHome: ah;
        place inpbuf: buf;
    }
```

```
        { Message msg = *buf; }
        out {
            place atHome: ah;
            place inpbuf: -buf;
        }
        gate   /buf > 0
        gate   msg.dst.net == home && msg.dst.host == host
        gate   msg.msg.kind == (is MsgKind m_data)
    ;

•trans junkM 3
        in {
            place resident: false;
            place inpbuf: buf;
        }
        { Message msg = *buf; }
        out {
            place resident: false;
            place inpbuf: -buf;
        }
        gate   /buf > 0
        gate   msg.dst.net == net && msg.dst.host == host
        gate   msg.msg.kind == (is MsgKind m_data)
    ;

    trans sendM
        in {
            place atHome: ah;
            place outbuf: buf;
            place partner: p;
        }
        { Message msg = {{net,host}, p,
                        {{net,host}, p, m_data}}; }
        out {
            place atHome: ah;
            place outbuf: buf+msg;
            place partner: p;
        }
//      gate   %buf > 0
        gate   /buf == 0
        gate   p.host != host
    ;
#endif

// Transitions for moving a mobile from one network to another

trans vacate
#ifdef mob3
: trans leave3
#endif
...
        in {
            place inpbuf: buf1;
            place outbuf: buf2;
            place resident: true;
            place atHome: ah;
        }
        out {
            place inpbuf: {};
            place outbuf: {};
            place resident: false;
        }
        gate   /buf1 == 0 && /buf2 == 0
    ;

trans occupy
#ifdef mob3
: trans arrive3
```

```
#endif
...
    in {
        place inpbuf: {};
        place outbuf: {};
        place resident: false;
    }
    out {
        place inpbuf: buf1;
        place outbuf: buf2;
        place resident: true;
        place atHome: ah;
    }
;
...
```

homeAgent.pn. The module *homeAgent.pn* corresponds to Fig. 12. It is reasonably straightforward and hence is not included.

foreignAgent.pn. The module *homeAgent.pn* corresponds to Fig. 13. Like the home agent, it is reasonably straightforward and is omitted.

6.4   Results

Preliminary results for the sizes of the state space are shown in Table 1. The model had two networks and was parameterised by the number of mobile nodes (given by the column labelled *Mobiles*) and the maximum queue length for the input and output buffers (indicated by the multiple columns labelled *Max queue length*). It was also possible to enable or disable the sending of data messages (as indicated by the column labelled *Data?*), and the movement of mobile nodes between networks (as indicated by the column labelled *Move?*). Then the sizes of the state space for each combination of parameters are given by the columns labelled *Nodes* and *Arcs*.

Not surprisingly, the size of the state space is quite small as long as the mobile nodes cannot send data or move from one network to another. Basically, the only things that can happen are that the foreign agents will generate advertisements and these will be discarded by the mobile nodes which are always known to be at home. In fact, it is only when mobile nodes move between networks that the registration of nodes occurs, together with the associated encapsulation and forwarding of messages. The careful allocation of priorities ensures that, as long as the mobile nodes cannot move, the maximum queue length does not affect the size of the state space.

As a result of the exercise of modelling and simulating Mobile IP, an error in the Mobile UNITY model was discovered. When a foreign agent receives an encapsulated message (destined for a mobile node which is away from home), the unpacked datagram is simply added to the local Ethernet. However, this unpacked datagram is destined for the mobile node, whose home network address does not match the current one. Thus, the datagram could be routed back to its home network. This is possible in the Mobile Unity model. It could be argued that this is not a major problem because RFC 3344 [22] allows for different data link implementations, and does not specify which is to be used. Nevertheless, the allocation and possible reuse of care-of addresses is an important issue for Mobile IP, and this problem does highlight the value of being able to simulate a model and even perform state space exploration. In our model, we have overcome

Table 1. State space sizes for Mobile IP

	Options		Max queue length 1		Max queue length 2	
Mobiles	Data?	Move?	Nodes	Arcs	Nodes	Arcs
2	–	–	7	8	7	8
3	–	–	9	11	9	11
4	–	–	11	14	11	14
2	✓	–	17	20	17	20
3	✓	–	35	43	35	43
4	✓	–	63	78	63	78
2	✓	✓	1668	2180	1668	2180
3	✓	✓	60708	85684	134508	187115
4	✓	✓	> 3.5 M	> 5.1 M		

this problem by the further use of priorities — mobile nodes access local messages at a higher priority than the agents and the exporting of those messages to the Internet.

7   Related Work

The Nets-within-Nets paradigm proposed by Valk has been the focus of a significant effort in terms of object-oriented design and mobility [8,9,10,25]. The fundamental notion is that there are (at least) two levels of Petri Nets — the system net provides the most abstract view of the system. The tokens resident in the places of the system net may be black tokens (with no associated data) or representations of object nets. The two-level hierarchy can be generalised to an arbitrary number of levels, but that is not necessary for our purposes. Three different semantics have been proposed for the nets-within-nets paradigm — a reference semantics (where tokens in the system net are references to common object nets), a value semantics (where tokens in the system net are distinct object nets), and a history process semantics (where tokens in the system net are object net processes) [25]. The reference semantics (as supported by the Renew tool [12]) has been used to model mobile agent systems [8,9]. However, a reference semantics provides a global reference scope, so that connectivity is enhanced but locality becomes meaningless. These authors have acknowledged that a value semantics is really required for mobility [10]. Then, locality is meaningful but connectivity is more constrained — an object net residing in a system place can primarily synchronise with the system transitions adjacent to the place. In other words, the object net token has to be removed from the place in order to interact with it. The interaction between co-resident tokens has more recently been added using another form of synchronous channel. However, the notation in Renew suggests that the interaction is achieved by the two object nets being accessed as side conditions of a system net transition.

Our proposal has certain points of contact with the nets-within-nets paradigm. The notation of having arcs incident on locations is akin to system net places containing object nets which can be removed (or added) by adjacent system net transitions. However, our locations have a more general fusion context. We have also refined the results of [10] in noting that if locations have bounded places, then we obviate the need for generalised clear and set arcs for shifting subsystem locations, and hence reachability and boundedness can remain decidable.

There have been a number of calculi proposed for working with mobility. Mobility was one of the key motivations behind the π-calculus [19]. However, the π-calculus did not explore the interplay between connectivity and locality — it had a flexible method for exchanging names and thus modifying connectivity, but there was no sense in which the connectivity was limited by locality. (The scope rules simply limited the accessibility to names.)

The ambient calculus [6] identifies *ambients* as a sphere of computation. They are properly nested which then determines locality. Capabilities are provided for entering, leaving and dissolving ambients. Movement across ambient boundaries can be *subjective* — the process in the ambient decides to employ the capability — or *objective* — the process outside the ambient dictates the move. As in the π-calculus, connectivity is provided by the ability to communicate capabilities or names over channels.

The seal calculus [26] identifies *seals* as agents or mobile computations. Here, seal boundaries are the main protection mechanism and seal communication is restricted to a single level in the hierarchy. Mobility is not under the control of a seal but of its parent — thus subjective moves of the ambient calculus are not supported.

The capabilities of the above calculi can be broadly mapped into the Mobile Petri Net formalism of this paper, which then makes it possible to specify and reason about causality and concurrency, as in [4]. Ambients and seals can be mapped to locations. We can cater for synchronous and asynchronous communication. Moves are objective, and fusion can be constrained to the enclosing location as in the Seal calculus.

8   Conclusions

This paper has reviewed a Petri Net formalism suitable for studying mobility, and specifically the interplay between locality and connectivity. It has extended Modular Petri Nets with the notion of nested modules called *locations*. The nesting of modules determines locality while the fusion context of each module determines connectivity. Locations are constrained so that the firing of their transitions depends on whether the locations are occupied. The identification of *vacate*, *occupy* and *regular* transitions helps determine the effect of each transition on the occupied status of its location. For notational convenience, we add broad arcs incident on locations to represent multiple vacate and occupy transitions.

This paper has assessed the proposed formalism by producing a model of the Mobile IP standard [22]. While the paper has not included the full annotations of the subnets, we feel that the graphical Petri Net notation makes it much easier to grasp the flow of information in comparison to the purely textual representation of the Mobile Unity model [18]. Further, the ability to execute the model greatly enhances the diagnostic possibilities for the model.

The proposed Petri Net formalism for mobility is quite simple and general even though the changes to Modular Petri Nets are not extensive. Consequently, it is relatively simple to study mobile systems using currently available modular analysis tools such as Maria [15], which already has support for nested modules. With a bit more effort, it is also possible to map these mobile systems into Hierarchical Coloured Petri Nets and analyse them in the tool CPNTools [7,11], but without support for transition fusion, it may not be able to reap the benefits of modular state spaces.

The model for Mobile IP was implemented in *Maria* and an interesting consequence was the identification of a problem in the Mobile Unity model. Ongoing work will consider more detailed state space exploration for the Mobile IP model, using techniques such as those applied to Object-Based Petri Nets in order to manage state space explosion [11].

Acknowledgements

The author gratefully acknowledges the helpful comments of Laure Petrucci and the anonymous referees which have significantly improved the quality of this paper.

References

1. Chiola, G., Dutheillet, C., Franceschinis, G., Haddad, S.: Stochastic Well-Formed Colored Nets and Symmetric Modeling Applications. IEEE Transactions on Computers 42(11), 1343–1360 (1993)
2. Christensen, S., Hansen, N.D.: Coloured Petri Nets Extended with Channels for Synchronous Communication. In: Valette, R. (ed.) ICATPN 1994. LNCS, vol. 815, pp. 159–178. Springer, Heidelberg (1994)
3. Christensen, S., Petrucci, L.: Modular analysis of Petri Nets. The Computer Journal 43(3), 224–242 (2000)
4. Devillers, R., Klaudel, H., Koutny, M.: Petri Net Semantics of the Finite π-Calculus. In: de Frutos-Escrig, D., Núñez, M. (eds.) FORTE 2004. LNCS, vol. 3235, pp. 309–325. Springer, Heidelberg (2004)
5. Dufourd, C., Finkel, A., Schnoebelen, P.: Reset nets between decidability and undecidability. In: Larsen, K.G., Skyum, S., Winskel, G. (eds.) ICALP 1998. LNCS, vol. 1443, pp. 103–115. Springer, Heidelberg (1998)
6. Cardelli, L., Gordon, A.: Mobile Ambients. In: Nivat, M. (ed.) Foundations of Software Science and Computational Structures. LNCS, vol. 1998, pp. 140–155. Springer, Heidelberg (1998)
7. Jensen, K., Kristensen, L., Wells, L.: Coloured Petri nets and CPN tools for modelling and validation of concurrent systems. Journal of Software Tools for Technology Transfer 9(3-4), 213–254 (2007)
8. Köhler, M., Moldt, D., Rölke, H.: Modelling the Structure and Behaviour of Petri Net Agents. In: Colom, J.-M., Koutny, M. (eds.) ICATPN 2001. LNCS, vol. 2075, pp. 224–241. Springer, Heidelberg (2001)
9. Köhler, M., Moldt, D., Rölke, H.: Modelling Mobility and Mobile Agents Using Nets within Nets. In: van der Aalst, W.M.P., Best, E. (eds.) ICATPN 2003. LNCS, vol. 2679, pp. 121–139. Springer, Heidelberg (2003)
10. Köhler, M., Rölke, H.: Properties of Object Petri Nets. In: Cortadella, J., Reisig, W. (eds.) ICATPN 2004. LNCS, vol. 3099, pp. 278–297. Springer, Heidelberg (2004)
11. Lakos, C.A., Kristensen, L.M.: State Space Exploration of Object-Based Systems using Equivalence Reduction and the Sweepline Method. In: Peled, D.A., Tsay, Y.-K. (eds.) ATVA 2005. LNCS, vol. 3707, pp. 187–201. Springer, Heidelberg (2005)

12. Kummer, O., Wienberg, F., Duvigneau, M., Schumacher, J., Köhler, M., Moldt, D., Rölke, H., Valk, R.: An extensible editor and simulation engine for Petri nets: Renew. In: Cortadella, J., Reisig, W. (eds.) ICATPN 2004. LNCS, vol. 3099, pp. 484–493. Springer, Heidelberg (2004)
13. Lakos, C., Christensen, S.: A General Systematic Approach to Arc Extensions for Coloured Petri Nets. In: Valette, R. (ed.) ICATPN 1994. LNCS, vol. 815, pp. 338–357. Springer, Heidelberg (1994)
14. Lakos, C.A.: A Petri Net View of Mobility. In: Wang, F. (ed.) FORTE 2005. LNCS, vol. 3731, pp. 174–188. Springer, Heidelberg (2005)
15. Mäkelä, M.: Maria: Modular Reachability Analyser for Algebraic System Nets. In: Esparza, J., Lakos, C.A. (eds.) ICATPN 2002. LNCS, vol. 2360, pp. 434–444. Springer, Heidelberg (2002)
16. Mäkelä, M.: Maria - Modular Reachability Analyzer for Algebraic System Nets (Version 1.3.4). Technical report, Helsinki University of Technology, Laboratory for Theoretical Computer Science (June 2003)
17. Mäkelä, M.: Model Checking Safety Properties in Modular High-Level Nets. In: van der Aalst, W.M.P., Best, E. (eds.) ICATPN 2003. LNCS, vol. 2679, pp. 201–220. Springer, Heidelberg (2003)
18. McCann, P.J., Roman, G.C.: Modeling Mobile IP in Mobile UNITY. ACM Transactions on Software Engineering and Methodology 8(2), 115–146 (1999)
19. Milner, R.: Elements of Interaction. Communications of the ACM 36(1), 78–89 (1993)
20. Milner, R.: The Flux of Interaction. In: Colom, J.-M., Koutny, M. (eds.) ICATPN 2001. LNCS, vol. 2075, pp. 19–22. Springer, Heidelberg (2001)
21. Perkins, C.: IP Mobility Support. Rfc 2002, Internet Engineering Task Force (IETF) (October 1996)
22. Perkins, C.: IP Mobility Support for IPv4. Rfc 3344, Internet Engineering Task Force (IETF) (August. 2002)
23. Reisig, W., Rozenberg, G. (eds.): Lectures on Petri Nets I: Basic Models. LNCS, vol. 1491. Springer, Heidelberg (1998)
24. Reisig, W., Rozenberg, G. (eds.): Lectures on Petri Nets II: Applications. LNCS, vol. 1492. Springer, Heidelberg (1998)
25. Valk, R.: Object Petri Nets — Using the Nets-within-Nets Paradigm. In: Desel, J., Reisig, W., Rozenberg, G. (eds.) Lectures on Concurrency and Petri Nets. LNCS, vol. 3098, pp. 819–848. Springer, Heidelberg (2004)
26. Vitek, J., Castagna, G.: Towards a Calculus of Secure Mobile Computations. In: IEEE Workshop on Internet Programming Languages, Chicago. IEEE, Los Alamitos (1998)

A Discretization Method
from Coloured to Symmetric Nets:
Application to an Industrial Example

Fabien Bonnefoi[1], Christine Choppy[2], and Fabrice Kordon[3]

[1] DSO/DSETI, Cofiroute, 92310 Sèvres, France
Fabien.Bonnefoi@cofiroute.fr
[2] LIPN - CNRS UMR 7030, Université Paris XIII, 93430 Villetaneuse, France
Christine.Choppy@lipn.univ-paris13.fr
[3] LIP6 - CNRS UMR 7606, Université P. & M. Curie, 75252 Paris, France
Fabrice.Kordon@lip6.fr

Abstract. Intelligent Transport Systems (ITS) tend to be more distributed and embedded. In many cases, continuous physical parameters are part of the systems description. Analysis techniques based on discrete models must integrate such constraints. In this paper, we propose a methodological way to handle such hybrid systems with model checking on Petri Nets and algebraic methods. Our methodology is based on transformations from Coloured Petri Nets (CPN) for their expressiveness to Symmetric Petri Nets (SN) to take advantage of their efficient verification techniques. Our methodology also addresses the impact of discretization on the precision of verification. In scientific computing, the discretization process introduces "error intervals" that could turn a given verified property into a false positive one. In that case, the property to be verified might have to be transformed to cope with such precision errors.

Keywords: Discretization, Symmetric Petri Nets, Coloured Petri Nets, Intelligent Transport Systems, Hybrid Systems.

1 Introduction

Future supervision systems tend to be more distributed and embedded. Parallelism brings a huge complexity and then, a strong need to deduce good and bad behaviours on the global system, from the known behaviour of its actors. This is crucial since safety critical missions can be supervised by such systems. Intelligent Transport Systems (ITS) are a typical example: many functions tend to be integrated in vehicles and road infrastructure. Moreover, in many cases physical constraints are part of such systems. Analysis techniques based on discrete models must integrate such constraints: we then speak of *hybrid* systems.

A major trend in formal analysis is to cope with such systems. This raises many issues in terms of analysis complexity. Some techniques are dedicated to continuous analysis such as algebraic approaches like B [1]. However, such approaches are difficult to set up and most industries prefer push-button tools.

K. Jensen, J. Billington, and M. Koutny (Eds.): ToPNoC III, LNCS 5800, pp. 159–188, 2009.
© Springer-Verlag Berlin Heidelberg 2009

Model checking easily offers such push-button tools but does not cope well with continuous systems. Most model checking techniques deal with discrete and finite systems. Thus, management of hybrid systems is not easy or leads to potentially infinite systems that are difficult to verify. For example, management of continuous time requires much care, even to only have decidable models. Hybrid Petri Nets [19] might be a solution to model and analyze hybrid systems but no tool is available to test neither safety nor temporal logic properties [40].

In this paper, we propose a methodology to handle hybrid systems with model checking on Petri Nets and algebraic methods. Our methodology is based on transformations from Coloured Petri Nets (CPN) [30] to Symmetric Petri Nets[1] (SN) [11,12]. CPN expressiveness allows an easy modelling of the system to be analyzed. SN are of interest for their analysis because of the symbolic state space that is efficient to represent the state space of large systems. Since SN offer a limited set of operations on colours, transformation from CPN requires much care from the designer as regards the types to be discretized.

Our methodology also addresses an important question: what is the impact of discretization on the precision of verification? As in scientific computing, the discretization process introduces "precision errors" that could turn a given verified property into a false positive one. In that case, the property to be verified might have to be transformed to take into consideration such precision errors.

Sect. 2 briefly recalls the notions of CPN and SN, as well as abstraction/refinement, type issues. Our methodology which involves modelling, discretization and verification is presented in Sect. 3, and we show in Sect. 4 how we model our Emergency Braking application. The various discretization concepts on our case study are detailed in Sect. 5, and our experiments on net analyses are presented in Sect. 6. Some perspectives on discretization are also discussed in Sect. 7 before a conclusion.

2   Building Blocks

This section presents the building blocks from the state of the art used in the discretization method.

2.1   Coloured Petri Nets

Coloured Petri nets [30] are high level Petri nets where tokens in a place carry data (or colours) of a given type. Since several tokens may carry the same value, the concept of multiset (or bag) is used to describe the marking of places.

In this paper, we assume the reader is familiar with the concept of multisets. We thus recall briefly the formal definition of coloured Petri nets as in [30]. It should be noted however that the types considered for the place tokens may be basic types (e.g. boolean, integers, reals, strings, enumerated types) or structured

---

[1] *Symmetric nets* were formerly known as *Well-Formed nets*, a subclass of *High-level Petri nets*. The new name appeared during the ISO standardisation of Petri nets [27].

types – also called compound colour sets – (e.g. lists, product, union, etc.). In both cases, the type definition includes the appropriate (or usual) functions.

Different languages were proposed to support the type definition for coloured Petri nets (e.g. algebraic specification languages as first introduced in [43], object oriented languages [8]), and an extension of the Standard ML language was chosen for CPN Tools [17]. As always, there may be a tradeoff between the expressivity of a specification language, and efficiency when tools are used to compute executions, state graphs, etc. If expressivity is favored, it could be desirable to allow any appropriate type and function, while when tools should be used to check the behaviour and the properties of the system studied, the allowed types and functions are restricted (as the language allowed for CPN Tools or as in Symmetric Nets presented in Sect. 2.2). Here, to start with we want to allow a specification language that fits as much as possible what is needed to describe the problem under study, and then to show how the specification is transformed so as to allow computations and checks by tools.

In the following, we refer to $EXPR$ as the set of expressions provided by the net inscription language (net inscriptions are arcs expressions, guards, colour sets and initial markings), and to $EXPR_V$ as the set of expressions $e \in EXPR$ such that $Var[e] \subseteq V$ ($Var[e]$ denotes the set of variables occurring in e).

Definition 2.1. *A non-hierarchical coloured Petri net CPN [30] is a tuple* $CPN = (P, T, A, \Sigma, V, C, G, E, I)$ *such that:*

1. P is a finite set of *places*.
2. T is a finite set of *transitions* such that $P \cap T = \emptyset$.
3. $A \subseteq P \times T \cup T \times P$ is a set of directed *arcs*.
4. Σ is a finite set of non empty *colour sets* (types).
5. V is a finite set of typed *variables* such that $\forall v \in V, Type[v] \in \Sigma$.
6. $C : P \to \Sigma$ is a *colour set function* assigning a colour set (or a type) to each place.
7. $G : T \to EXPR_V$ is a *guard* function assigning a guard to each transition such that $Type(G(t)) = Bool$, and $Var[G(t)] \subseteq V$, where $Var[G(t)]$ is the set of variables of $G(t)$.
8. $E : A \to EXPR_V$ is an *arc expression function* assigning an arc expression to each arc such that $Type(E(a)) = C(p)_{MS}$, where p is the place connected to the arc a.
9. $I : A \to EXPR_V$ is an *initialisation function* assigning an initial marking to each place such that $Type(I(p)) = C(p)_{MS}$.

As explained in Sect. 3, the first step of our methodology is to produce a CPN model for the application. The next step is a transformation motivated by the discretization of continuous functions to obtain a symmetric net.

2.2 Symmetric Nets

Symmetric nets were introduced in [11,12], with the goal of exploiting symmetries in distributed systems to get a more compact representation of the state space.

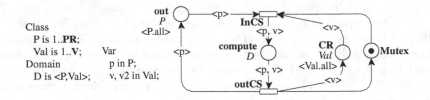

Fig. 1. Example of Symmetric Net

The concept of symmetric nets is similar to the coloured Petri net one. However, the allowed types for the places as well as allowed colour functions are more restricted. These restrictions allow us to compute symmetries and obtain very compact representations of the state space, enabling the analysis of complex systems as in [28].

Basically, types must be finite enumerations and can only be combined by means of cartesian products. Allowed functions in arc expressions are: identity, successor, predecessor and broadcast (that generates one copy of any value in the type). These constraints affect points 4, 6, 7, 8, 9 in Definition 2.1.

The Symmetric net in Fig. 1 models a class of threads (identified by type P) accessing a critical resource **CR**. Threads can get a value within the type Val from **CR**. Constants **PR** and **V** are integer parameters for the system. The class of threads is represented by places **out** and **compute**.

Place **compute** corresponds to some computation on the basis of the value provided by **CR**. At this stage, each thread holds a value that is replaced when the computation is finished. Place **Mutex** handles mutual exclusion between threads and contains token with no data ("black tokens" in the sense of the Petri Net standard [29]). Place **out** initially holds one token for each value in P (this is denoted <P.all>) and place **CR** holds one value for each value in type Val.

The main interest of SN is the possibility to generate a *symbolic state space*. A symbolic state (in the symbolic state space) represents a set of concrete states with a similar structure. This is well illustrated in Fig. 3 where the upper symbolic state represents the four upper states in Fig. 2.

In Fig. 3, the bottom state corresponds to the initial marking where **out** contains one token per value in type P and **CR** one token per value in type Val.

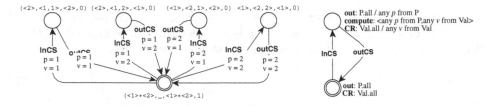

Fig. 2. State space of the model in Fig. 1   **Fig. 3.** Symbolic state space

The top state represents a set of states with permutations on binding variables p and v. There, place **compute** holds one token only while places **out** and **CR** hold one token per value in the place type minus the value used to build the token in place **compute**.

Let us note that in some case (like here), the symbolic[2] state space does not change with types P and *Val* or initial markings, which yields a very compact representation of the system behaviour. The symbolic state space may even be exponentially smaller than the explicit state space.

Verification of properties can be achieved either by a structural analysis, on the symbolic state space (model checking), or on the unfolded associated Place/Transition (P/T) net (essentially to compute structural properties).

2.3 Transformation, Abstraction and Refinement

Abstraction and refinement are part of the use of formal specifications. While abstraction is crucial to concentrate on essential aspects of the problem to be solved (or the system to be built), and to reason about them, more elaborate details need to be further introduced in the refinement steps. A similar evolution is taking place when a general pattern or template is established to describe the common structure of a family of problems, and when this template is instantiated to describe a single given problem.

Three kinds of refinement for coloured Petri nets are introduced in [34,35], the type refinement, the node refinement and the subnet refinement. These refinements are correct if behaviours are preserved and if, to any behaviour of a refined net it is possible to match a behaviour of the abstract net.

Another motivation is raised by the use of tools to check the behaviour and properties of the model, since it may involve the discretization of some domains so as to reduce the number of possible values to consider in the state space. It thus involves a simplification of some domains that may be considered as an abstraction.

3 Methodology for Discretization

This section presents our methodology to model and analyse an hybrid system. We give an overview and then detail its main steps and the involved techniques.

3.1 Overview of the Methodology

Fig. 4 sketches our methodology. It takes as input a set of requirements. It is thus divided in two parts:

---

[2] The word *symbolic* has several meanings in model checking. Here, it refers to the symbolic state space, which is a set-based representation of the state space. It is also used in *symbolic model checking* to indicate the encoding of explicit states by means of decision diagrams (such as BDD). This term is also used later in the paper.

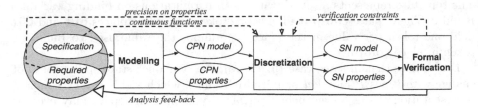

Fig. 4. Overview of our methodology

- the *specification* describes the system (we only consider in this work the behavioural aspects),
- the *required properties* establish a set of assertions to be verified by the system.

Once the specification written using "classical" techniques, the system is modelled using high-level Petri Nets (CPN) that allow one to insert complex colour functions such as ones involving real numbers. These functions come from the specifications of the system (in Intelligent Transport Systems, numerous behaviours are described by means of equations describing physical models). These functions are inserted in arc labels into the CPN-model produced by the **Modelling** step. Required properties are also set in terms of CPN.

However, the CPN system cannot be analysed in practice since the system is too complex (due to the data and functions involved). So, the **Discretization** step is dedicated to the generation of an associated system expressed using Symmetric Nets. Symmetric Nets are well suited to specify such systems that are intrinsically symmetric [5]. Operations such as structural analysis or model checking can be achieved for much larger systems. Formal analysis of the system is performed at the **Formal Verification** step.

Let us note that a similar transformation is achieved in [3] instead from coloured Petri nets to counter systems. The goal in this work is similar, being able to analyze a CPN specification, but not in the domain of hybrid systems.

The following sections present the three main steps of our methodology and especially focus on the **Discretization** step that is the most delicate one as well as the main contribution of this paper.

3.2   Modelling

There are heterogeneous elements to consider in Intelligent Transport Systems (ITS): computerized actors (such as cars or controllers in a motorway infrastructure) have to deal with physical variables such as braking distances, speed and weight.

If those continuous variables were not modelled, only a subpart of the "required properties" of the system could be checked. Especially, it is not possible to verify properties related to quantitative variables.

Our work aims at providing a more precise representation of the system in the Petri net models by representing those quantitative variables. To design the CPN model we used a template adapted to the case study presented in Sect. 4. The "interfaces" of the Petri net model, presented in Sect. 5, were already identified. The main task was to identify control and data flows that are involved in this subpart of the system, and that must be modelled to allow formal verification. Also, operations made on those flows were identified.

Then, the different selected variables of the system were represented using equivalent types in CPN. For example, continuous variables of the system were modelled with the real type of CPN formalism. The functions of the system that manipulate the continuous variables were represented using arc expressions.

3.3 Discretization

The discretization step takes CPN with their properties as inputs, and produces SN with their properties as outputs. To achieve this goal, a discretization of the real data and functions involved is performed. As a result, the types involved in the CPN are abstracted, and the real functions are represented by a place providing tuples of appropriate result values.

We propose different steps to manage the discretization of continuous functions in Symmetric Nets:
- Step 1: Continuous feature discretization
- Step 2: Error propagation computing
- Step 3: Type transformation and modelling of complex functions in Symmetric Nets.

Step 1 - Continuous feature discretization. Discretization is the process of transforming continuous models and equations into discrete counterparts. Depending on the domain to which this process is applied we use also the words "digitizing", "digitization", "sampling", "quantization" or "encoding". Techniques for discretization differ according to application domains and objectives.

Let us introduce the following definitions to avoid ambiguity in this paper:

Definition 3.1. *A region is a n-dimensional polygon (i.e. a polytope) made by adjacent points of an n-dimensional discretized function.*

Definition 3.2. *A mesh is a set of regions used to represent a n-dimensional discretized function for modeling or analysis.*

There exist many discretization methods that can be classified between global or local, supervised or unsupervised, and static or dynamic methods [20].

- **Local methods** produce partitions that are applied to localized regions of the instance space. Those methods usually use decision trees to produce the partitions (i.e. the classification).
- **Global methods** (like binning) [20] produce a mesh over the entire n-dimensional continuous instance space, where each feature is partitioned into regions. The mesh contain $\prod_{i=1}^{n} k_i$ regions, where k_i is the number of partitions of the ith feature.

In our study we consider the **equal width interval binning method** [20] as a first approach to discretize the continuous features. Equal width interval binning is a global unsupervised method that involves dividing the range of observed values for the variable into k equally sized intervals, where k is a parameter provided by the user. If a variable x is bounded by x_{min} and x_{max}, the interval width is:

$$\Delta = \frac{x_{max} - x_{min}}{k} \tag{1}$$

Step 2 - Error propagation computing. To model a continuous function in Symmetric Nets it is necessary to convert it into an equivalent discrete function. This operation introduces inaccuracy (or error) which must be taken into account during the formal verification of the model. This inaccuracy can be taken into account in the Symmetric Net properties in order to keep them in accordance with the original system required properties. The other solution is to change the original required properties taking into account the introduced inaccuracy.

The issues are well expressed in [10]: *"in science, the terms uncertainties or errors do not refer to mistakes or blunders. Rather, they refer to those uncertainties that are inherent in all measurements and can never be completely eliminated. (...) A large part of a scientist's effort is devoted to understanding these uncertainties (error analysis) so that appropriate conclusions can be drawn from variable observations. A common complaint of students is that the error analysis is more tedious than the calculation of the numbers they are trying to measure. This is generally true. However, measurements can be quite meaningless without knowledge of their associated errors."*

There are different methods to compute the error propagation in a function [36,10]. The most current one is to determine the separate contribution due to errors on input variables and to combine the individual contributions in a quadrature.

$$\Delta_{f(x,y,..)} = \sqrt{\Delta_{fx}^2 + \Delta_{fy}^2 + ...} \tag{2}$$

Then, different methods to compute the contribution of input variables to the error in the function are possible, like the "derivative method" or the "computational method".

– The derivative method evaluates the contribution of a variable x to the error on a function f as the product of the error on x (i.e. Δ_x) with the partial derivative of $f(x, y, ..)$:

$$\Delta_{fx} = \frac{\partial f(x, y, ..)}{\partial x} \Delta_x \tag{3}$$

– The computational method computes the variation by a finite difference:

$$\Delta_{fx} = | f(x + \Delta_x, y, ..) - f(x, y, ..) | \tag{4}$$

The use of individual contributions in a quadrature relies on the assumption that the variables are independent and that they have a Gaussian distribution for their mean values. This method is interesting as it gives a good evaluation of the error. But we do not have a probabilistic approach, and we do not have a Gaussian distribution of the "measured" values.

In this paper, we prefer to compute the maximum error bounds on f due to the errors on variables as it gives an exact evaluation of the error propagation. Let $f(x)$ be a continuous function, x be the continuous variable, and x_{disc} the discrete value of x. If we choose a discretization step of $2 \cdot \Delta_x$ we can say that for each x_{disc} image of x by the discretization process, $x \in [x_{disc} - \Delta_x, x_{disc} + \Delta_x]$ (which is usually simplified by the expression $x = x_{disc} \pm \Delta_x$). We can compute the error $\Delta_{f(x)}$ introduced by the discretization:

$$f(x) = f(x_{disc}) \pm \Delta_{f(x)} \qquad \Delta_{f(x)} = f(x \pm \Delta_x) - f(x) \qquad (5)$$

We can also say that the error on $f(x)$ is inside the interval :

$$\Delta_{f(x)} \in [Min(f(x \pm \Delta_x) - f(x)), Max(f(x \pm \Delta_x) - f(x))] \qquad (6)$$

This method can also be applied with functions of multiple variables. In this case, for a function f of n variables $f(x \pm \Delta_x, y \pm \Delta_y, ..)$ has 2^n solutions. The maximum error bounds on f are:

$$\Delta_f \in [Min(f(x \pm \Delta_x, y \pm \Delta_y, ..) - f(x, y, ..)), Max(f(x \pm \Delta_x, y \pm \Delta_y, ..) - f(x, y, ..))]$$
$$(7)$$

An example of this method applied to an emergency braking function is presented in Sect. 5.2.

Step 3a - Type transformation. Once the best discretization actions are decided with regards to our goals, the CPN model may be transformed into a symmetric net.

Let us first note that some types do not need to be transformed because they are simple enough (e.g. enumerated types) and do not affect the state space complexity.

When the types are more complex, two kinds of transformation are involved in this process, that concern the value set (also called carrier set), and the complex functions. The value set transformation results from the discretization of all infinite domains into an enumerated domain.

A node refinement is applied to transitions that involve a complex function on an output arc expression. As explained below and in Fig. 5, there are two possibilities to handle this. In our method, such functions are represented by tuples of discrete values (values of the function arguments and of the result) that are stored in a **values** place. The **values** place is both input and output of the refined transition, thus for any input data provided by the original input arc(s), the **values** place yields the appropriate tuple with the function result.

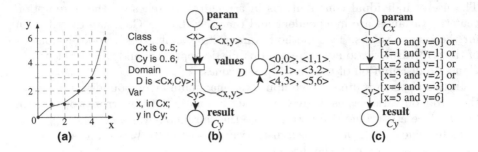

Fig. 5. Example of function discretization using a place or a transition guard

Step 3b - Modelling of complex functions in Symmetric Nets. To cope with the modelling of complex functions in Symmetric Nets (for example, the computation of braking distance according to the current speed of a vehicle), we must discretize and represent them either in a specific place or as a guard of a transition. When a place is used, it can be held in an SN-module ; it then represents the function and can be stored in a dedicated library.

Fig. 5 represents an example of function discretization. The left side (a) shows a function that is discretized, and the right side shows the corresponding Petri net models : in model (b), the function is discretized by means of a place, in model (c), it is discretized by mean of a transition guard. In both cases, correct associations between x and y are the only ones to be selected when the transition fires. Note that in model (b) **values** markings remain constant.

This technique can be generalized to any function $x = f(x_1, x_2, ..., x_n)$, regardless of its complexity. Non deterministic functions can also be specified in the same way (for example, to model errors in the system). Let us note that:

- the discretization of any function becomes a modelling hypothesis and must be validated separately (to evaluate the impact of imprecision due to discretization),
- given a function, it is easy to automatically generate the list of values to be stored in the initial marking of the place representing the function, or to be put in the guard of the corresponding transition.

The only drawback of this technique is a loss in precision compared to continuous systems that require appropriate hybrid techniques [14]. Thus, the choice of a discretization schema must be evaluated, for example to ensure that uncertainty remains in a safe range.

It is also possible to model functions by using inequations in the guard of Fig. 5(c). However, comparison between free variables break the model symmetries if there are any. This is why, in model of Fig. 5(c) the guard only use comparisons between a free variable and a constant.

3.4 Verification

Our models are analysed using:

- *Structural techniques* (invariant computation, structural bounds, etc) on P/T nets. Since our nets are coloured, an unfolding tool able to cope with large systems [33] is used to derive the corresponding P/T net to compute structural properties.
- *Model checking*, there exist efficient model checking techniques that are dedicated to this kind of systems and make intensive use of symmetries as well as of decision diagrams. Such techniques revealed to be very efficient for this kind of systems by exploiting their regularity [28,5].

However, due to the complexity of such systems, discretization is a very important point. If symmetric net coloured classes are too large (i.e. the discretization interval is too small), we face a combinatorial explosion (for both model checking or structural analysis by unfolding). On the other hand, if the error introduced by the discretization is too high, the property loses its "precision" and the verification of properties may lose its significance.

This is why in Fig. 4, the discretization step needs *verification constraints* as inputs from the verification step. A compromise between combinatorial explosion and precision in the model must be found.

4 Modelling the Emergency Braking Problem

The case study presented in this paper is a subpart of an application from the "Intelligent Road Transport System" domain. It is inspired from the European project SAFESPOT [7]. This application is called "Hazard and Incident Warning" (H&IW), and its objective is to warn the driver when an obstacle is located on the road. Different levels of warning are considered, depending on the criticality of the situation. This section presents the "Emergency Braking module" of the application and how it can be specified using the CPN formalism.

4.1 Presentation of the Case Study

SAFESPOT is an Integrated Project funded by the European Commission, under the strategic objective "Safety Cooperative Systems for Road Transport". The Goal of SAFESPOT is to understand how "intelligent" vehicles and "intelligent" roads can cooperate to produce a breakthrough in road safety. By combining data from vehicle-side and road-side sensors, the SAFESPOT project will allow to extend the time in which an accident is foreseen. The transmission of warnings and advices to approaching vehicles (by means of vehicle-to-vehicle and vehicle-to-infrastructure communications [39,38,18]), will extend in space and time the driver's awareness of the surrounding environment.

Functional Architecture. The SAFESPOT applications [4] rely on a complex functional architecture. If the sensors and warning devices differ between

SAFESPOT vehicles and SAFESPOT infrastructure, the functional architecture is designed to be almost the same for these two main entities of the system providing a peer-to-peer network architecture. It enables real-time exchange of vehicles' status and of all detected events or environmental conditions from the road. This is necessary to take advantage of the cooperative approach and thus enable the design of effective safety applications.

As presented in Fig. 6, information measured by sensors is provided to the "Data Processing / Fusion" module or transmitted through the network to the "Data Processing / Fusion" module of other entities. This module analyses and processes arriving data to put them on the "Local Dynamic Map" (LDM) of the system. The "Local Dynamic Map" enables the cooperative applications of the system to retrieve relevant variables and parameters depending on their purpose. The applications are then able to trigger relevant warnings to be transmitted to appropriate entities and displayed via an onboard Human Machine Interface (HMI) or road side Variable Message Signs (VMS). In SAFESPOT, five main infrastructure-based applications were defined: "Speed Alert", "Hazard and Incident Warning", "Road Departure Prevention", "Co-operative Intersection Collision Prevention" and "Safety Margin for Assistance and Emergency Vehicles". These applications are designed to provide the most efficient recommendations to the driver.

Hazard and incident application. The aim of the "Hazard and Incident Warning" application is to warn the drivers in case of dangerous events on the road. Selected events are: accident, presence of unexpected obstacles on the road, traffic jam ahead, presence of pedestrians, presence of animals and presence of a vehicle driving in the wrong direction or dangerously overtaking. This application also analyses all environmental conditions that may influence the road friction or decrease the drivers' visibility. Based on the cooperation of vehicles and road side sensors, the "Hazard and Incident Warning" application provides warnings to the drivers and feeds the SAFESPOT road side systems and vehicles with information on new driving situations. This application is essential to provide other applications with the latest relevant road description.

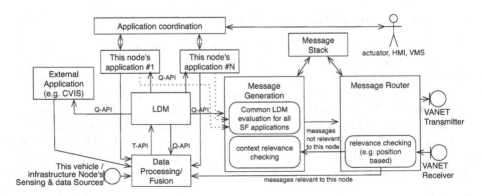

Fig. 6. SAFESPOT High Level Architecture

The emergency braking module. The emergency braking module is one sub-system in the "Hazard and Incident Warning" distributed application. It communicates with other subsystems. The behavior of this subsystem is significant in the SAFESPOT system and must be analyzed.

In the case of an obstacle on the road, the emergency braking module receives/retrieves the speed, deceleration capability and the relative distance to a static obstacle for the monitored vehicle. With these data, it will compute a safety command to be transmitted to the driver and to other applications of the system. Those commands represent the computed safety status of a vehicle. The three commands (or warnings) issued by this module are "Comfort" if no action is required from the driver, "Safety" if the driver is supposed to start decelerating, and "Emergency" if the driver must quickly start an emergency braking. This is illustrated in Fig. 7. Note that if a driver in an "Emergency" status does not brake within one second, an automated braking should be triggered by the "Prevent" system (which is another European project).

Petri nets are well suited to describe and analyse this type of application. However, a part of the "Hazard and Incident Warning" application algorithm is based on the analysis of continuous variables like vehicle speed or position of an obstacle. Those data are part of the data flow of the system ; they are also determinant for the control flow of the system.

Many properties can be verified using Petri nets without modelling continuous variables. However, some properties require continuous variables to be modelled, like those presented Sect. 4.3. Then we face a combinatorial explosion and have to enhance the Petri net formalism as well as the modelling methodology to enable the verification of such systems.

4.2 Mathematical Model of the Emergency Braking Module

The *emergency braking module* implements a strategy function to determine the safety status of a given vehicle. This function computes the *braking distance* of a vehicle from its speed and deceleration capabilities.

Let $v \in V$ be the velocity (speed) of a vehicle with $V \subset \mathbb{R}^+$. Let also $b \in B$ be the braking capability of the vehicle with $B \subset \mathbb{R}^+$. The braking distance function is then:

$$f(v, b) = \frac{v^2}{2b} \tag{8}$$

Let then $d \in D$ be the relative distance of the obstacle to the vehicle with $D \subset \mathbb{R}^+$. The main algorithm of the "Emergency braking module" defines two

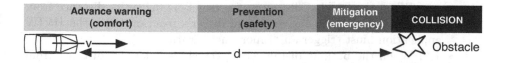

Fig. 7. Emergency braking safety strategy

thresholds to determine when a vehicle goes from a "Comfort state" to a "Safety state", and from a "Safety state" to an "Emergency state". Those thresholds are based on the time left for the driver to react. According to the application specification, if the driver has more than three seconds to react he is in a "Comfort state", then if he has less than three seconds but more than one second he is in the "Safety state", if he has less than one second to react, he is in the "Emergency State". The values of those thresholds are expressed as follows:

$$EB\_Safety = \frac{v^2}{2b} + v * 3 - d \tag{9}$$

$$EB\_Emergency = \frac{v^2}{2b} + v * 1 - d \tag{10}$$

The resulting algorithm of the strategy function can be represented as follows:

```
function Eb_Strategy(d,v,b){
   Eb_Safety = (v^2)/(2b) + v * 3 - d;
   Eb_Emergency = (v^2)/(2b) + v * 1 - d;
   if (Eb_Safety < 0) then  Command = 'Comfort';
   else if (Eb_Emergency < 0) then Command = 'Safety';
      else  Command = 'Emergency'  endif
   return Command;}
```

In SAFESPOT, v values are considered to be in $[0, 46]m/s$, b in $[3, 9]m/s^{-2}$ and d in $[0, 500]m$. If variables are outside those sets, other applications are triggered (this becomes out of the scope of the emergency braking module). For example, speeds above $46m/s$ are managed by the "Speed Alert" application.

4.3   Required Properties

The SAFESPOT and H&IW application specifications are completed with required properties, structured following the FRAME method [22], to be satisfied by the system. An analysis of the H&IW required properties shows that of the 47 main requirements, 18 (i.e. 38%) involve continuous space and/or time constraints. The method presented in this paper focuses on those properties. Here are examples of this kind of properties for the emergency braking module:

- **Property 1:** commands must be appropriately activated. This can be refined as follows.
  - **1.1:** When the braking distance of a vehicle is below its distance from a static obstacle plus one second of driver's reaction time, the H&IW application must trigger an "Emergency" warning.
  - **1.2:** When the braking distance of a vehicle is below its distance from a static obstacle plus three seconds of driver's reaction time, the H&IW application must trigger a "Safety" warning.

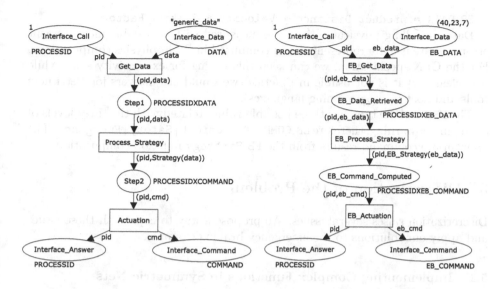

Fig. 8. Template Coloured Petri net for the H&IW applications

Fig. 9. Coloured Petri net instantiated for the Emergency Braking application

- **1.3:** When multiple obstacles are present on the road, the H&IW application must trigger the associated multiple warnings[3].
- **Property 2:** commands must be issued progressively stronger
  - When a vehicle is approaching an obstacle, the H&IW application must trigger the different warnings *Confort, Safety, Emergency* in the order corresponding to the danger faced by the vehicle. Therefore, if a driver does not react to the first *Comfort* warning, a *Safety* warning must follow, and then the *Emergency* one.

4.4   The Coloured Petri Net Specification

Several modules in the H&IW application share the same architecture, namely for a given process, data is retrieved from the interface. Then, a command is computed, and sent to appropriate modules in the system. The CPN of Fig. 8 exhibits this generic behaviour (i.e. the template in Sect. 3.2). Transition Get_Data has two input arcs from places Interface_Call and Interface_Data. Place Interface_Call is typed with PROCESSID which may be an integer subset (here the marking is a token with value 1). Once a process is called and data is retrieved, place Step1 carries tokens that are couples (pid,data). Transition Process_Strategy provides a command resulting from computations on data.

In Fig. 9 this generic schema is instantiated for the Emergency Braking Application (so, generic_data and generic_command become resp. of type EB_DATA and EB_COMMAND). Since data are Distance, Velocity, and Braking_Factor:

---

[3] When multiple warnings are triggered, they are filtered by the Application Coordinator like shown in Fig. 6.

EB_DATA = product Distance * Velocity * Braking_Factor.

Data modelling physical entities are measured with a possible measurement error and are usually represented and computed in $\mathbb{R}^*$ in physics computations. For the CPN specification, we can keep this typing for expressivity sake, while it is clear that it is not usable in practice (we would use integers for Petri nets tools and float in programming languages).

The EB_COMMAND type has three possible values related with the three levels of command or warning, therefore EB_COMMAND = Comfort | Safety | Emergency. The appropriate command results from the EB_Strategy function computation.

5   Discretization of the Problem

Discretization raises several issues. We propose a way to cope with these issues and apply our solutions to the emergency braking example.

5.1   Implementing Complex Functions in Symmetric Nets

Starting from the CPN model we use the methodology presented in Sect. 3.

First, CPN types must be transformed into discrete types. Using the equal width interval binning discretization method (presented in Sect. 3.3) with a number of k_v, k_b and k_d intervals for each variable we obtain a mesh of $k_v \times k_b \times k_d$ regions (as defined in definitions 3.2 and 3.1) in the resulting discretized function. The resulting sets for variables d, v and b are then composed of k ordered elements. For example, with $k = k_v = k_b = k_d = 10$ the resulting discretized type of v is $[0, 4.6, 9.2, ..., 46]$ and the discretized braking function contains 10^3 regions. With $k = 100$, the domain of v is $[0, 0.46, 0.92, ..., 46]$ and the mesh is composed of 10^6 regions.

Sect. 3.3 presents two solutions to model complex functions in Symmetric Nets. We select solution b in Fig. 5 because it is more efficient when computing the symbolic state space. Therefore we add place **EB_Strategy_Table** in the Symmetric Petri net (Fig. 10), and the initial associated marking that is presented in Sect. 5.4.

We chose a simple and generic discretization method that does not take into account the specificity of functions to be discretized. Other discretization methods like those using variable intervals can reduce the number of markings with the same level of accuracy in the resulting discretized function. These aspects are discussed in sections 7.1 and 7.2.

Finally, depending on the analyzed properties, it is also possible to compute and use the equivalence classes. This aspect is discussed in Sect. 6.2.

5.2   Computation of the Error Propagation in Symmetric Nets

As presented in Sect. 3, we compute the precision error introduced by the discretization operation. The resulting error in the computation of the "Threshold" is:

Table 1. Error bounds for different discretization parameters

Discretization par. k / $card(EBData)$	$v = 13m/s,\ b = 8m/s^{-2},$ $d = 500m$	$v = 36m/s,\ b = 4m/s^{-2},$ $d = 100m$
10 / 10^3	$\Delta_{Eb\_Saf} \in [-70.83m, 74.84m]$ $\Delta_{Eb\_Emerg} \in [-61.64m, 65.64m]$	$\Delta_{Eb\_Saf} \in [-118.9m, 144.5m]$ $\Delta_{Eb\_Emerg} \in [-109.7m, 135.3m]$
20 / $8 * 10^3$	$\Delta_{Eb\_Saf} \in [-35.87m, 36.81m]$ $\Delta_{Eb\_Emerg} \in [-31.27m, 32.26m]$	$\Delta_{Eb\_Saf} \in [-61.97m, 68.28m]$ $\Delta_{Eb\_Emerg} \in [-57.37m, 63.68m]$
50 / $12.5 * 10^3$	$\Delta_{Eb\_Saf} \in [-14.45m, 14.61m]$ $\Delta_{Eb\_Emerg} \in [-12.62m, 12.77m]$	$\Delta_{Eb\_Saf} \in [-25.47m, 26.47m]$ $\Delta_{Eb\_Emerg} \in [-23.63m, 24.63m]$
100 / 10^6	$\Delta_{Eb\_Saf} \in [-7.25m, 7.29m]$ $\Delta_{Eb\_Emerg} \in [-6.33m, 6.37m]$	$\Delta_{Eb\_Saf} \in [-12.85m, 13.10m]$ $\Delta_{Eb\_Emerg} \in [-11.93m, 12.19m]$

$$\Delta_{Eb\_Safety} = Eb\_Safety(v \pm \Delta_v, b \pm \Delta_b, d \pm \Delta_d) - Eb\_Safety(v, b, d) \qquad (11)$$

$$\Delta_{Eb\_Safety} = (\frac{(v \pm \Delta_v)^2}{2(b \pm \Delta_b)} + 3(v \pm \Delta_v) - (d \pm \Delta_d)) - (\frac{v^2}{2b} + 3v - d) \qquad (12)$$

$$\Delta_{Eb\_Safety} = \frac{(v \pm \Delta_v)^2}{2(b \pm \Delta_b)} - \frac{v^2}{2b} \pm 3\Delta_v \pm \Delta_d \qquad (13)$$

For example, let us consider (cf. Table 1) a classic private vehicle driving at $v = 13m/s$ (i.e. $50km/h$), on a dry road (i.e. $b = 8m/s^2$), at $d = 500m$ from an obstacle. If we consider $k = 100$ intervals and an error of respectively $\pm0.45m/s$ for v, $\pm0.06m/s^2$ for d and $\pm5m$ for p. Then we obtain:[4]

$\Delta_{Eb\_Safety} \in [-7.25m, +7.29m]$ $\qquad$ $\Delta_{Eb\_Emergency} \in [-6.33m, +6.37m]$

For the same vehicle at 100 meters from the obstacle, driving at $v = 36m/s$ (i.e. $130km/h$), on a wet road (i.e. $b = 4m/s^2$), we obtain:

$\Delta_{Eb\_Safety} \in [-12.85m, +13.10m]$ $\qquad$ $\Delta_{Eb\_Emergency} \in [-11.93m, +12.19m]$

Those results provide an information on the precision of the Symmetric Net properties. Table 1 gives some error bounds computed from four values for parameter k. As expected, precision of computed thresholds depends on k. However, precision also depends on the values of variables. For example, values of v and b are a determinant for error bound computation. Exploiting those precisions, to validate the Symmetric Net model and its properties, requires to consider carefully those values.

5.3 Validating the Discretization in Symmetric Nets

Discretization of variables and function in the Symmetric Net model in Fig. 10 introduces imprecision. Depending on properties that need to be verified, this imprecision must be considered. For example, properties presented Sect. 4.3 can be verified using CTL (Computation Tree Logic) [21] formulae. With a discretization factor of $k = 100$ values on input variables (cf last line of Table 1),

---

[4] In Table 1, Δ_{Eb_Saf} and Δ_{Eb_Emerg} stand for Δ_{Eb_Safety} and $\Delta_{Eb\_Emergency}$ respectively.

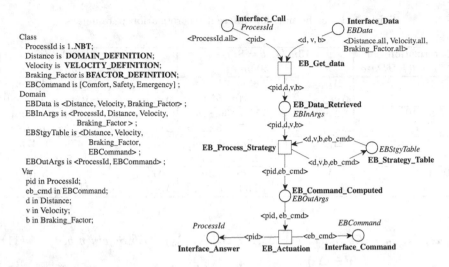

Fig. 10. Symmetric Petri net of the Emergency Braking module (marking of place **EB_Strategy_Table** is not displayed for sake of place)

Property 1 (in Sect. 4.3) can be verified with an accuracy smaller than $\pm7,3m$ on a relative distance, for a velocity of $13m/s$ on a dry road ($b = 8m/s^{-2}$).

If the introduced imprecision is acceptable with regards to the properties to be verified, then the system designer can state that the discretization is valid for those properties. Otherwise, a better accuracy may be required and a new discretization must be done.

It is also possible to integrate the imprecision in the CTL formulae. To do so, more constraining values of input variables must be chosen (i.e. a higher speed, a lower braking factor or a closer obstacle) in the CTL formulae. In our case, the simplest way is to choose a lower value of obstacle position that takes into account the discretization error. For example, the CTL formula:

`AG((EB_Data_Retrieved == <13,8,500>) => AX(EB_Cmd_Cpt==<Safety>))`

becomes:

`AG((EB_Data_Retrieved == <13,8,(500-7.29)>) => AX(EB_Cmd_Cpt==<Safety>)).`

In some cases, it is possible to compute the discretization of input variables depending on the required precision on the function (see Sect. 7.2).

5.4   Transformation to Obtain the Symmetric Net

This section deals with the transformation of the CPN into a Symmetric Net. First, we present the general principles that are then applied on models dedicated to the verification of the two properties defined in Sect. 4.3.

Computing Discretization. Our objective is to get a dedicated Symmetric net from the CPN of Fig. 9. To do so, we must: *i)* discretize continuous types and, *ii)* generate the tables corresponding to the functions to be modeled. The result is presented in Fig. 10.

The type **EB_DATA** in Fig. 9 is associated to *EBData* in Fig. 10, that is a cartesian product of three discrete types: **Distance**, **Velocity** and **Braking_Factor**. In Fig. 10, the definition of these types is not represented because it depends on the discretization criteria (e.g. the number of values one can handle).

As presented in Sect. 3.3, the complex function **EB_Strategy** of Fig. 9 is associated to the place **EB_Strategy_Table** in Fig. 10. Its initial marking is a conversion table for the discretized function. The binding between variables d, v, b in inputs arcs of transition **EB_Process_Strategy** enables the selection of the appropriate command in variable eb_cmd.

The marking stored in **EB_Strategy_Table** is generated from the discretized values of domains **Distance**, **Velocity** and **Braking_Factor** by means of the **EB_Strategy** function presented in Sect. 4.2.

The model to verify property 1 (Sect. 4.3). There is no need to change anything to verify property 1 in the model of Fig. 10.

The initial marking of place **Interface_Data** must provide any possible value for inputs since we aim at verifying that all conditions lead to appropriate commands as stated in properties 1.1 and 1.2. This is generated by the tuple of broadcast functions <Distance.all,Velocity.all,Braking_Factor.all>. If we consider several parallel threads to handle several obstacles (see property 1.3 in Sect. 4.3), it is of interest to consider a value greater than one for constant **NBT** in the declaration, thus generating several tokens in place **Interface_Call**. The initial marking of place **Interface_Call** is a set of **ProcessId** (obtained by the broadcast constant <ProcessId.all>). This does not add much complexity in the model and does not invalidate the computation of error propagation. Thus, multiple parallel accesses stated in property 1.3 can be concretely verified.

The model to verify property 2 (Sect. 4.3). This property involves interactions with a functional environment (runtime), and a person (a passive driver). Fig. 11 presents a model on which a passive driver has been added on the right, and part of the functional environment (the runtime is reduced to a feedback loop) on the left. This model allows to verify properties on the module dynamicity, like the fact that, if a driver does not react, the different warning levels (Comfort Safety Emergency) will be issued while the situation becomes more and more dangerous (see Sect. 4.3).

Conclusion. This section shows the modelling of some properties to be checked on the emergency braking application. Our models are based on the template proposed in Fig. 9 and we introduce some variations to adapt the template to the verification of properties 1 and 2.

These variations have an impact on the verification complexity. For example, the reachability graph for the second model (with the passive driver) grows linearly with the discretization of the **Distance** colour domain. On the contrary, the complexity of the first model grows much faster.

In these examples, the feedback loops only contain discrete variables (i.e. a command or the PID), but no discretized continuous variables. This avoids the

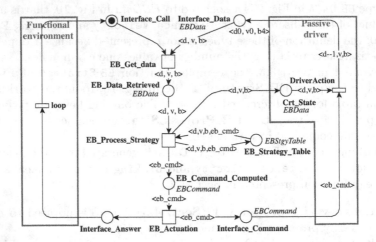

Fig. 11. Model of Fig. 10 enriched with a passive driver (declaration part and marking of place **EB_Strategy_Table** is the one of the model in Fig. 10)

accumulation of errors introduced by the discretization. If a discretized variable is involved in a feedback loop, this entails more constraints on the discretization, and the model should be made in such a way that the errors are not infinitely accumulated.

The next section provides some analysis results. We focus on the first model since it is more challenging with regards to the combinatorial explosion problem.

6   Net Analysis

This section briefly presents the analysis experimentation performed on the symmetric net of Fig. 10 with various configurations.

The use of a discretization method with symmetric nets generates complex models with large markings. It is important to know the consequences on the net analysis and model checking tools.

Objectives. The objectives of this analysis are to analyse the net properties as well as to overcome sources of combinatorial explosion.

Experimental method. Fig. 12 shows the various techniques that can be used for checking properties. Structural analysis (place invariants, bounds, etc.) can

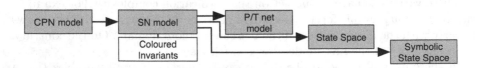

Fig. 12. Overview of the analyses

be performed on the unfolded place/transitions nets as proposed in Sect. 2.2. Coloured invariants can also be computed directly from the coloured net. Model checking can be performed either on the state space or the symbolic state space according to the tool chosen.

Technical aspects. The analysis of the Petri Net is a complex operation that requires different transformations of the model like unfolding or reduction. Various tools were used for the analysis. First CPN models were designed with CPN-Tools [17], then symmetric nets were designed using Coloane, an interface for CPN-AMI [37] and PetriScript (a script language to design SN) to generate initial markings. CPN-AMI is used for the analysis of symmetric nets.

6.1   Structural Analysis

We first used structural analysis techniques because they do not require the state space construction and are thus of more reduced complexity.

Symmetric net analysis. We computed coloured invariant on the SN, experimenting with various discretizations. The only computed invariant involves place "EB_Strategy_Table", as expected (its marking is stable by definition). Discretization has no significant impact on the memory used for the computation.

Structural Reduction. By applying structural reduction on Petri nets [2,23,25], it is possible to reduce the net of Fig. 10 into the one of Fig. 13. They are structurally equivalent when we want to check that all situations lead to a planned situation in the system with regards to property 1. The new net avoids most of the useless interleaving when the controller is simultaneously executed by several threads.

The reduced net of Fig. 13 has the same declaration as the one of Fig. 10.

Analysis on unfolded nets. Discretization has an impact on the memory required to compute unfolded nets. In fact, the size of the **EB_Data** domain has a cubic growth and thus impacts the resulted P/T net. For example, the number of P/T (np) places grows with $np = 5 * (k)^3 + 8$, where k is the discretization parameter (Sect. 3.3). The CPN-AMI unfolder to P/T nets confirms this formula.

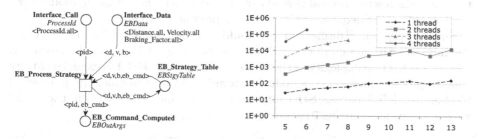

Fig. 13. Reduced net from the one of Fig. 10 (without declaration and marking for **EB_Strategy_Table**)

Fig. 14. Evolution of the symbolic state space for the net of Fig. 13 when discretization and involved threads varies

We also proved by unfolding that both nets (Fig. 10 and 13) are bounded. Since the marking has no impact on the validity of structural properties, the analysis is relevant and shows the interest of our methodology.

6.2   Behavioural Analysis

PROD [42] and GreatSPN [13], two model checkers integrated in CPN-AMI [37], were used to complete the behavioural analysis.

State Space Computation. According to our tests, the complexity of state space generation is similar to that of unfolding for both memory and time. The symbolic state space generated with GreatSPN shows some (minor) optimization compared to explicit model checking generated by PROD. This is probably due to the lack of symmetries in the marking of place EB_Strategy_Table. This should be overcome using two techniques: *i)* the introduction of test arc in SN, *ii)* a dedicated algorithm to enable the model checker to detect places with a stable marking.

The complexity of the binding for transition **EB_Process_Strategy** is:

$$\sum_{i=1..NbP} C_{IDM}^i \tag{14}$$

where $NbP = |ProcessId|$ (the number of values in type $ProcessId$) and $IDM = |M_0(\textbf{Interface\_Data})|$ (the initial number of tokens in place **Interface_Data**). Since $M_0(\textbf{Interface\_Data})$) contains $|Distance| \times |Velocity| \times |Braking\_Factor|$ tokens, this net quickly generates a large state space, requiring dedicated model checking based analysis.

Therefore, we were not surprised when PROD could generate the state space only for a limited number of threads or a very small number of values in Distance and Velocity types (for our experiments, we fixed Brakinf_Factor to 2 values, corresponding to average braking factors for dry and wet roads). This is due to the fact that an explicit representation of state spaces is stored in memory.

Some measures provided by experimentations on GreatSPN are provided in Fig. 14. No value is presented when calculus reached a time-out of 24 hours.

Fig. 14 shows that GreatSPN copes better with the complexity induced by the parallelism when several threads access the Emergency Braking module (we use the automatic detection of symmetries presented in [41]).

Surprisingly, some discretizations could not be computed for more than 2 threads (and with more values, for 2 threads) because of CPU more than memory. This is due to the complexity of the binding of transition **EB_Process_Strategy** that is only symmetric for domain ProcessId. As an example, the number of bindings for **EB_Process_Strategy** (computed using formula 14) in the model with 7 values for domains Distance and Velocity is:

$$C_{98}^1 + C_{98}^2 + C_{98}^3 + C_{98}^4 = 3\,769\,227 \tag{15}$$

since places **Interface_Data** and **Interface_Call** initially holds 98 and 4 tokens. The excessive execution time is due to the canonization function that is

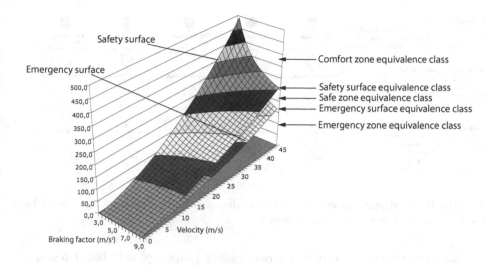

Fig. 15. Semantic equivalence classes and surfaces from equations 9 and 10

required by the generation of the symbolic state space. Memory consumption never exceeded 100Mbyte.

6.3 Coping with the Complexity Problem

Our discretization approach allows us to reduce infinite problems to discrete ones. However, verification for a reasonable discretization still raises some problems. This section proposes some hints to cope with such problems.

The first solution we propose consists in a specific modelling technique that can be used for reachability analysis only. The second one resides in the integration of dedicated techniques in model checkers.

Semantic Behavioural Equivalence Classes. In SN, the symbolic state space is computed when equivalence classes are provided. They are either defined by the system modeller in colour domains or computed automatically from the SN structure [41]. Here, the structure of our models does not comply with the computation of such equivalence classes.

However, solutions of the equations 9 and 10 define equivalence classes in the system by defining limits between the different situations of the system: "Comfort", "Safety", "Emergency". Fig. 15 shows the five equivalence classes we deduce from the two surfaces computed from the equations of the system.

Thus, we may consider one element per equivalence class in the state space. Projection of these elements on the involved dimensions allows us to reduce colour domains to a very small set of values. In the net of Fig. 10, we may consider only five points in the state space of the system, thus leading to five values in the *distance*, *velocity* and *braking factor* colour domains.

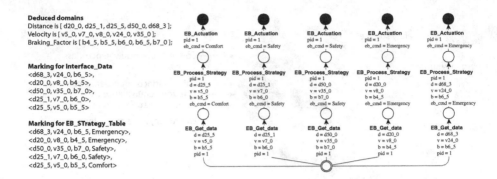

Fig. 16. Reduced state space for the colour domains and place marking defined by means of the semantic equivalent classes

This discretization is relevant for reachability properties only but it remains correct since all the possible equivalence classes in the state space are reachable. Fig. 16 shows the reduced state space for the model of Fig. 10.

We name such equivalence classes *semantic* because, contrary to the classical equivalence classes in SN, they are computed from the definition of the systems and from its structure.

Building Dedicated Model Checkers. It is clear that for such systems dedicated model checkers should be designed. Based on our experience, let us identify some useful feature that could cope with the combinatory explosion problem:

1. The use of partial order techniques like in [6]: this is a way to avoid the exploration of redundant paths.
2. Check for places with a stable marking: this is a typically useful optimization when techniques such as decision diagrams like BDD [9] are used. If variables encoding places with a stable marking are place on top of the decision diagram, then the marking is represented only once. For example, PROD stores place **EB_STrategy_Table** and its huge marking was stored for each state in the state space.
3. The use of high-level decision diagrams such as DDD [15], MDD [32] or SDD [16] can also help. These decision diagrams directly encode discrete data (for DDD and MDD) and can handle hierarchy (for SDD). These enable very efficient state space encoding techniques[5] by means of high-level decision diagrams. In particular, this is a way to cope with the previous optimization.
4. The use of recent extension to symmetric Petri Nets where tokens can hold bags themselves [31]. Such an extension allows to fire symbolically transitions with a similar structure than the one of **EB_Process_Strategy** [24].
5. It was shown that GreatSPN could be parallelized in an efficient way [26]. Then, generation of the state space is distributed over a set of machines, thus

---

[5] The term *symbolic model checking* is used to refer to this diagram-decision based technique. This is different from the symbolic state space.

Table 2. Discretization with optimized criteria

Discretization par.	$v = 13m/s, b = 8m/s^{-2},$ $d = 500m$	$v = 36m/s, b = 4m/s^{-2},$ $d = 100m$
$k_v = 114, k_b = 73$ $k_d = 120$	$\Delta_{Eb\_Saf} \in [-6.167m, 6.203m]$	$\Delta_{Eb\_Saf} \in [-12.09m, 12.40m]$
$card(EBData) < 10^6$	$\Delta_{Eb\_Emerg} \in [-5.367m, 5.403m]$	$\Delta_{Eb\_Emerg} \in [-11.29m, 11.60m]$

allowing the use of more CPU (of interest since canonization in GreatSPN is CPU consuming) and memory.

Unfortunately, these techniques are not yet implemented, or implemented as prototypes only. However, we think they should allow the analysis of discretized systems in the future.

7   Some Perspectives about Discretization

We have described and applied a discretization method to cope with hybrid systems and handle continuous variables in a safe and discrete manner. In this section, we open a discussion on several aspects.

7.1   Optimized Discretization Parameters

The methodology presented in this paper is based on the use of a discretization algorithm to discretize continuous variables. In Sect. 5.2, we used "equal width interval binning" algorithm because it is simple to implement. This algorithm, like many others, relies on discretization parameters that can be optimized for a given set of continuous variables and functions.

However, in the emergency braking module example, we may study the partial derivates of the error on the two thresholds (Δ_{EB_Safety} and $\Delta_{EB\_Emergency}$). We then find that variables v and d are more influent than b. For example, the partial derivate of the error on the EB_Safety threshold (equation 13) with respect to the variable v is:

$$\frac{\partial \Delta_{Eb\_Safety}}{\partial v} = \frac{v \pm \Delta_v}{b \pm \Delta_b} - \frac{v}{b} \tag{16}$$

This allows to find optimized discretization parameters considering the respective influence of each involved variable. To do so, parameters for each discretized variable are considered depending on its influence on error propagation.

Table 2 presents the resulting error when discretization parameters are optimized using partial derivatives[6]. It shows that we can reduce the resulting error of about 10% with discretization parameters based on partial derivatives.

---

[6] Δ_{Eb_Saf} and Δ_{Eb_Emerg} stand for Δ_{Eb_Safety} and $\Delta_{Eb\_Emergency}$ respectively.

The study of the best discretization method and parameters for a given set of continuous variable and function is a complex problem that may give interesting results. It is a promising field for future work on optimization of the methodology presented in this paper.

7.2   Tuning the Discretization

It is of interest to compute the discretization intervals of discretized types (here k_b, k_v and k_d) according to the maximum error tolerated on one type involved in a property where error must be bounded *a priori*.

Let us consider as an example the braking distance function (8) presented Sect. 4.2. It is possible to compute the discretization intervals of variables v and b, based on the accuracy required for the function Δ_f. Let $\pm\Delta_f$ be the tolerated error on f, and $\pm\Delta_v$, $\pm\Delta_b$ be the resulting errors on v and b. Using the error bounds propagation as presented Sect. 3.3 we get:

$$\pm\Delta_f = \frac{(v \pm \Delta_v)^2}{2 * (b \pm \Delta_b)} - \frac{v^2}{2 * b} \tag{17}$$

We then obtain[7]:

$$\Delta_b = -\frac{2 * b^2 * \Delta_f - 2 * b * \Delta_v * v - b * \Delta_v^2}{2 * b * \Delta_f + v^2} \tag{18}$$

and two solutions for Δ_v that are a little bit more complex.

Let v_{min}, v_{max}, b_{min} and b_{max} be the bounds of v and b. The cardinality of V and B sets are:

$$Card(V) = \frac{v_{max} - v_{min}}{2 * \Delta_v} \qquad Card(B) = \frac{b_{max} - b_{min}}{2 * \Delta_b} \tag{19}$$

Now, consider that we want the same cardinalities for V and B colour sets ($k_b = k_v$). We obtain[8]:

$$\Delta_v = \frac{(v_{max} - v_{min})\Delta_b}{b_{max} - b_{min}} \tag{20}$$

Using the value of Δ_v of equation (20) in equation (18), it is now possible to compute Δ_b from the desired Δ_f.

For only two variables, this method is complex as it gives multiple solutions that need to be analyzed to choose the appropriate solutions. However, it provides a way to compute the discretization intervals of input variables depending on the desired output error.

---

[7] We intentionally removed the $\pm$ operator to increase readability.

[8] Note that it is possible to choose another factor between $Card(V)$ and $Card(B)$ as explained in Sect. 7.1.

8   Conclusion

This paper proposes a way to integrate continuous aspects of complex specifications into a discretized Petri Net model for model checking purpose. Our approach takes place in the context of Intelligent Transport Systems and, more precisely, the management of emergency braking when an obstacle is identified on the road.

Discretization methods rely on the equations describing the problem. In our work, these equations come from the physical model interacting with the system. It is crucial for engineers to evaluate the quality of the proved properties and their impact on the system verification.

To this end, we compute a discretized abstraction from these equations. The abstraction quality is then evaluated with regards to the properties to be checked This is a key point in modelling and evaluating a system by means of formal specification. Typically, imprecision raised by discretization is corrected by either applying a more precise discretization or adding constants in formulas expressing properties to be checked.

These equations are attached to a Coloured Petri net (CPN) template. Discretization is proposed to provide a finite model, even if large. This CPN is then transformed into a symmetric net (SN) to take advantage of the dedicated verification techniques. Analysis is performed on the SN.

Although the analysis performed on our example remains limited by the combinatorial explosion, we can extract the main properties (boundness, reachability properties) when colour domains are of reasonable size. Let us note that without discretization, evaluation of such continuous systems by means of model checking techniques is still an open issue.

In the context of a SAFESPOT application, several modules run in parallel and may introduce more continuous types and variables. In particular, experimenting propagation of discretization constraints between different modules need a particular attention. Future work will then have to evaluate how a larger number of variables and constraints could be managed. We must also investigate to cope with several modules at a time.

In our methodology, several discretization algorithms can be applied. As a first trial, we experimented a simple algorithm but other ones based on non-uniform discretization intervals are promising. These others discretization techniques might introduce new constraints in formal verification and in error propagation computation but it is an interesting field in the future.

Finally, the notion of semantic behavioural equivalences extracted from the equations and injected in the specification by means of initial marking is of interest. It may help in tackling the combinatorial explosion when several modules are considered for reachability analysis. Exploitation of such information from the problem, when possible, seems the most interesting perspective for future work on this methodology.

Acknowledgements. We thank the anonymous referees for their detailed and fruitful comments.

References

1. Abrial, J.-R.: The B book - Assigning Programs to meanings. Cambridge Univ. Press, Cambridge (1996)
2. Berthelot, G.: Transformations and decompositions of nets. In: Brauer, W., Reisig, W., Rozenberg, G. (eds.) APN 1986. LNCS, vol. 254, pp. 359–376. Springer, Heidelberg (1987)
3. Billington, J., Gallasch, G.E., Petrucci, L.: Verification of the class of stop-and-wait protocols modelled by coloured Petri nets. Nord. J. Comput. 12(3), 251–274 (2005)
4. Bonnefoi, F., Bellotti, F., Scendzielorz, T., Visintainer, F.: SAFESPOT Applications for Infrasructure-based Co-operative Road Safety. In: 14th World Congress and Exhibition on Intelligent Transport Systems and Services (2007)
5. Bonnefoi, F., Hillah, L., Kordon, F., Renault, X.: Design, modeling and analysis of ITS using UML and Petri Nets. In: 10th International IEEE Conference on Intelligent Transportation Systems (ITSC 2007), pp. 314–319. IEEE Press, Los Alamitos (2007)
6. Brgan, R., Poitrenaud, D.: An efficient algorithm for the computation of stubborn sets of well formed Petri nets. In: DeMichelis, G., Díaz, M. (eds.) ICATPN 1995. LNCS, vol. 935, pp. 121–140. Springer, Heidelberg (1995)
7. Brignolo, R.: Co-operative road safety - the SAFESPOT integrated project. In: APSN - APROSYS Conference. Advanced Passive Safety Network (May 2006)
8. Buchs, D., Guelfi, N.: A formal specification framework for object-oriented distributed systems. IEEE Trans. Software Eng. 26(7), 635–652 (2000)
9. Burch, J.R., Clarke, E.M., McMillan, K.L., Dill, D.L., Hwang, L.J.: Symbolic model checking: 10^{20} states and beyond. Inf. Comput. 98(2), 142–170 (1992)
10. Covault, R.B.C., Driscoll, D.: Uncertainties and Error Propagation - Appendix V of Physics Lab Manual. Case Western Reserve University (2005)
11. Chiola, G., Dutheillet, C., Franceschinis, G., Haddad, S.: Stochastic well-formed colored nets and symmetric modeling applications. IEEE Trans. Computers 42(11), 1343–1360 (1993)
12. Chiola, G., Dutheillet, C., Franceschinis, G., Haddad, S.: A symbolic reachability graph for coloured Petri nets. Theoretical Computer Science 176(1–2), 39–65 (1997)
13. Chiola, G., Franceschinis, G., Gaeta, R., Ribaudo, M.: GreatSPN 1.7: Graphical Editor and Analyzer for Timed and Stochastic Petri Nets Performance Evaluation. Special issue on Performance Modeling Tools 24(1&2), 47–68 (1995)
14. Christofides, P., El-Farra, N.: Control Nonlinear and Hybrid Process Systems: Designs for Uncertainty, Constraints and Time-delays. Springer, Heidelberg (2005)
15. Couvreur, J.-M., Encrenaz, E., Paviot-Adet, E., Poitrenaud, D., Wacrenier, P.-A.: Data decision diagrams for Petri net analysis. In: Esparza, J., Lakos, C.A. (eds.) ICATPN 2002. LNCS, vol. 2360, pp. 101–120. Springer, Heidelberg (2002)
16. Couvreur, J.-M., Thierry-Mieg, Y.: Hierarchical Decision Diagrams to Exploit Model Structure. In: Wang, F. (ed.) FORTE 2005. LNCS, vol. 3731, pp. 443–457. Springer, Heidelberg (2005)

17. The CPN Tools Homepage (2007), http://www.daimi.au.dk/CPNtools
18. Daniel, J., Luca, D.: IEEE 802.11p: Towards an International Standard for Wireless Access in Vehicular Environments. In: Proceedings of Vehicular Technology Conference, VTC Spring, pp. 2036–2040. IEEE, Los Alamitos (2008)
19. David, R., Alla, H.: On Hybrid Petri Nets. Discrete Event Dynamic Systems: Theory and Applications 11(1-2), 9–40 (2001)
20. Dougherty, J., Kohavi, R., Sahami, M.: Supervised and unsupervised discretization of continuous features. In: Int. Conf. on Machine Learning, pp. 194–202 (1995)
21. Emerson, E.A., Halpern, J.Y.: Decision procedures and expressiveness in the temporal logic of branching time. J. Comput. Syst. Sci. 30(1), 1–24 (1985)
22. Frame Forum. The FRAME forum home page, http://www.frame-online.net
23. Haddad, S.: A reduction theory for coloured nets. In: Rozenberg, G. (ed.) APN 1989. LNCS, vol. 424, pp. 209–235. Springer, Heidelberg (1990)
24. Haddad, S., Kordon, F., Petrucci, L., Pradat-Peyre, J.-F., Trèves, N.: Efficient State-Based Analysis by Introducing Bags in Petri Net Color Domains. In: 28th American Control Conference (ACC 2009), St-Louis, USA. IEEE, Los Alamitos (2009)
25. Haddad, S., Pradat-Peyre, J.-F.: New efficient Petri nets reductions for parallel programs verification. Parallel Processing Letters 16(1), 101–116 (2006)
26. Hamez, A., Kordon, F., Thierry-Mieg, Y., Legond-Aubry, F.: dmcG: a distributed symbolic model checker based on GreatSPN. In: Kleijn, J., Yakovlev, A. (eds.) ICATPN 2007. LNCS, vol. 4546, pp. 495–504. Springer, Heidelberg (2007)
27. Hillah, L., Kordon, F., Petrucci, L., Trèves, N.: PN standardisation: A survey. In: Najm, E., Pradat-Peyre, J.-F., Donzeau-Gouge, V.V. (eds.) FORTE 2006. LNCS, vol. 4229, pp. 307–322. Springer, Heidelberg (2006)
28. Hugues, J., Thierry-Mieg, Y., Kordon, F., Pautet, L., Baarir, S., Vergnaud, T.: On the Formal Verification of Middleware Behavioral Properties. In: FMICS 2004, pp. 139–157. Elsevier, Amsterdam (2004)
29. ISO/IEC-JTC1/SC7/WG19. International Standard ISO/IEC 15909: Software and Systems Engineering - High-level Petri Nets, Part 1: Concepts, Definitions and Graphical Notation (December 2004)
30. Jensen, K., Kristensen, L.M.: Coloured Petri Nets, Modelling and Validation of Concurrent Systems. Monograph. Springer, Heidelberg (2008)
31. Junttila, T.: On the symmetry reduction method for Petri Nets and similar formalisms. PhD thesis, Helsinki University of Technology, Espoo, Finland (2003)
32. Kam, T., Villa, T., Brayton, R., Sangiovanni-Vincentelli, A.L.: Multi-Valued Decision Diagrams: Theory and Applications. International Journal on Multiple-Valued Logic 4(1-2), 9–62 (1998)
33. Kordon, F., Linard, A., Paviot-Adet, E.: Optimized Colored Nets Unfolding. In: Najm, E., Pradat-Peyre, J.-F., Donzeau-Gouge, V.V. (eds.) FORTE 2006. LNCS, vol. 4229, pp. 339–355. Springer, Heidelberg (2006)
34. Lakos, C., Lewis, G.: Incremental State Space Construction for Coloured Petri Nets. In: Colom, J.-M., Koutny, M. (eds.) ICATPN 2001. LNCS, vol. 2075, pp. 263–282. Springer, Heidelberg (2001)
35. Lewis, G.: Incremental specification and analysis in the context of coloured Petri nets. PhD thesis, University of Hobart, Tasmania (2002)
36. Lindberg, V.: Uncertainties and Error Propagation - Part I of a manual on Uncertainties, Graphing, and the Vernier Caliper. Rochester Inst. of Technology (2000)
37. LIP6/MoVe. The CPN-AMI home page, http://www.lip6.fr/cpn-ami/

38. IEEE 802.11 Working Group for WLAN Standards. IEEE 802.11 tm Wireless Local Area Networks. IEEE, Los Alamitos (2008)
39. ISO TC204 WG-16. CALM architecture. ISO (2007)
40. Petri Nets Steering Committee. Petri Nets Tool Database: quick and up-to-date overview of existing tools for Petri Nets,
    http://www.informatik.uni-hamburg.de/TGI/PetriNets/tools/db.html
41. Thierry-Mieg, Y., Dutheillet, C., Mounier, I.: Automatic Symmetry Detection in Well-Formed Nets. In: van der Aalst, W.M.P., Best, E. (eds.) ICATPN 2003. LNCS, vol. 2679, pp. 82–101. Springer, Heidelberg (2003)
42. Varpaaniemi, K., Halme, J., Hiekkanen, K., Pyssysalo, T.: Prod reference manual. Technical report, Helsinki University of Technology (1995)
43. Vautherin, J.: Parallel systems specifications with coloured Petri nets and algebraic specifications. In: Rozenberg, G. (ed.) APN 1987. LNCS, vol. 266, pp. 293–308. Springer, Heidelberg (1987)

The ComBack Method Revisited:
Caching Strategies and Extension with
Delayed Duplicate Detection*

Sami Evangelista[1], Michael Westergaard[1], and Lars M. Kristensen[2]

[1] Computer Science Department, Aarhus University, Denmark
{evangeli,mw}@cs.au.dk
[2] Department of Computer Engineering, Bergen University College, Norway
lmkr@hib.no

Abstract. The ComBack method is a memory reduction technique for explicit state space search algorithms. It enhances hash compaction with state reconstruction to resolve hash conflicts on-the-fly thereby ensuring full coverage of the state space. In this paper we provide two means to lower the run-time penalty induced by state reconstructions: a set of strategies to implement the caching method proposed in [20], and an extension through delayed duplicate detection that allows to group reconstructions together to save redundant work.

Keywords: explicit state model checking, state explosion problem, state space reduction, hash compaction, delayed duplicate detection.

1 Introduction

Model checking is a formal method used to detect defects in system designs. It consists of a systematic exploration of the reachable states of the system whose behavior can be formally represented as a directed graph. Its nodes are system states and its arcs are possible transitions from one state to another. This principle is simple, can be easily automated, and, in case of errors, a counter-example is provided to the user.

However, even simple systems may have an astronomical or even infinite number of states. This state explosion problem is a severe obstacle for the application of model checking to industrial size systems. Numerous possibilities are available to alleviate, or at least delay, this phenomenon. One can for example exploit the redundancies in the system description that often induce symmetries [4], exploit the independence of some transitions to reduce the exploration of redundant interleavings [8], or encode the state graph using compact data structures such as binary decision diagrams [2].

Hash compaction [16,21] is a graph storage technique that reduces the amount of memory used to store states. It uses a hash function h to map each encountered state s into a fixed-size bit-vector $h(s)$ called the *compressed state descriptor* which is stored in memory as a representation of the state. The *full state*

---

* Supported by the Danish Research Council for Technology and Production.

K. Jensen, J. Billington, and M. Koutny (Eds.): ToPNoC III, LNCS 5800, pp. 189–215, 2009.

descriptor is not stored in memory. Thus, each discovered state is represented compactly using typically 32 or 64 bits. The disadvantage of hash compaction is that two different states may be mapped to the same compressed state descriptor which implies that the hash compaction method may not explore all reachable states. The probability of *hash collisions* can be reduced in several ways, e.g., by using multiple hash functions [12,16], but the method still cannot guarantee full coverage of the state space. Partial coverage of the state space is acceptable if the intent is to find errors, but not sufficient if the goal is to prove the correctness of a system specification.

The ComBack method [20] extends hash compaction with a backtracking mechanism that allows reconstruction of full state descriptors from compressed ones and thus resolves conflicts on-the-fly to guarantee full coverage of the state space. Its underlying principle is to store for each state a sequence of events whose execution can generate this state. Thus, when the search algorithm checks if it already visited a state s, it can reconstruct full state descriptors for states mapped to the same hash value as s and compare them to s. Only if none of the states reconstructed is equal to s does the algorithm consider it as a new state.

The ComBack method stores a small amount of information per state, typically between 16 and 24 bytes depending on the system being analyzed. Thus it is especially suited to industrial case studies for which the full state descriptor stored by a classical search algorithm can be very large (from 100 bytes to 10 kilo-bytes). This important reduction, however, has a cost in time: a ComBack based algorithm will explore many more arcs in order to reconstruct states. As the graph is given implicitly, visiting an arc consists of applying a successor function that can be arbitrarily complex, especially for high-level languages such as Promela [10] or Coloured Petri nets [11]. Experiments made in [20] report an increase in run-time up to more than 600% for real-life protocols.

The goal of the work presented in this paper is to propose solutions to tackle this problem. Firstly, starting from the proposal of [20] to use a cache of full state descriptors to shorten sequences, we first propose different caching strategies. Secondly, we extend the ComBack method with delayed duplicate detection, a technique widely used by disk-based model checkers [17]. The principle is to delay the instant we check if a state has already been visited from the instant of its generation. Any state reached is put into a set of candidates and only occasionally is this set compared to the set of already visited states in order to identify new ones. The underlying idea of this operation is that comparing these two sets may be much cheaper than checking separately if each candidate has already been visited. Applied to the ComBack method, this results in saving the exploration of transitions that are shared by different sequences. For instance if sequences $a.b.c$ and $a.b.d$ reconstruct respectively states s and s', we may group the reconstructions of s and s' in order to execute sequence $a.b$ only once instead of twice. This will result in the execution of 4 events instead of 6 events.

This article has the following structure. The basic elements of labeled transition systems and the ComBack method are recalled in Section 2. In Section 3, different caching strategies are proposed. An algorithm that combines the

ComBack method with delayed duplicate detection is presented in Section 4. Section 5 reports on experiments made with the ASAP tool [19] which implements the techniques proposed in this paper. Finally, Section 6 concludes this paper.

2   Background

In this section we give the basic ingredients required for understanding the rest of this paper and provide a brief overview of the ComBack method [20].

2.1   Transition Systems

As the methods proposed in this paper are not linked to a specific formalism they will be developed in the framework of labeled transition systems, the most low-level representation of concurrent systems.

Definition 1 (Labeled Transition System). *A labeled transition system is a tuple* $S = (S, E, T, s_0)$, *where* S *is a set of* ***states***, E *is a set of* ***events***, $T \subseteq S \times E \times S$ *is the* ***transition relation***, *and* $s_0 \in S$ *is the* ***initial state***.

In the rest of this paper we assume that we are given a labeled transition system $S = (S, E, T, s_0)$. Let $s, s' \in S$ be two states and $e \in E$ an event. If $(s, e, s') \in T$, then e is said to be *enabled* in s and the *occurrence* (execution) of e in s leads to the state s'. This is also written $s \xrightarrow{e} s'$. An *occurrence sequence* is an alternating sequence of states s_i and events e_i written $s_1 \xrightarrow{e_1} s_2 \cdots s_{n-1} \xrightarrow{e_{n-1}} s_n$ and satisfying $s_i \xrightarrow{e_i} s_{i+1}$ for $1 \leq i \leq n - 1$. For the sake of simplicity, we assume that events are *deterministic*[1], i.e., if $s \xrightarrow{e} s'$ and $s \xrightarrow{e} s''$ then $s' = s''$.

We use $\rightarrow^*$ to denote the transitive and reflexive closure of T, i.e., $s \rightarrow^* s'$ if and only if there exists an occurrence sequence $s_1 \xrightarrow{e_1} s_2 \cdots s_{n-1} \xrightarrow{e_{n-1}} s_n$, $n \geq 1$, with $s = s_1$ and $s' = s_n$. A state s' is *reachable* from s if and only if $s \rightarrow^* s'$. The *state space* of a system is the directed graph (V, E) where $V = \{ s' \in S \mid s_0 \rightarrow^* s' \}$ is the set of nodes and $E = \{(s, e, s') \in T \mid s, s' \in V\}$ is the set of edges.

2.2   The ComBack Method

A classical state space search algorithm (Algorithm 1) operates on a set of visited states V and a queue of states to visit Q. An iteration of the algorithm (lines 4–7) consists of removing a state from the queue, generating its successors and inserting the successor states that have not been visited so far into both the visited set and the queue for later exploration. We use the term of *state expansion* to refer to this process.

---

[1] The ComBack method can be extended to support non-deterministic transition systems. The interested reader may consult Section 5 of [20] that describes such an extension.

Algorithm 1. A classical search algorithm.

1: $\mathcal{V} \leftarrow$ **empty** ; $\mathcal{V}.insert\ (s_0)$
2: $\mathcal{Q} \leftarrow$ **empty** ; $\mathcal{Q}.enqueue\ (s_0)$
3: **while** $\mathcal{Q} \neq$ **empty do**
4:      $s \leftarrow \mathcal{Q}.dequeue\ ()$
5:      **for** $e, s' \mid (s, e, s') \in T$ **do**
6:          **if** $s' \notin \mathcal{V}$ **then**
7:              $\mathcal{V}.insert\ (s')$; $\mathcal{Q}.insert\ (s')$

Algorithm 2. A search algorithm based on hash compaction.

1: $\mathcal{V} \leftarrow$ **empty** ; $\mathcal{V}.insert\ (\underline{h(s_0)})$
2: $\mathcal{Q} \leftarrow$ **empty** ; $\mathcal{Q}.enqueue\ (s_0)$
3: **while** $\mathcal{Q} \neq$ **empty do**
4:      $s \leftarrow \mathcal{Q}.dequeue\ ()$
5:      **for** $e, s' \mid (s, e, s') \in T$ **do**
6:          **if** $\underline{h(s')} \notin \mathcal{V}$ **then**
7:              $\mathcal{V}.insert\ (\underline{h(s')})$; $\mathcal{Q}.insert\ (s')$

Using hash compaction [21], items stored in the visited set are not actual state descriptors but compressed descriptors, typically 32-bit integers, obtained through a hash function h. Algorithm 2 uses this technique. The few differences with Algorithm 1 have been underlined. This storage scheme is motivated by the observation that full state descriptors are often large for realistic systems, i.e., typically between 100 bytes and 10 kilo-bytes, which drastically limits the size of state spaces that can be explored. Though hash compaction considerably reduces memory requirements, it comes at the cost of possibly missing some parts of the state space (and therefore potentially some errors). Indeed, as h may not be injective, two different states may erroneously be considered the same if they are mapped to the same hash value. Hence, hash compaction is preferably used at the early stages of the development process for its ability to quickly discover errors rather than proving the correctness of the system.

The ComBack method extends hash compaction with a backtracking mechanism that allows it to retrieve actual states from compressed descriptors in order to resolve hash collisions on-the-fly and guarantee full coverage of the state space. This is achieved by modifying the hash compaction algorithm as follows:

- A *state number*, or identifier, is assigned to each visited state s.
- A *state table* stores for each compressed state descriptor a *collision list* of state numbers for visited states mapped to this compressed state descriptor.
- A *backedge table* is maintained which for each state number of a visited state s stores a *backedge* consisting of an event e and a state number of a visited predecessor s' such that $s' \xrightarrow{e} s$.

The key algorithm of the ComBack method is the insertion procedure that checks whether a state s is already in the visited set and inserts it into the visited set if needed. The insertion procedure can be illustrated with the help of Fig. 1, which

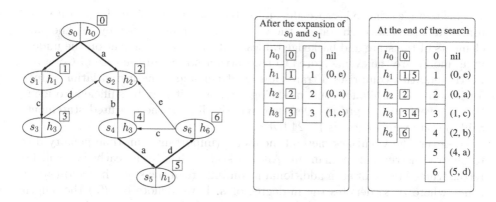

Fig. 1. A state space and the state and backedge tables at two stages

depicts a simple state space. Each ellipse represents a state. The hash value of each state is written in the right part of the ellipse. The state and backedge tables used to resolve hash conflicts have been depicted to the right of the figure for two different steps of the search. For the sake of clarity, we have also depicted on the state space the identifier of each state (the square next to the ellipse) and highlighted (using thick arcs) the transitions that are used to backtrack to the initial state, i.e., the edges constituting the backedge table. Note that these identifiers also coincide with the expansion order of states.

After the expansion of s_0 and s_1, the set of visited states is $\{s_0, s_1, s_2, s_3\}$. As no hash conflict is detected, a single state is associated with each hash value in the state table (the left table of the first rounded box). In the backedge table (the right table of the first rounded box) a nil value is associated with state 0 (the initial state) as any backtracking will stop here. The table also indicates that the actual value of state 1 (s_1) is retrieved by executing event e on state 0 and so on for the other entries of the table. After the execution of event b on state s_2 we reach s_4. Algorithm 2 would claim that s_4 has already been visited—as $h(s_3) = h(s_4)$—and stop the search at this point, missing states s_5 and s_6. Using the two tables the hash conflict between s_3 and s_4 can be handled as follows. The insertion procedure first looks in the state table if any state has already been mapped to $h(s_4) = h_3$ and finds value 3. The comparison of state 3 (of which we no longer have the actual state descriptor) to s_4 is first done by recursively following the pointers of the backedge table until the initial state is reached, i.e., 3 then 1 and then 0. Then the sequence of events associated with the entries of the table that have been met during backtracking, i.e., $e.c$, is executed on the initial state. Finally, a comparison between s_3 and s_4 indicates that s_4 is new. We therefore assign a new identifier (4) to s_4, insert it in the collision list of hash value h_3 and insert the entry $4 \to (2, b)$ in the backedge table.

Throughout this article the term *state reconstruction* (or simply *reconstruction*) is used to refer to the process of backtracking to the initial state and then executing a sequence of events to retrieve a full state descriptor. The sequence executed will be called the *reconstructing sequence*.

This storage scheme is especially suited to systems exhibiting large state vectors as it represents each state in the visited set with only a few bytes. The only elements of the state and backedge tables that are still dependent on the underlying model are the events stored to reconstruct states. In the case of Coloured Petri Nets, this comprises a transition identifier and some instantiation values for its variables while for some modeling languages it may be sufficient to identify an event with a process identifier and the line of the executed statement. Still, a state rarely exceeds 16–24 bytes.

However, the ComBack method incurs an (important) run-time penalty due to the reconstruction mechanism. After a state s has been reached it will be reconstructed once for each additional incoming arc of the state, hence $in(s) - 1$ times where $in(s)$ denotes the in-degree of s. If we denote by $d(s)$ the length of the shortest path from s_0 to s, the number of event executions due to state reconstructions is lower bounded by:

$$\sum_{s \in S} (in(s) - 1) \cdot d(s)$$

Note that in Breadth-First Search (BFS) each sequence executed to reconstruct a state s is exactly of length $d(s)$ while it may be much longer in Depth-First Search (DFS). Evidence of this is shown in Table 1 of [20] showing that the ComBack method combined with DFS is in some cases much slower than with BFS while the converse is not true.

In addition, the time spent on reconstructing states depends, to a large extent, on the complexity of executing an event that ranges from trivial (e.g., for Place/Transition Nets) to high, e.g., for Promela or Coloured Petri Nets for which executing an event may include the execution of embedded code.

3   Caching Strategies

A cache that maps state identifiers to full descriptors is a good way to reduce the cost of state reconstructions. The purpose of such a cache is twofold. Firstly, the reconstruction of a state identified by i may be avoided if i is cached. Secondly, if a state has to be reconstructed we may stop backtracking as soon as we encounter a state belonging to the cache and thus execute a shorter reconstruction sequence from this state. As an example, consider again the configuration of Fig. 1. Caching the mapping $2 \rightarrow s_2$ may be useful in two ways:

– To avoid the reconstruction of state 2. A lookup in the cache directly returns state s_2, which saves the backtrack to s_0 and the execution of event a.
– For the reconstruction of state 4. During the backtrack to s_0 the algorithm finds out that state 2 is cached, retrieves its descriptor and only executes event b from s_2 to obtain s_4, once again saving the execution of event a.

We now propose four strategies to implement such a cache. We focus on strategies based on BFS as the traversal order it induces enables to take advantage of some typical characteristics of state spaces [14].

Random cache. The simplest way is to implement a randomized cache. This gives us the first following strategy.

> **Strategy R:** *When a new state is put in the visited set, it is inserted in the cache with probability p (1 if the cache is not full) and the state to replace (if needed) is chosen randomly.*

Fifo cache. A common characteristic of state spaces is the high proportion of forward transitions[2], typically around 80%. This has a significant consequence in BFS in which levels are processed one by one: most of the transitions outgoing from a state will lead to a new state or to a state that has been recently generated from the same level. Hence, a good strategy in BFS seems to be to use a fifo cache, since when a new state at level $l+1$ is reached from level l it is likely that one of the following states of level l will also reach it. If the cache is large enough to contain any level of the graph, only backward transitions will generate reconstructions as forward transitions will always result in a cache hit. This strategy can be implemented as follows.

> **Strategy F:** *When a new state is put in the visited set, insert it unconditionally into the cache. If needed, remove the oldest state from the cache.*

Heuristic based cache. Obviously, the benefit we can obtain from caching a state may largely differ from one state to another. For instance, it is pointless to cache a state s that does not have any successor state pointing to it in the backedge table as it will not shorten any reconstruction sequence, but only avoid the reconstruction of s.

To evaluate the interest of caching some state s we propose to use the following caching heuristic H.

$$H(s) = d(s) \cdot p(s), \text{ with } p(s) = \frac{r(s)}{L(d(s))}$$

where

- $d(s)$ is the distance of s to the initial state in the backedge table
- $r(s)$ is the number of states that reference s in the backedge table
- $L(n)$ is the number of states at level n, i.e., with a distance of n from the initial state

A cache hit is more interesting if it occurs early during the backtrack as it will shorten the sequence executed. Thus the benefit of caching a state s increases with its distance $d(s)$. Through rate $p(s)$ we estimate the probability that s belongs to some reconstructing sequence. This increases if many states point to

---

[2] If we define *level l* as the set of states that are reachable from s_0 in l steps (and not less), a transition that has its source in level l and its target in level $l+1$ is called a *forward transition*. Any other transition is called a *backward transition*.

s in the backedge table and decreases with the number of states on the same level as s. The distance of s could also be considered in the computation of $p(s)$ as s cannot appear in a reconstructing sequence of a length less than $d(s)$. Our choice is based on another typical characteristic of state spaces [18]: backward transitions are usually short in the sense that the levels of its destination and source are often close. Thus, in BFS, if a state has to be reconstructed, it is likely that the length of its reconstructing sequence is close to the current depth which is an upper bound of the length of a reconstructing sequence. Hence, assuming that the state space has this characteristic, the distance will only have a small impact on $p(s)$.

Our third strategy is based on this heuristic.

> **Strategy H:** *After all outgoing transitions of state s have been visited compute $H(s)$. Let s' be the state that minimizes H in the cache. If $H(s') < H(s)$ replace s' by s in the cache.*

Note that after the visit of s, all necessary information to compute $H(s)$ is available since all its successors have been generated and the BFS search order implies that $L(d(s))$ is known.

Distance based cache. Heuristic H does not take into account the presence of already cached states. Yet it may be useless to cache a state with a high value of H if, for example, the state it points to in the backedge table is itself in the cache. The last strategy we propose is a slight variation of strategy H. It is also based on heuristic H but it is parameterized by an integer k that specifies the shortest possible sequence between two cached states.

> **Strategy D:** *Apply strategy H. Do not cache a state if, in the backedge table, one of its ancestors of degree k or less is already cached.*

Other possibilities are available. In [6] a reduction technique also based on state reconstruction is proposed. The algorithm is parameterized by an integer k and only caches states at levels $0, k, 2 \cdot k, 3 \cdot k \ldots$. The motivation of this strategy is to bound the length of reconstructing sequences to $k - 1$. As presented, the strategy in [6] does not bound the size of the cache but k could be dynamically increased to solve this problem.

Different strategies may also be combined. We can for example cache recently inserted states following strategy F and when a state leaves this cache it can be inserted into a second level cache maintained with strategy H. Thus we will keep some recently visited states in the cache and some old strategic states.

4   Combination with Delayed Duplicate Detection

Duplicate detection consists of checking the presence of a newly generated state in the set of visited states. If the state has not been visited so far, it must be included in the visited set and later expanded. With delayed duplicate detection (DDD), this check is delayed from the instant of state generation by putting the state reached in a *candidate set* that contains potentially new states, i.e., states reached via event executions but not checked to be in the visited set. In this scheme, duplicate detection consists of comparing the visited and candidate sets to identify new states. This is motivated by the fact that this comparison may be much cheaper than checking individually for the presence of each candidate in the visited set.

Algorithm 3 is a generic algorithm based on DDD. Besides the usual data structures, we find a candidate set *candidates* filled with states reached through event execution (lines 7–8). An iteration of the algorithm (lines 4–9) consists of expanding all queued states and inserting their successors in the candidate set. Once the queue is empty duplicate detection starts. We identify new states by removing visited states from candidate states (line 11). States remaining after this procedure are put in the visited set and in the queue (lines 12–14).

The key point of this algorithm is the way the comparison at line 11 is conducted. In the disk-based algorithm of [17], the candidate set is kept in a memory hash table and visited states are stored sequentially in a file. New states are detected by reading states one by one from the file and deleting them from the table implementing the candidate set. States remaining in the table at the end of this process are therefore new. Hence, in this context, DDD replaces a large number of individual disk look-ups — that each would likely require reading a disk block — by a single file scan. It should be noted that duplicate detection may also be performed if the candidate set fills up, i.e., before an iteration (lines 4–9) of the algorithm has been completed.

4.1   Principle of the Combination

The underlying idea of using DDD in the ComBack method is to group state reconstructions to save the redundant execution of events shared by different reconstruction sequences. This is illustrated by Fig. 2. The search algorithm first visits states s_0, s_1, s_2, s_3 and s_4 each mapped to a different compressed state descriptor. Later, state s is processed. It has two successors: s_4 (already met) and s_5 mapped to h_3 which is also the compressed state descriptor of s_3. With the basic reconstruction mechanism we would have to first backtrack to s_0, execute sequence $a.b.d$ to reconstruct state 4 and find out that e does not, from s, generate

---

Algorithm 3. A generic search algorithm using delayed duplicate detection

```
1: V ← empty ; V.insert (s0)          10: proc duplicateDetection () is
2: Q ← empty ; Q.enqueue (s0)         11:     new ← candidates \ V
3: while Q ≠ empty do                 12:     for s ∈ new do
4:     candidates ← empty             13:         V.insert (s)
5:     while Q ≠ empty do             14:         Q.enqueue (s)
6:         s ← Q.dequeue ()
7:         for e, s' | (s, e, s') ∈ T do
8:             candidates.insert (s')
9:     duplicateDetection ()
```

a new state, and then execute $a.b.c$ from s_0 to discover a conflict between s_5 and s_3 and hence that f generates a new state. Nevertheless, we observe some redundancy in these two reconstructions: as sequences $a.b.c$ and $a.b.d$ share a common prefix $a.b$, we could group the two reconstructions together so that $a.b$ is executed once for both s_3 and s_4. This is where DDD can help us. As we visit s, we notice that its successors s_4 and s_5 are mapped to hash values already met. Hence, we put those in a candidate set and mark the identifiers of states that we have to reconstruct in order to check whether s_4 and s_5 are new or not, i.e., 3 and 4.

Duplicate detection then consists of reconstructing marked states and to delete them from the candidate set. This can be done by conducting a DFS starting from the initial state in search of marked states. However, as we do not want to reconstruct the whole search tree, we have to keep track of the sub-tree that we are interested in. Thus, we additionally store for each identifier the list of its successors in the backedge table that have to be visited. The DFS then prunes the tree by only visiting successors included in this list. On our example this will result in the following traversal order: s_0, s_1, s_2, s_3 and finally s_4.

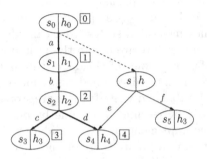

Fig. 2. The prefix $a.b$ of the reconstructing sequences of s_3 and s_4 can be shared

4.2   The Combined Algorithm

We now propose Algorithm 4 that combines the ComBack method with DDD. As it is straightforward to extend the algorithm with a full state descriptor cache (discussed in Section 3) we only focus on the basic combination here.

The two main data structures in the algorithm are the queue $\mathcal{Q}$ containing full descriptors of states to visit together with their identifiers and the visited set $\mathcal{V}$. The latter comprises three structures: a *stateTable* as in the basic ComBack method, a *backedgeTable* mapping state numbers to predecessors and some auxiliary information used by the algorithm, and a *candidates* set consisting of states that may or may not be new states.

First, consider again the LTS from Fig. 2. In Fig. 3, we see the example annotated using the same notation used in Fig. 1 at the left. That is, each ellipse represents a state and the hash value of the state is written in the right part of the ellipse. The state number is shown in a square next to the ellipse. The search has investigated states s_0, s_1, s_2, s_3 and s_4, and is about to investigate s. We have not shown the state table, which simply maps h_i to i for $i = 0 \ldots 4$. We have just picked s from the queue, which is now empty. At the right of Fig. 3, we show the backedge table before executing any event from s, after executing

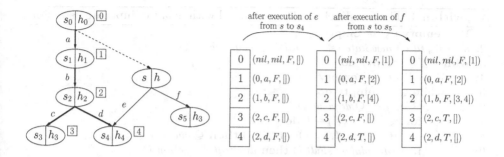

Fig. 3. Evolution of the backedge table after the execution of e and f from s

the event e, and after executing the event f. The first two components of the tuples in the backedge table are the state number of the predecessor and the event executed to reach the state as in Fig. 1. The third component is a boolean telling whether the state should be compared to the states in the candidate set. Before executing any events, this boolean is set to False for all states as we have just performed duplicate detection. When executing event e, we reach state s_4, which has hash value h_4 corresponding to state 4. We therefore add s_4 to the candidate set and set the boolean value to True for state 4 in the backedge table (the middle one in Fig. 3). The last component of the tuple in the backedge is a list of interesting successors, i.e., successor states that should be investigated when performing duplicate detection. We backtrack from state 4 to the initial state. For state number 2 we add the successor 4 as interesting, for state 1 we add state 2, and for the initial state 0 we add state 1 as can be seen in the middle backedge table in Fig. 3. After executing f from s, we obtain s_5, which has hash value h_3 corresponding to state 3. We add s_5 to the candidate set (which now consists of s_4 and s_5), and flags that state 3 should be compared against the candidate set. We also update the lists of interesting successors, adding 3 at state 2. As state number 2 is already marked as interesting successor of state 1, we stop backtracking at this point.

We can now perform duplicate detection by starting in the initial state. We see that we need to expand state 1, so we execute the event a. In state 1, we execute b to reach 2. In state 2, we must execute events to reach states 3 and 4. Executing c to reach state number 3, we obtain state s_3. As we have set the boolean to True, we attempt to remove s_3 from the candidate set. The candidate set did not contain s_3, so it still contains states s_4 and s_5. The successor list of state 3 is empty, so we execute d from state 2 to reach state 4. This also has the boolean set to True, so we remove s_4 from the candidate set, which now only contains the state s_5. Duplicate detection is then done, and we continue the search from s_5.

To sum up, the *stateTable* of the visited set $\mathcal{V}$ maps hash values to state numbers exactly like in the basic ComBack method, the *backedgeTable* maps a state number id to a tuple $(id_{pred}, e, check, succs)$ where

Algorithm 4. The ComBack method extended with delayed duplicate detection

1: $\mathcal{V} \leftarrow$ **empty** ; $\mathcal{Q} \leftarrow$ **empty**
2: $n \leftarrow 0$; $id \leftarrow newState\,(s_0, nil, nil)$; $\mathcal{Q}.enqueue\,(s_0, id)$
3: **while** $\mathcal{Q} \neq$ **empty do**
4:      $\mathcal{V}.candidates \leftarrow$ **empty**
5:      **while** $\mathcal{Q} \neq$ **empty do**
6:          $(s, s_{id}) \leftarrow \mathcal{Q}.dequeue\,()$
7:          **for** $e, s' \mid (s, e, s') \in T$ **do**
8:              **if** $insert\,(s', s_{id}, e) = NEW(s'_{id})$ **then** $\mathcal{Q}.enqueue\,(s', s'_{id})$
9:          **if** $\mathcal{V}.candidates.isFull\,()$ **then** $duplicateDetection\,()$
10:      $duplicateDetection\,()$

11: **proc** $newState\,(s, id_{pred}, e)$ **is**
12:      $id \leftarrow n$; $n \leftarrow n + 1$
13:      $\mathcal{V}.stateTable.insert\,(id, h(s))$
14:      $\mathcal{V}.backedgeTable.insert\,(id \rightarrow (id_{pred}, e, false, []))$
15:      **return** id

16: **proc** $insert\,(s, id_{pred}, e)$ **is**
17:      $ids \leftarrow \{id \mid (h(s), id) \in \mathcal{V}.stateTable\}$
18:      **if** $ids = \emptyset$ **then**
19:          $id \leftarrow newState\,(s, id_{pred}, e)$
20:          **return** $NEW(id)$
21:      **else**
22:          $\mathcal{V}.candidates.insert\,(s, id_{pred}, e)$
23:          **for** id **in** ids **do**
24:              $\mathcal{V}.backedgeTable.setCheckBit\,(id)$
25:              $backtrack\,(id)$
26:          **return** $MAYBE$

27: **proc** $backtrack\,(id)$ **is**
28:      $id_{pred} \leftarrow \mathcal{V}.backedgeTable.getPredecessorId\,(id)$
29:      **if** $id_{pred} \neq nil$ **then**
30:          **if** $id \notin \mathcal{V}.backedgeTable.getSuccessorList\,(id_{pred})$ **then**
31:              $\mathcal{V}.backedgeTable.addSuccessor\,(id_{pred}, id)$
32:              $backtrack\,(id_{pred})$

33: **proc** $duplicateDetection\,()$ **is**
34:      $dfs\,(s_0, 0)$
35:      **for** (s, id_{pred}, e) **in** $\mathcal{V}.candidates$ **do**
36:          $id \leftarrow newState\,(s, id_{pred}, e)$
37:          $\mathcal{Q}.enqueue\,(s, id)$
38:      $\mathcal{V}.candidates \leftarrow$ **empty**

39: **proc** $dfs\,(s, id)$ **is**
40:      $check \leftarrow \mathcal{V}.backedgeTable.getCheckBit\,(id)$
41:      **if** $check$ **then** $\mathcal{V}.candidates.delete\,(s)$
42:      **for** $succ$ **in** $\mathcal{V}.backedgeTable.getSuccessorList\,(id)$ **do**
43:          $e \leftarrow \mathcal{V}.backedgeTable.getReconstructingEvent\,(succ)$
44:          $dfs\,(s.exec\,(e), succ)$
45:      $\mathcal{V}.backedgeTable.unsetCheckBit\,(id)$
46:      $\mathcal{V}.backedgeTable.clearSuccessorList\,(id)$

- id_{pred} and e are the identifier of the predecessor and the reconstructing event as in the basic ComBack method;
- *check* is a boolean specifying if the duplicate detection procedure must verify whether or not the state is in the candidate set;
- *succs* is the identifier list of its successors which must be generated during the next duplicate detection as previously explained.

The *candidates* set is a set of triples (s, id_{pred}, e) where s is the full descriptor of a candidate state. In case duplicate detection reveals that s does not belong to the visited set, id_{pred} and e comprise the reconstruction information that will be associated with the state in *backedgeTable*.

Consider again Algorithm 4 which shows the ComBack method extended with delayed duplicate detection. The main procedure (lines 1–10) works basically as Algorithm 3. A notable difference is that procedure *insert* (see below) may return a two-valued answer:

NEW - if the state is surely new. In this case, the identifier assigned to the inserted state is also returned by the procedure. The state can be unconditionally inserted in the queue for a later expansion.

MAYBE - if we cannot answer without performing duplicate detection.

Procedure *newState* inserts a new state in the visited set together with its reconstruction information. It computes a new identifier for s, a state to insert, and update the *stateTable* and *backedgeTable*.

Procedure *insert* receives a state s, the identifier id_{pred} of one of its predecessors s' and the event used to generate s from s'. It first performs a lookup in the *stateTable* for identifiers of states mapped to the same hash value as s (line 17). If this search is unsuccessful (lines 18–20), this means that s has definitely not been visited before. It is unconditionally inserted in $\mathcal{V}$, and its identifier is returned by the procedure. Otherwise (lines 21–26), the answer requires the reconstruction of states whose identifiers belong to set *ids*. We thus save s in the candidate set for a later duplicate detection, set the check bit of all identifiers in *ids* to true so that the corresponding states will be checked against candidate states during the next duplicate detection, and backtrack from these states.

The purpose of the *backtrack* procedure is, for a given state s with identifier id, to update the successor list of all the states on the path from s_0 to s in the backedge table so that s will be visited by the DFS performed during the next duplicate detection. The procedure stops as soon as a state with no predecessor is found, i.e., s_0, or if id is already in the successor list of its predecessor, in which case this also holds for all its ancestors.

Duplicate detection (lines 33–38) is conducted each time the candidate set is full (line 9), i.e., it reaches a certain peak size, or the queue is empty (line 10). Using the successor lists constructed by the backtrack procedure, we initiate a depth-first search from s_0 (see procedure *dfs*). Each time a state with its check bit set to true is found (line 41) we delete it from the candidate set if needed. When a state leaves the stack we set its check bit to false and clear its successor list (lines 45–46). Once the search finishes (lines 35–37) any state remaining in the candidate set is new and can be inserted into the queue and the visited set.

4.3 Additional Comments

We discuss below several issues regarding the proposed algorithm.

Memory issues. Our algorithm requires the storage of additional information used to keep track of states that must be checked against the candidate set during duplicate detection. This comprises, for each state, a boolean value (the *check* bit) and a list of successors that must be visited. As any state may belong to the successor list of its predecessor in the backedge table, the memory overhead is theoretically one bit plus one integer per state. However, our experiments reveal (see Section 5) that even very small candidate sets show good memory performance. Therefore, successor lists are usually short and the extra memory consumption low. We did not find any model for which our algorithm aborted due to a lack of memory, but where the one of [20] did terminate.

Grouping reconstructions of queued states. In [20] the possibility to reduce memory usage by storing identifiers instead of full state descriptors in the queue (Variant 4 in Section 5) was mentioned. This comes at the cost of an additional reconstruction per state required to get a full descriptor for the state that can be used to generate its successors. The principle of grouping state reconstructions can also be applied to the states waiting in the queue. The idea is to dequeue blocks of identifiers from the queue instead of individual ones and reconstruct those in a single step using a procedure similar to *dfs* given in Algorithm 4.

Compatibility with depth-first search. A nice characteristic of the basic ComBack method is its total decoupling from the search algorithm thereby making it fully compatible with, e.g., LTL model checking [3,9]. Delaying detection from state generation makes an algorithm implicitly incompatible with a depth-first traversal in which the state processed is always the most recent state generated. At first glance, the algorithm proposed in this section also belongs to that category. However, we can exploit the fact that the insertion procedure can decide if a state is new without actually putting it in the candidate set (if the hash value of the state has never been met before). The idea is that the search can progress as long as new states are met. If some state is then put in the candidate set, the algorithm puts a marker on the stack to remember that a potentially new state lies here. Finally, when a state is popped from the stack, duplicate detection is performed if markers are present on top of the stack. If we find out that some of the candidate states are new, the search can continue from these ones. This makes delayed detection compatible with depth-first search at the cost of performing additional detections, during the backtrack phase of the depth-first search algorithm.

5 Experimental Results

We report in this section the data we collected during several experiments with the proposed techniques[3]. We used the ASAP verification tool [19] where we

---

[3] Additional data on these experiments may be found in [7].

have implemented the algorithms described in this article. A nice characteristic of ASAP is its independence from the description language of the model. This allows us to perform experiments on CPN (Coloured Petri net) and DVE (the input language of the DiVinE verification tool [1]) models. All CPN models come from our own collection. DVE models were taken from the BEEM database [15] though we did not consider models belonging to the families "Planning and scheduling" and "Puzzles" as these are mostly very simple models, e.g., the towers of Hanoi, that have nothing to do with real life examples.

For CPN models, the hash function used is defined inductively on the state of the model. In CPNs, a state is a *marking* of a set of *places*. Each marking is a *multi-set* over a given *type*. We use a standard hash function for each type. We extend this hash function to multi-sets by using a *combinator function*, which takes two hash values and returns a new hash value. We extend the hash functions on markings of places to a hash function of the entire model by using the combinator function on the place hash functions. We proceed exactly the same way for DVE instances, except that the components of the system are now variables, channels, and process states rather than places. The performance of these functions in term of conflicts is usually very good. Typically with a 32 bit hash signature there is generally no collision for small instances (100,000 states or less) and for larger instances (e.g., up to 10^6–10^7 states), we can reasonably expect to cover more than 95% of the state space using a hash compaction based algorithm. The quality of this function is evidenced by Table 1 in [20] that reports the number of collisions for several CPN instances.

This section uses several abbreviations. Table 1 lists all abbreviations used in this section and it provides a short description of all selected models. Each instance name consists of the concatenation of a model name followed by an instance number (i.e., an instantiation of the model parameters). For instance `firewire_tree.5` is the 5th instance (in the BEEM database) of the model named `firewire_tree`. The instantiation values of parameters are not relevant for this study.

5.1   Experiment 1: Evaluation of Caching Strategies

In this first experiment we evaluated the different strategies proposed in Section 3. We selected 143 DVE instances having from 10,000 to 10,000,000 states and ran the ComBack algorithm of [20] using BFS with 10 caching strategies and 4 sizes of cache expressed as a fraction of the state space size: { 10^{-4}, 10^{-3}, 10^{-2}, 10^{-1} }. Of these 10 strategies 4 are simple: R (with a replacement probability $p = 0.5$), F, H, D (with a minimal distance $k = 5$ between cached states); and 6 are combinations of the first ones: F_{20}-H_{80}, F_{50}-H_{50}, F_{80}-H_{20}, F_{20}-D_{80}, F_{50}-D_{50} and F_{80}-D_{20}. We measured after each run the number of event executions that were due to state reconstructions. The results are summarized in Table 2. On the top four rows we give, for each strategy, the average over all instances of the number of event executions due to state reconstruction with this strategy divided by the same number obtained with strategy R. The bottom four rows report for each strategy the number of instances for which it performed best.

Table 1. List of abbreviations and short description of models used in Section 5

Abbreviations	
BFS	Breadth-First Search
CPN	Colored Petri Net
DDD	Delayed Duplicate Detection
DVE	Input language of the DiVinE toolset
Caching strategies	
D	Distance based caching strategy
F	Fifo caching strategy
H	Heuristic based caching strategy
R	Random caching strategy
F_X-D_Y	Combination of caching strategies F and D where X% (resp. Y%) of the cache is allocated to a Fifo sub-cache (resp. Distance based sub-cache)
F_X-H_Y	Combination of caching strategies F and H where X% (resp. Y%) of the cache is allocated to a Fifo sub-cache (resp. Heuristic based sub-cache)
Introduced in Experiment 2	
Std.	Execution time using a standard storage method, i.e., without any reduction technique, or using hash compaction if the classical search ran out of memory
DDD(X)	Characterize a search using the ComBack method with delayed duplicate detection. X denotes the size of the candidate set as a fraction of the cache size.
$T\uparrow$	Denote, for a search using the ComBack method, the increase of the execution time as the ratio $\frac{\text{execution time of this run}}{\text{execution time with a standard search}}$.
$E\uparrow$	Denote, for a search using the ComBack method, the increase of the number of event executions as the ratio $\frac{\text{event executions during this run}}{\text{transitions of the graph}}$.
Introduced in Experiment 3	
S	Caching strategy used
F_C	Proportion of full state descriptors allocated to the cache
F_{CS}	Proportion of full state descriptors allocated to the candidate set
F_Q	Proportion of full state descriptors allocated to the queue

Models	
CPN models	
dymo	Dynamic MANET on-demand routing protocol (from [5])
erdp	Edge router discovery protocol (from [13])
protocol	Simplified stop and wait protocol
telephones	Telecommunication service
DVE models	
brp	Bounded retransmission protocol
brp2	Timed version of the bounded retransmission protocol
cambridge	Cambridge ring protocol
firewire_link	Layer link protocol of the firewire protocol (IEEE 1394)
firewire_tree	Tree identification protocol of the firewire protocol (IEEE 1394)
needham	Needham-Schroeder public key authentication protocol
synapse	Synapse cache coherence protocol

These latter are only given for information and our interpretations will be based on the values of the top rows.

Strategy F performs well compared to strategy R but its performance degrades as we allocate more states to the cache. This is also confirmed by the fact that combinations F-H and F-D seem to perform better for a large cache when the proportion of states allocated to the fifo sub-cache is low. Apparently with this strategy, we quickly reach a limit where all (or most of) forward transitions lead to a cached (or new) state and most backward transitions lead to a non cached state. Such a cache failure always implies backtracking to the initial state (the fifo strategy implies that if a state is not cached, none of its ancestors in the backedge table is cached) which can be quite costly. Beyond this point, allocating more states to the cache is almost useless. We even see that strategy F is largely outperformed by strategy R for the largest cache size we experimented with.

The performance of strategy H is poor for small caches but progresses well compared to strategy F. With this strategy, most transitions will be followed by a state reconstruction. However, our heuristic works rather well and reconstructing sequences are usually much shorter than with strategy F. Still, strategy H is usually outperformed by strategy F for small cache sizes due to a high presence of forward transitions in state spaces [14]. To sum up, strategy F implies few reconstructions but long reconstructing sequences and strategy H has the opposite characteristics.

As expected, strategy D improves strategy H although only slightly. It prevents the algorithm from caching two states that are linked by too short a path in the backedge table. However, for the largest cache size we experimented with, the algorithm could not use all the memory allocated to the cache due to this restriction. This is the only case where strategy D performed worse than H.

From all these observations it is not surprising to see that the best strategy is to always keep a small fifo cache and allocate remaining memory to a second level cache maintained with strategy D, that is, to keep a small number of recently visited states to limit the number of reconstructions and many strategic states from previous levels that will help us shorten reconstructing sequences.

Note that for a large cache holding 10% of the state space, the strategy used impacts less than for small caches. This is evidenced by the fact that the values observed for strategy H and D approach 1 and more generally that all values of line 10^{-2} are smaller than those of line 10^{-1}.

Out of these 143 instances we selected 4 instances that have some specific characteristics (brp2.6, cambridge.6, firewire_tree.4 and synapse.6) and evaluated strategies F, D and F_{20}-D_{80} with different sizes of cache ranging from 1,000 to 10,000. The collected data is plotted in Fig. 4. On the x-axis is the different cache sizes used. For each run we recorded the number of event executions due to reconstructions and compared it to the same number obtained with strategy F. For instance, with brp2.6 and a cache of 2,000 states, reconstructions generated approximately three times more event executions with strategy D than with strategy F. We also provide the characteristics of these graphs in terms of number of states and transitions, average degree, i.e., average number

Table 2. Evaluation of caching strategies on 143 DVE instances

Cache	Strategy									
size	R	F	H	F_{20} H_{80}	F_{50} H_{50}	F_{80} H_{20}	D	F_{20} D_{80}	F_{50} D_{50}	F_{80} D_{20}
10^{-4}	1.000	0.397	1.123	0.365	0.339	0.339	1.057	0.356	**0.332**	0.337
10^{-3}	1.000	0.452	0.940	0.292	0.293	0.317	0.763	**0.272**	0.278	0.308
10^{-2}	1.000	0.587	0.479	0.187	0.208	0.261	0.385	**0.162**	0.188	0.247
10^{-1}	1.000	2.090	0.670	0.258	0.348	0.538	0.771	**0.229**	0.280	0.472
10^{-4}	3	**55**	3	9	18	28	2	18	21	15
10^{-3}	0	12	3	25	26	19	2	**50**	22	13
10^{-2}	0	4	2	42	12	6	4	**85**	9	8
10^{-1}	0	0	4	53	10	6	0	**66**	19	14

of transitions per state, number of levels and number of forward transitions as a proportion of the overwhole number of transitions.

The graph of `firewire_tree.4` only has forward transitions, which is common for leader election protocols. Therefore, a sufficiently large fifo cache is the best solution. This is one of the few instances where increasing the cache size benefits strategy F more than D. Moreover, its average degree is rather high, which leads to a huge number of reconstructions with strategy D. On the contrary the graph of `cambridge.6` has a relatively large number of backward transitions. Increasing the fifo cache did not bring any substantial improvement: from 262,260,647 executions with a cache size of 1,000 it went down to 260,459,235 executions with a cache size of 10,000. Strategy D is especially interesting for `synapse.6` as its graph has a rather unusual property: a low fraction of its states have a high number of successors (from 13 to 18). These states are thus shared by many reconstructing sequences and, using our heuristic, they are systematically kept in the cache. Thus, strategy D always outperforms strategy F even for small caches. The out-degree distribution of the graph of `brp2.6` has the opposite characteristics: 49% of its states have 1 successor, 44% have 2 successors and other states have 0 or 3 successors. Therefore, there is no state that is really interesting to keep in the cache. This is evidenced by the fact that the progressions of distance based strategies (relative to strategy F) are not so good. It goes from 3.157 to 2.200 for strategy D and from 0.725 to 0.537 for strategy F_{20}-D_{80}.

5.2 Experiment 2: Evaluation of Delayed Duplicate Detection

To experiment with delayed detection we picked out 63 DVE instances having from 1,000,000 to 60,000,000 states and 12 CPN instances having from 100,000 to 5,000,000 states. The ComBack method was especially helpful for the `dymo` and `erdp` models which are models of two industrial protocols—a routing protocol [5] and an edge router discovery protocol [13]—and have large descriptors (1,000–5,000 bytes).

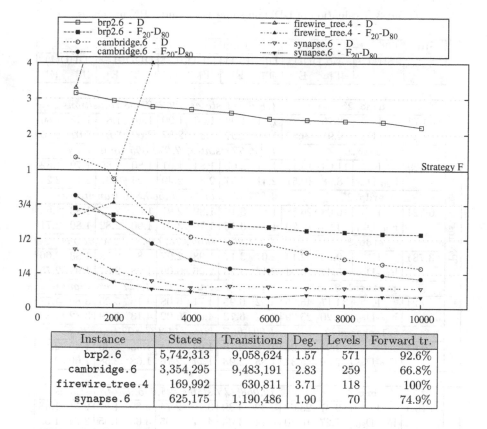

Instance	States	Transitions	Deg.	Levels	Forward tr.
brp2.6	5,742,313	9,058,624	1.57	571	92.6%
cambridge.6	3,354,295	9,483,191	2.83	259	66.8%
firewire_tree.4	169,992	630,811	3.71	118	100%
synapse.6	625,175	1,190,486	1.90	70	74.9%

Fig. 4. Evolution of strategies F, D and F_{20}-D_{80} on some selected instances

Table 3 summarizes our observations. We only report data for 6 instances of each family but still provide the average increases on all instances (see Table 4(a)). We used caching strategies F and F_{20}-D_{80} with a cache size corresponding to 1% of the state space for DVE instances and 0.1% for CPN instances. This choice is motivated by the observation that state vectors of Coloured Petri Nets (100–10,000 bytes) are usually much larger than those of DVE models (10–500 bytes). For each instance we performed 6 tests: one with a standard storage method, i.e., full state descriptors are kept in the visited set, (column Std.), one with the ComBack method without delaying detection (column No DDD) and four with delayed detection enabled with different candidate set sizes expressed as a fraction the memory given to the cache (columns DDD(0.1), DDD(0.2), DDD(0.5) and DDD(1)). For CPN instances, we kept identifiers in the queue rather than full state descriptors—as described in [20], Variant 4 of Section 5. Most of the instances studied have rather large levels (typically more than 10% of the state space) which prevented us from keeping full state descriptors. The optimization described in Section 4.3 that consists of grouping the reconstruction of queued identifiers was turned on and each block of states reconstructed

Table 3. Evaluation of DDD on some DVE and CPN instances

Std.		ComBack								
	Strat.	No DDD		DDD(0.1)		DDD(0.2)		DDD(0.5)		DDD(1)	
		T↑	E↑	T↑	E↑	T↑	E↑	T↑	E↑	T↑	E↑

CPN instances

`dymo.2`[*]		*4,196,714 states, 29,227,638 transitions*									
15,073	F	8.75	17.83	1.31	1.94	1.36	1.90	1.40	1.87	1.45	1.86
	F_{20}-D_{80}	4.97	9.05	1.88	2.97	1.92	2.87	2.00	2.85	2.02	2.75
`dymo.6`		*1,256,773 states, 7,377,095 transitions*									
1,429	F	11.64	15.17	1.91	2.00	1.84	1.91	1.80	1.85	1.79	1.82
	F_{20}-D_{80}	8.98	10.51	2.41	2.57	2.32	2.49	2.24	2.37	2.14	2.24
`erdp.2`[*]		*4,277,126 states, 30,503,876 transitions*									
6,324	F	10.08	26.58	1.85	3.16	1.89	2.97	1.84	2.77	1.83	2.65
	F_{20}-D_{80}	4.84	10.80	1.75	3.16	1.79	3.01	1.90	2.82	1.89	2.71
`erdp.3`[*]		*2,344,208 states, 18,739,842 transitions*									
3,784	F	9.99	22.13	2.07	3.17	1.98	2.97	1.89	2.78	1.83	2.66
	F_{20}-D_{80}	5.44	10.04	2.14	3.23	2.06	3.06	1.98	2.87	1.92	2.75
`protocol.3`[*]		*2,130,381 states, 11,584,421 transitions*									
424	F	34.17	45.85	5.32	6.95	4.71	6.08	4.16	5.25	3.80	4.77
	F_{20}-D_{80}	18.20	25.89	5.12	6.73	4.61	6.02	4.13	5.29	3.82	4.82
`telephones.2`		*1,004,967 states, 11,474,892 transitions*									
894	F	3.90	8.06	1.76	2.78	1.64	2.54	1.53	2.30	1.45	2.13
	F_{20}-D_{80}	2.90	5.32	1.85	2.93	1.77	2.76	1.68	2.58	1.60	2.39

DVE instances

`brp.4`		*12,068,447 states, 25,085,950 transitions*									
16.4	F	9.37	23.02	4.42	3.39	3.76	2.74	2.89	1.87	2.51	1.48
	F_{20}-D_{80}	7.27	10.80	5.14	3.04	4.57	2.55	3.65	1.85	3.22	1.53
`brp2.6`		*5,742,313 states, 9,058,624 transitions*									
6.5	F	8.20	16.79	7.86	5.64	6.68	5.16	6.62	5.26	6.41	5.27
	F_{20}-D_{80}	5.59	7.06	8.55	5.31	6.03	3.88	7.16	4.85	7.07	4.91
`cambridge.7`		*11,465,015 states, 54,850,496 transitions*									
197	F	23.12	65.43	1.92	3.12	1.90	3.05	1.88	2.97	1.83	2.83
	F_{20}-D_{80}	2.83	5.63	1.79	2.40	1.79	2.38	1.78	2.36	1.75	2.30
`firewire_link.5`[*]		*18,553,032 states, 59,782,059 transitions*									
358	F	3.34	3.54	1.30	1.18	1.26	1.14	1.23	1.11	1.22	1.10
	F_{20}-D_{80}	1.45	1.32	1.32	1.15	1.29	1.12	1.26	1.10	1.26	1.09
`needham.4`		*6,525,019 states, 22,203,081 transitions*									
20.2	F	2.29	2.41	1.73	1.29	1.74	1.29	1.75	1.28	1.75	1.27
	F_{20}-D_{80}	2.08	1.72	1.98	1.33	2.00	1.33	2.01	1.33	2.01	1.32
`synapse.7`		*10,198,141 states, 19,893,297 transitions*									
24.4	F	9.11	13.53	2.02	1.42	2.02	1.41	2.01	1.39	2.00	1.36
	F_{20}-D_{80}	2.43	1.94	2.21	1.31	2.21	1.31	2.22	1.31	2.23	1.30

Parameters for CPN instances	cache size 0.1% of the state space
	items in the queue state identifiers
Parameters for DVE instances	cache size 1% of the state space
	items in the queue full state descriptors

from the queue had the same size as the candidate set. Hence, when DDD was not used this optimization was turned off. In column Std. we provide the execution time in seconds using a standard search algorithm. In columns T↑ we measure the run-time increase (compared to the standard search) as the ratio $\frac{\text{execution time of this run}}{\text{T (with standard search)}}$ and in column E↑ the increase of the number of event executions as the ratio $\frac{\text{event executions during this run}}{\text{transitions of the graph}}$. Hence, a value of 1 for E↑ means that we executed exactly the same number of events as the basic algorithm and that no state reconstruction occurred. Some runs using standard storage ran of out memory. This is indicated by a ⋆ following the instance name. For these, we provide the time obtained with hash compaction as a lower approximation.

We notice that DDD is indeed useful to save the redundant exploration of transitions during reconstruction, and this, even with small candidate sets. Typically, the number of executions is reduced by a factor of 3–5 or even more if we keep identifiers in the queue and group their reconstruction. It seems that, using BFS, states generated successively are "not so distant" in the graph so their reconstructing sequences are quite similar, which allows many sharings. This especially holds during the reconstruction of queued states. Two states reconstructed this way often have the same predecessor in the backedge table (since they have been generated from that state) or at least a common ancestor of low degree. Hence, the average decrease of executions is more sensible for CPNs.

However, although DDD saves many redundant operations in state reconstructions, this does not always impact on the time saved as we would have expected. Indeed DDD is much more interesting when analyzing CPN instances rather than DVE instances. If we consider for example the average increase in the number of events and in the run-time of the 63 DVE instances with caching strategy F_{20}-D_{80} (see Table 4(a)) we could reduce the number of events executed by more than a factor of 2 ($4.00 \rightarrow 1.63$) whereas the average time ($3.45 \rightarrow 2.65$) did not decrease in a comparable way. The reason is that executing an event is much faster for DVE models than for CPN models. Events are typically simple in the DVE language, e.g., a variable incrementation, whereas they can be quite complex with CPNs and include the execution of code embedded in the transition. Therefore, the fact of maintaining the candidate set or successors lists has a non negligible impact for DVE models which means that DDD reduces time only if the number of executions decreases in a significant way, e.g., for instance brp.4.

We saw in the previous experiment that model characteristics largely impact on the performance of caching strategies and, hence, that a significant increase in the size of the cache did not necessarily lead to the expected decrease of event executions. This assertion is less valid when duplicate detection is used as we can see from Table 4(b). This one gives, for each configuration (caching strategy and candidate set size), the distribution of values in column E↑ for all 63 DVE instances (rather than the 6 selected). We notice that, by using a very small candidate set (0.1% of the state space), the number of instances in the last range, i.e., with E↑ > 4, is drastically reduced and, more generally, that allocating more states to the candidate set always brings some improvement.

Table 4. Summary of data for Experiment 2

(a) Average on all instances of time and event executions increase

Strat.	No DDD		DDD(0.1)		DDD(0.2)		DDD(0.5)		DDD(1)	
	T↑	E↑	T↑	E↑	T↑	E↑	T↑	E↑	T↑	E↑
Average on 63 DVE instances										
F	7.15	13.40	2.70	2.12	2.59	1.99	2.43	1.81	2.36	1.66
F_{20}-D_{80}	3.45	4.00	2.92	1.96	2.81	1.86	2.71	1.74	2.65	1.63
Average on 12 CPN instances										
F	16.25	24.03	4.05	4.98	3.45	4.18	2.97	3.52	2.71	3.20
F_{20}-D_{80}	9.59	13.61	3.77	4.81	3.36	4.24	3.01	3.72	2.83	3.44

(b) Distribution of the values in column E↑ for the 63 DVE instances

Range of E↑		Strat. F					Strat. F_{20}-D_{80}				
		Candidate set size					Candidate set size				
		0	0.1	0.2	0.5	1	0	0.1	0.2	0.5	1
1	– 1.25	3	9	9	13	14	3	8	8	11	14
1.25	– 1.50	2	8	8	7	15	6	9	10	12	17
1.50	– 1.75	0	0	1	6	9	6	5	7	9	9
1.75	– 2	1	11	15	18	15	2	15	15	18	13
2	– 4	5	33	28	18	9	24	24	22	12	9
	> 4	52	2	2	1	1	22	2	1	1	1

With the largest candidate set size (1% of the state space), duplicate detection generated less event executions than the exploration algorithm, i.e., E↑ < 2, for 53 instances out of the 63 selected, regardless the caching strategy. Instances cambridge.7 and brp2.6 (see Table 3) belong to the 10 instances that do not have this property. To sum up, increasing the candidate set size is (almost) always useful; and the benefit of DDD is hence more predictable than the benefit of state caching (which depends to a large extent on the characteristics of model).

A somewhat negative observation is that the relative advantage of strategy F_{20}-D_{80} previously observed in Experiment 1 is lost when DDD is used. We indeed observe that numbers reported in Table 4(a) are roughly the same for both strategies: the caching strategy impacts less when delayed detection is used. For DVE instances the algorithm executed slightly more events with strategy F but was also faster. This is due to the time overhead induced by the distance based strategy: each time a state is likely to enter the cache we have to verify in the backedge table that none of its ancestors of degree k (5 in our case) is itself cached. For CPN instances, strategy F generally lead to fewer event executions. Since we stored identifiers instead of full state descriptors in the queue, a fifo cache was more appropriate: by keeping states of the last BFS levels, the probability that an unprocessed state that has to be reconstructed is cached greatly increases. Hence, the grouped reconstruction of queued states occurs less frequently.

These results are best explained in light of the observations made in the first experiment. The goal of strategies F and D are indeed different. The first one reduces the number of reconstructions while the second one reduces the number of events executed per reconstruction. Thus, delayed duplicate detection and caching strategy D work at the same level and may be redundant whereas DDD is fully complementary to strategy F. Hence, it is not surprising that DDD brings better results when combined with a fifo caching strategy.

5.3   Experiment 3: Optimal Use of Available Memory

In this last experiment we look at a more practical problem. We basically want to know how to tune the different parameters of the ComBack algorithm in order to guarantee a good reduction of the execution time, whatever the input model. This is of interest since the previous experiments stressed that huge variations can be observed for one model when using different combinations of the various extensions we proposed.

First, let us note that by keeping state identifiers in the queue we bound the number of full state descriptors used by the algorithm. Each of these descriptors may be kept in three places: in the queue (if we group the reconstruction of states waiting to be processed), in the cache or in the candidate set. Now, let us suppose that the user has an idea of the available memory and the full state descriptor size[4]. From this user supplied information, we can roughly decide on a maximal number of full state descriptors that may reside in memory. Since there is an obvious conflict between the data structures used by the algorithm, i.e., a full state descriptor may only be kept in one of these, we have to decide which portion should be allocated to the queue, to the cache and to the candidate set. Hence, the problem considered is the following. Given a maximal number of full state descriptors F given to the algorithm, find the configuration that minimizes the extra time introduced by the ComBack method. By configuration we mean here a tuple (S, F_C, F_{CS}, F_Q) where S is a caching strategy and F_C, F_{CS} and F_Q are respectively the proportion of the F states allocated to the cache, to the candidate set and to the queue.

To try to answer this question we performed numerous runs using the different instances and parameters listed by Table 5(a) (naturally, under the constraint that $F_C + F_{CS} + F_Q = 1$). We voluntarily selected small values for parameter F and large state spaces (with respect to F) to simulate verification runs where the state space is large, e.g., 10^9 states, and the state descriptor is so large that only a small fraction of them can be kept in main memory. We did not experiment with values greater than 0.4 for parameter F_Q as beyond this bound the performance of the algorithm quickly degraded. Table 5(b) summarizes our results. The performance reported in column E↑ is the average over all instances of event executions increase (see Table 1). Note that the table has been sorted according to the values reported in that column. For the sake of clarity we only report the performance of a few configurations including the ones that performed the best and the worst. We can make the following observations regarding this data.

---

[4] Otherwise, a partial search exploring a sample of the state space can provide a rather good approximation of this size.

- First, for each of the three tables, the top part is occupied by configurations using the three techniques (i.e., for which $F_C > 0$, $F_{CS} > 0$ and $F_Q > 0$) while the bottom part is almost only composed of configurations for which one of these parameters is set to 0. Hence, none of these techniques should be disabled.
- All other things being equal, strategy F_{20}-D_{80} is outperformed by strategies F and F_{80}-D_{20}. It is therefore preferable to allocate more states to a fifo cache although a small distance based cache can bring some improvements compared to a pure fifo cache.
- It seems that the best way to proceed is, whatever the value of F, to attribute the majority of the memory to the cache (50–60%), a small amount to group the reconstruction of queued states (10–20%) and the remainder to the candidate set for delayed duplicate detection (20–40%). Although this

Table 5. Summary of data for Experiment 3

(a) Parameters used for experimentation

Parameters	$F \in \{ 100, 1{,}000, 10{,}000 \}$ $F_C \in \{ 0, 0.1, 0.2, \ldots, 1 \}$ $F_{CS} \in \{ 0, 0.1, 0.2, \ldots, 1 \}$ $F_Q \in \{ 0, 0.1, 0.2, 0.3, 0.4 \}$ $S \in \{ F, F_{20}\text{-}D_{80}, F_{80}\text{-}D_{20} \}$
Selected instances	all DVE instances with $[100 \cdot F - 1{,}000 \cdot F]$ states

(b) Results

$F=10^2$ (58 instances)

Configuration				$E\uparrow$
S	F_C	F_{CS}	F_Q	
F_{80}-D_{20}	0.5	0.3	0.2	6.51
F	0.4	0.3	0.3	6.64
F_{20}-D_{80}	0.9	0	0.1	8.91
F_{20}-D_{80}	0.5	0.5	0	13.7
F_{80}-D_{20}	0.8	0.2	0	16.5
F	1	0	0	25.9
-	0	1	0	30.9

$F=10^3$ (46 instances)

Configuration				$E\uparrow$
S	F_C	F_{CS}	F_Q	
F_{80}-D_{20}	0.6	0.2	0.2	5.45
F	0.4	0.3	0.3	5.54
F_{20}-D_{80}	0.9	0	0.1	9.51
F_{20}-D_{80}	0.8	0.2	0	13.3
F_{80}-D_{20}	0.8	0.2	0	18.4
F	1	0	0	34.6
-	0	1	0	74.4

$F=10^4$ (41 instances)

Configuration				$E\uparrow$
S	F_C	F_{CS}	F_Q	
F_{80}-D_{20}	0.6	0.3	0.1	3.59
F	0.4	0.3	0.3	3.72
F_{80}-D_{20}	0.8	0.2	0	6.58
F_{20}-D_{80}	0.9	0	0.1	7.42
F_{80}-D_{20}	0.1	0.9	0	13.7
-	0	1	0	20.7
F	1	0	0	29.5

information is not provided by the table, these are also the configurations for which the standard deviation is the lowest, meaning that the performance of the algorithm is more predictable and depend less on the specific characteristics of the state space.

6   Conclusion

The ComBack method has been designed to explicitly store large state spaces of models with complex state descriptors. The important memory reduction factor it may provide is however counterbalanced by an increase in run-time due to the on-the-fly reconstruction of states. We proposed in this work two ways to tackle this problem. First, strategies have been devised in order to efficiently maintain a full state descriptor cache, used to perform less reconstructions and shorten the length of reconstructing sequences. Second, we combined the method with delayed duplicate detection to group reconstructions and save the execution of events that are shared by multiple sequences. We have implemented these two extensions in ASAP and performed extensive experimentation on both DVE models from the BEEM database and CPN models from our own collection. These experiments validated our proposals on many models. Compared to a random replacement strategy, a combination of our strategies could, on an average made over a hundred of DVE instances, decrease the number of transitions visited by a factor of five. We also observed that delayed duplicate detection is efficient even with very small candidate sets. In the best cases, we could even approach the execution time of a hash compaction based algorithm. Moreover, by storing identifiers instead of full descriptors in the queue we bound the number of full state descriptors that reside in memory. Hence, our data structures can theoretically consume less memory during the search than hash compaction structures. We experienced this situation on several occasions.

In this work, we mainly focused on caching strategies for a breadth-first search. BFS is helpful to find short error-traces for safety properties, but not if we are interested in the verification of linear time properties, which is inherently based on a depth-first search. The design of strategies for other types of searches is thus a future research topic. In addition, the combination with delayed duplicate detection opens the way to an efficient multi-threaded algorithm based on the ComBack method. The underlying principle would be to have some threads exploring the state space and visiting states while others are responsible for performing duplicate detection. We are currently working on such an algorithm.

Acknowledgments. We thank the anonymous reviewers for their detailed comments that helped us to improve this article.

References

1. Barnat, J., Brim, L., Cerná, I., Moravec, P., Rockai, P., Simecek, P.: DiVinE - A Tool for Distributed Verification. In: Ball, T., Jones, R.B. (eds.) CAV 2006. LNCS, vol. 4144, pp. 278–281. Springer, Heidelberg (2006)

2. Burch, J.R., Clarke, E.M., Dill, D.L., Hwang, L.J., McMillan, K.: Symbolic Model Checking: 10^{20} States and Beyond. In: LICS 1990, pp. 428–439 (1990)
3. Couvreur, J.-M.: On-the-Fly Verification of Linear Temporal Logic. In: Wing, J.M., Woodcock, J.C.P., Davies, J. (eds.) FM 1999. LNCS, vol. 1708, pp. 253–271. Springer, Heidelberg (1999)
4. Emerson, E.A., Sistla, A.P.: Symmetry and Model Checking. Formal Methods in Systems Design 9(1-2), 105–131 (1996)
5. Espensen, K.L., Kjeldsen, M.K., Kristensen, L.M.: Modelling and Initial Validation of the DYMO Routing Protocol for Mobile Ad-Hoc Networks. In: van Hee, K.M., Valk, R. (eds.) PETRI NETS 2008. LNCS, vol. 5062, pp. 152–170. Springer, Heidelberg (2008)
6. Evangelista, S., Pradat-Peyre, J.-F.: Memory Efficient State Space Storage in Explicit Software Model Checking. In: Godefroid, P. (ed.) SPIN 2005. LNCS, vol. 3639, pp. 43–57. Springer, Heidelberg (2005)
7. Evangelista, S., Westergaard, M., Kristensen, L.M.: The ComBack Method Revisited: Caching Strategies and Extension with Delayed Duplicate Detection. Technical report, DAIMI, Aarhus University, Denmark (2008), http://www.cs.au.dk/~evangeli/doc/comback-extensions.pdf
8. Godefroid, P.: Partial-Order Methods for the Verification of Concurrent Systems – An Approach to the State-Explosion Problem. LNCS, vol. 1032. Springer, Heidelberg (1996)
9. Godefroid, P., Holzmann, G.J.: On the Verification of Temporal Properties. In: Danthine, A.A.S., Leduc, G., Wolper, P. (eds.) PSTV 1993. IFIP Transactions, vol. C-16, pp. 109–124. North-Holland, Amsterdam (1993)
10. Holzmann, G.J.: The Model Checker SPIN. IEEE Transactions on Software Engineering 23(5), 279–295 (1997)
11. Jensen, K., Kristensen, L.M.: Coloured Petri Nets – Modelling and Validation of Concurrent Systems. Springer, Heidelberg (2009)
12. Knottenbelt, W., Mestern, M., Harrison, P., Kritzinger, P.: Probability, Parallelism and the State Space Exploration Problem. In: Puigjaner, R., Savino, N.N., Serra, B. (eds.) TOOLS 1998. LNCS, vol. 1469, pp. 165–179. Springer, Heidelberg (1998)
13. Kristensen, L.M., Jensen, K.: Specification and Validation of an Edge Router Discovery Protocol for Mobile Ad-hoc Networks. In: Ehrig, H., Damm, W., Desel, J., Große-Rhode, M., Reif, W., Schnieder, E., Westkämper, E. (eds.) INT 2004. LNCS, vol. 3147, pp. 248–269. Springer, Heidelberg (2004)
14. Pelánek, R.: Typical Structural Properties of State Spaces. In: Graf, S., Mounier, L. (eds.) SPIN 2004. LNCS, vol. 2989, pp. 5–22. Springer, Heidelberg (2004)
15. Pelánek, R.: BEEM: Benchmarks for Explicit Model Checkers. In: Bošnački, D., Edelkamp, S. (eds.) SPIN 2007. LNCS, vol. 4595, pp. 263–267. Springer, Heidelberg (2007), http://anna.fi.muni.cz/models/
16. Stern, U., Dill, D.L.: Improved Probabilistic Verification by Hash Compaction. In: Camurati, P.E., Eveking, H. (eds.) CHARME 1995. LNCS, vol. 987, pp. 206–224. Springer, Heidelberg (1995)
17. Stern, U., Dill, D.L.: Using Magnetic Disk Instead of Main Memory in the Murφ Verifier. In: Vardi, M.Y. (ed.) CAV 1998. LNCS, vol. 1427, pp. 172–183. Springer, Heidelberg (1998)
18. Tronci, E., Della Penna, G., Intrigila, B., Venturini Zilli, M.: Exploiting Transition Locality in Automatic Verification. In: Margaria, T., Melham, T.F. (eds.) CHARME 2001. LNCS, vol. 2144, pp. 259–274. Springer, Heidelberg (2001)

19. Westergaard, M., Evangelista, S., Kristensen, L.M.: ASAP: An Extensible Platform for State Space Analysis. In: Franceschinis, G., Wolf, K. (eds.) PETRI NETS 2009. LNCS, vol. 5606, pp. 303–312. Springer, Heidelberg (2009)
20. Westergaard, M., Kristensen, L.M., Brodal, G.S., Arge, L.: The ComBack Method – Extending Hash Compaction with Backtracking. In: Kleijn, J., Yakovlev, A. (eds.) ICATPN 2007. LNCS, vol. 4546, pp. 445–464. Springer, Heidelberg (2007)
21. Wolper, P., Leroy, D.: Reliable Hashing without Collision Detection. In: Courcoubetis, C. (ed.) CAV 1993. LNCS, vol. 697, pp. 59–70. Springer, Heidelberg (1993)

Comparison of Different Algorithms to Synthesize a Petri Net from a Partial Language

Robin Bergenthum, Jörg Desel, and Sebastian Mauser

Department of Applied Computer Science, Catholic University of Eichstätt-Ingolstadt
firstname.lastname@ku-eichstaett.de

Abstract. In this paper we present two new algorithms that effectively synthesize a finite place/transition Petri net (p/t-net) from a finite set of labeled partial orders (a finite partial language). Either the synthesized p/t-net has exactly the non-sequential behavior specified by the partial language, or there is no such p/t-net. The first algorithm is an improved version of a synthesis algorithm presented in [14], which uses the classical theory of regions applied to the set of step sequences generated by the given partial language. Instead of computing all step sequences, the new algorithm directly works on appropriate prefixes specified in the partial language. The second algorithm is based on the theory of token flow regions for partial languages developed in [16,15,14]. While in [15,14] a so called basis representation is applied, the new algorithm combines the concepts of separation representation and token flows. We implemented both synthesis algorithms in our framework VipTool. A comparison of the two new algorithms with the two predecessor algorithms presented in [14,15] shows that both perform better than their respective predecessor. Therefore, a detailed comparison of these two synthesis algorithms is presented.

Keywords: Petri Net, Synthesis Algorithm, Partial Order Behavior, Region Theory.

1 Introduction

Synthesis of Petri nets from behavioral descriptions has been a successful line of research since the 1990s. There is a rich body of nontrivial theoretical results, and there are important applications in industry, in particular in hardware system design [5,11], and recently also in workflow design [21]. Moreover, there are several synthesis tools that are based on the theoretical results. The most prominent one is Petrify [4].

Originally, synthesis meant algorithmic construction of an unlabeled Petri net from sequential observations. It can be applied to various classes of Petri nets, including elementary nets [7] and place/transition nets (p/t-nets) [1]. Synthesis can start with a transition system representing the sequential behavior of a system or with a step transition system which additionally represents steps of concurrent events [1]. Synthesis can also be based on a language (originally a set of occurrence sequences or step sequences [6]). The *synthesis problem* is to decide whether, for a given behavioral specification, there exists an unlabeled Petri net of the respective class such that the behavior of this net coincides with the specified behavior.

K. Jensen, J. Billington, and M. Koutny (Eds.): ToPNoC III, LNCS 5800, pp. 216–243, 2009.

In [14,15], together with R. Lorenz, we developed two algorithms to solve the synthesis problem for p/t-nets where the behavior is given in terms of a finite partial language L, i.e. as a finite set of labeled partial orders (LPOs – also known as *partial words* [8] or *pomsets* [18]). Partial order semantics of Petri nets truly represents the concurrency of events and is often considered the most appropriate representation of behavior of Petri net models. Both algorithms apply the so called *theory of regions* in the setting of partial languages. A region defines a possible place of the synthesized net structurally w.r.t. the given specification.

All approaches to Petri net synthesis based on regions roughly follow the same idea:

– Instead of solving the synthesis problem (is there a net with the specified behavior?) and then – in the positive case – synthesizing the net, first a net is constructed from the given specification.
– The construction starts with the transitions taken from the behavioral specification.
– The behavior will be restricted by the addition of places and according initial markings and arcs.
– Each region yields a corresponding place in the constructed net. A region is defined in such a way that the behavior of the net with its corresponding place still includes the specified behavior.
– When all, or sufficiently many, regions are identified, all places of the synthesized net are constructed.
– If the behavior of the synthesized net coincides with the specified behavior, then the synthesis problem has a positive solution; otherwise there is no Petri net with the specified behavior and therefore the synthesis problem has a negative solution.
– In the latter case the synthesized net still includes all specified behavior, i.e. it represents an upper approximation of the specified behavior.

Each region yields a place. The crucial point is that the set of all regions is infinite in our setting of p/t-nets (e.g. for each place with no outgoing arcs there is a region defining this place). The aim is to find a finite set of regions which suffices to represent all relevant dependencies. In region theory there are basically two different approaches to calculate such a finite set of regions, called *basis representation* and *separation representation*. Both methods can be adapted to various region definitions as described in [17].

The two existing algorithms presented in [14,15] to synthesize a p/t-net from a finite partial language use the following principles: The first one, Alg. I, uses a region definition based on a translation of the given partial language into a language of step sequences, namely so called *transition regions* of the set of step sequences generated by the partial language are applied (called step transition regions). This approach is combined with the principle of separation representation. The second algorithm, Alg. II, uses a region definition given directly on the partial language, so called *token flow regions*, combined with the basis representation approach. We implemented both algorithms as plug-ins of our toolset VipTool [3] and compared their runtime in different small examples [14]. The first algorithm, Alg. I, using the transition regions is faster than Alg. II in examples which do not contain a lot of concurrency. The separation representation used in this algorithm leads to solution nets containing far less places than applying the basis representation used in the second approach. In examples containing a lot of concurrency, the translation of the partial language into a language of step sequences is too costly. In these examples, Alg. II, defining regions directly on the given partial language, is superior to Alg. I, but using the basis representation approach seems to worsen the overall runtime. Altogether, computing the complete basis seems to be inefficient so that in a practical approach only the separation representation is applicable. While it turned out that tokenflow regions are superior to transition regions in settings containing a lot of concurrency, it seems that transition regions are very fast in cases of few concurrency.

The aim of this paper is to deduce two new algorithms from the experiences of the results in [14]. The first one, Alg. III, is an advanced version of Alg. I using so called *LPO transition regions*. LPO transition regions are very similar to the transition regions of Alg. I, but it is not necessary to translate the given partial language into a language of step sequences. Instead, appropriate prefixes of the partial language are considered. Therefore, this new algorithm performs better than Alg I. This leads to a very fast algorithm solving the synthesis problem in the case of a partial language containing not too much concurrency. But if there is a lot of concurrency, the number of prefixes is very high making this algorithm problematic in runtime. The second new algorithm, Alg. IV, uses the token flow regions known from Alg II. The difference is that we now use the principle of separation representation, which proved to be superior to the basis representation (applied in Alg. II) in practical use. This leads to an algorithm which performs better than Alg. II. The algorithm is a very fast algorithm solving the synthesis problem in the case of a partial language containing much concurrency. The performance of the algorithm worsens if the partial language exhibits less concurrency.

Figure 1 shows a systematic classification of the algorithms discussed above. The two empty cells are not considered here, because the basis representation was in general inferior to the separation representation.

We implemented both new algorithms Alg. III and Alg. IV in VipTool [3] and performed experimental tests. As expected from complexity considerations, each of the two new algorithms is superior to one of the algorithms presented in [14,15] (Alg. I resp. Alg II). Thus, they are the most promising algorithms to synthesize a p/t-net from a finite partial language developed so far. To examine their relation in detail, we present a complexity comparison and experimental results covering both toy examples and

	Transition regions using step sequences	Transition regions using prefixes	Token flow regions
Separating rep.	Alg. I [14]	Alg. III	Alg. IV
Basis rep.	-	-	Alg. II [14,15]

Fig. 1. Overview of algorithms to synthesize a Petri net from a finite partial language

realistic settings. While the main focus is on performance we also discuss the size, i.e. the number of places, of the generated nets. Since both algorithms apply separation representation, this can be seen as an unbiased comparison between the concept of transition regions and the concept of token flow regions.

The remainder of the paper is organized as follows: We start with a brief introduction to p/t-nets, LPOs, partial languages and enabled LPOs in Section 2. In Section 3 we consider the problem of synthesizing a p/t-net from a given partial language. In Section 4 we first describe transition regions as developed in [14] and second improve those transition regions to a new advanced region definition of so called LPO transition regions. The definition of LPO transition regions leads to a new algorithm (Alg. III) solving the synthesis problem. In Section 5 we recall the definitions and main results from [16] about tokenflow regions. We develop a second new algorithm (Alg. IV) to solve the synthesis problem using token flow regions as described in [14,15] together with separation representation. Section 6 shows a comparison of the two new synthesis algorithms supported by experimental results. Finally, in Section 7 tool support is presented and practical applicability is discussed.

2 Preliminaries

In this section we introduce the definitions of *place/transition nets* (p/t-nets), *labeled partial orders* (LPOs), *partial languages* and LPOs *enabled w.r.t. p/t-nets*. We start with basic mathematical notations: by $\mathbb{N}$ we denote the *nonnegative integers*. $\mathbb{N}^+$ denotes the positive integers. The set of all *multi-sets* over a set A is the set $\mathbb{N}^A$ of all functions $f :$ $A \to \mathbb{N}$. Addition $+$ on multi-sets is defined as usual by $(m + m')(a) = m(a) + m'(a)$. We consider a componentwise $\leq$-relation on multi-sets and write $m < m'$ in the case $m \leq m'$, $m \neq m'$. We also write $\sum_{a \in A} m(a)a$ to denote a multi-set m over A and $a \in m$ to denote $m(a) > 0$.

Definition 1 (Place/transition net). *A place/transition-net (p/t-net) N is a quadruple (P, T, F, W), where P is a (possibly infinite) set of places, T is a finite set of transitions satisfying $P \cap T = \emptyset$, $F \subseteq (P \times T) \cup (T \times P)$ is the flow relation and $W : F \to \mathbb{N}^+$ is a weight function.*

We extend the weight function W. For pairs of net elements $(x, y) \in (P \times T) \cup (T \times P)$ with $(x, y) \notin F$ we define $W(x, y) = 0$. A *marking* of a p/t-net $N = (P, T, F, W)$ is a multi-set $m : P \to \mathbb{N}$ assigning $m(p)$ tokens to a place $p \in P$. A *marked p/t-net* is a pair (N, m_0), where N is a p/t-net, and m_0 is a marking of N, called *initial marking*. Figure 2 shows a marked p/t-net (N, m_0). Places are drawn as circles including tokens representing the initial marking, transitions as rectangles and the flow relation as arcs annotated by the values of the weight function (the weight 1 is not shown).

A multi-set of transitions $\tau \in \mathbb{N}^T$ is called a *step* (of transitions). A step τ is *enabled to occur* (concurrently) in a marking m if and only if $m(p) \geq \sum_{t \in \tau} \tau(t) W(p, t)$ for each place $p \in P$. In this case, its occurrence leads to the marking $m'(p) = m(p) + \sum_{t \in \tau} \tau(t)(W(t, p) - W(p, t))$. We write $m \xrightarrow{\tau} m'$ to denote that τ is enabled to occur in m and that its occurrence leads to m'. A finite sequence of steps $\sigma = \tau_1 \ldots \tau_n$, $n \in \mathbb{N}^+$, is called a *step occurrence sequence enabled in a marking m and leading to m_n*,

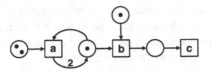

Fig. 2. A marked p/t-net (N, m_0)

denoted by $m \xrightarrow{\sigma} m_n$, if there exists a sequence of markings $m_1, \ldots, m_n$ such that $m \xrightarrow{\tau_1} m_1 \xrightarrow{\tau_2} \ldots \xrightarrow{\tau_n} m_n$. In the marked p/t-net (N, m_0) from Figure 2 only the step a and the step b are enabled to occur in the initial marking. In the marking reached after the occurrence of a, the step $a + b$ is enabled to occur. Steps containing transition c are only enabled after the occurrence of transition b.

We use partial orders with nodes (called events) *labeled by transition names* to specify scenarios describing the behavior of Petri nets.

Definition 2 (Labeled partial order). *A finite labeled partial order (LPO) is a triple $lpo = (V, <, l)$, where V is a finite set of events, $< \subseteq V \times V$ is an irreflexive and transitive relation over V, called the* set of edges *and $l : V \to T$ is a labeling function with set of labels T.*

Two events $v, v' \in V$ of an LPO $(V, <, l)$ are called *independent* if $v \neq v'$, $v \not< v'$ and $v' \not< v$. By $co \subseteq V \times V$ we denote the set of all pairs of independent events of V. A *co-set* is a subset $C \subseteq V$ fulfilling: $\forall v, v' \in C, v \neq v' : v \, co \, v'$. A *cut* is a maximal co-set. For a co-set C of an LPO $(V, <, l)$ and an event $v \in V \setminus C$ we write $v < (>) C$, if $v < (>) s$ for an element $s \in C$. We write $v \, co \, C$, if $v \, co \, s$ for all elements $s \in C$. A partial order $(V', <', l')$ is a *prefix* of another partial order $(V, <, l)$ if $V' \subseteq V$, $(v' \in V' \wedge v < v') \implies (v \in V')$, $<' = < \cap V' \times V'$ and $l' = l|_{V'}$. If $(V', <', l')$ is a strict prefix, i.e. $V' \neq V$, then we write $(V', <', l') \prec (V, <, l)$. The prefix relation $\prec$ defines a partial order on the set of prefixes of a given LPO. Given a co-set C of $(V, <, l)$, a *prefix of C* is a prefix $(V', <', l')$ of $(V, <, l)$ fulfilling $\{v \in V \mid v < C\} \subseteq V'$ and $C \cap V' = \emptyset$. Note that a co-set may have many prefixes. In particular, any prefix is a prefix of the empty set, any prefix of a co-set is also a prefix of any subset of the latter and for any co-set there is a unique minimal prefix. A cut has only one prefix.

An LPO $lpo = (V, <, l)$ is called *stepwise linear* if $co \cup id_V$ (id_V is the identity relation on V) is transitive. Given LPOs $lpo_1 = (V, <_1, l)$ and $lpo_2 = (V, <_2, l)$, lpo_2 is a *sequentialization* of lpo_1 if $<_1 \subseteq <_2$. If lpo_2 is stepwise linear, it is called *step linearization* of lpo_1.

As an example consider lpo_2 in Figure 3. It is not stepwise linear. Adding an arc between the second a-labeled and the c-labeled event yields a step linearization of lpo_2. Omitting the c-labeled event yields a prefix of lpo_2 which is a prefix of the co-set consisting only of the c-labeled event.

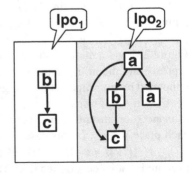

Fig. 3. A partial language.

We consider LPOs only up to isomorphism. Two LPOs $(V, <, l)$ and $(V', <', l')$ are called *isomorphic*, if there is a bijective mapping $\psi : V \to V'$ such that $l(v) = l'(\psi(v))$ for $v \in V$, and $v < w \iff \psi(v) <' \psi(w)$ for $v, w \in V$.

Definition 3 (Partial language). *Let T be a finite set. A set $L \subseteq \{\text{lpo} \mid \text{lpo is an LPO}$ with set of labels $T\}$ is called* partial language *over T.*

We always assume that each label from T occurs in a partial language over T. Figure 3 shows a partial language given by the set of LPOs $L = \{\text{lpo}_1, \text{lpo}_2\}$, which we will use as a running example.

There are different equivalent formal notions of runs of p/t-nets defining non-sequential semantics [12,23,14]. We only give the notion of enabled LPOs here: an LPO is enabled w.r.t. a marked p/t-net if, for each cut of the LPO, the marking reached by firing all transitions corresponding to events smaller than the cut enables the step (of transitions) given by the cut.

Definition 4 (Enabled LPO). *Let (N, m_0) be a marked p/t-net, $N = (P, T, F, W)$. An LPO* lpo $= (V, <, l)$ *with $l : V \to T$ is called* enabled (to occur) *in (N, m_0) if $m_0(p) + \sum_{v \in V \wedge v < C}(W(l(v), p) - W(p, l(v))) \geq \sum_{v \in C} W(p, l(v))$ for every cut C of* lpo *and every $p \in P$. Its occurrence leads to the final marking m' given by $m'(p) = m_0(p) + \sum_{v \in V}(W(l(v), p) - W(p, l(v)))$.*

Enabled LPOs are also called runs. *The set of all LPOs enabled in (N, m_0) is denoted by $\mathfrak{Lpo}(N, m_0)$. $\mathfrak{Lpo}(N, m_0)$ is called the* partial language of runs *of (N, m_0).*

There is an equivalent characterization of enabledness using step sequences and their correspondence to stepwise linear LPOs [14]: a stepwise linear LPO lpo' $= (V, <', l)$ can be represented by the step sequence $\sigma_{lpo'} = \tau_1 \dots \tau_n$ defined by $V = V_1 \cup \dots \cup V_n$, $<' = \bigcup_{i<j} V_i \times V_j$ and $\tau_i(t) = |\{v \in V_i \mid l(v) = t\}|$. An LPO lpo $= (V, <, l)$ is enabled in (N, m_0) if and only if, for each step linearization $lpo' = (V, <', l)$ of lpo, the step sequence $\sigma_{lpo'}$ is enabled in (N, m_0).

Observe that $\mathfrak{Lpo}(N, m_0)$ is always sequentialization and prefix closed, i.e. every sequentialization and every prefix of an enabled LPO is again enabled w.r.t. (N, m_0). Moreover, the set of labels of $\mathfrak{Lpo}(N, m_0)$ is always finite. Therefore, when specifying the non-sequential behavior of a searched p/t-net by a partial language, this partial language must necessarily be sequentialization and prefix closed and must have a finite set of labels. On the contrary, a user typically specifies whatever LPOs he likes, in particular he will not specify all prefixes and sequentializations but often mostly LPOs with maximal length and minimal dependencies. Such set of concrete LPOs L (which can neither be assumed to be sequentialization and prefix closed nor to include no sequentializations or prefixes) then defines a partial language $\mathcal{L}$ which emerges by adding all prefixes of sequentializations of LPOs in L. While for algorithmic purposes we have to deal with the (arbitrary) user specification L, to define the synthesis problem we consider $\mathcal{L}$. In this context, the partial language L in Figure 3 specifies the non-sequential behavior of a searched p/t-net by extending it to its prefix and sequentialization closure $\mathcal{L}$. Both LPOs shown in this figure are enabled w.r.t. the marked p/t-net (N, m_0) shown in Figure 2. It can be verified that (N, m_0) solves the synthesis problem w.r.t. L, which means $\mathfrak{Lpo}(N, m_0) = \mathcal{L}$.

3   Synthesis of P/T-Nets

We consider the problem of synthesizing a p/t-net from a finite partial language L specifying its non-sequential behavior.

Definition 5 (Synthesis problem)
Given: *A finite partial language L.*
Searched: *A marked p/t-net* (N, m_0) *fulfilling* $\mathfrak{Lpo}(N, m_0) = \mathcal{L}$ $(\mathcal{L} = \{lpo \mid lpo$ *is a prefix of a sequentialization of an LPO in L}) if such a net exists.*

We develop two different algorithms (Alg. III, Alg. IV) to compute a marked p/t-net (N, m_0) from a given set of LPOs L such that the prefix and sequentialization closure $\mathcal{L}$ of L satisfies $\mathcal{L} = \mathfrak{Lpo}(N, m_0)$ if this is possible. The idea to construct such a net (N, m_0) is as follows: The set of transitions of the searched net is given by the finite set of labels of L. Then, each LPO in L is enabled w.r.t. the marked p/t-net consisting only of these transitions (having an empty set of places). We then restrict the behavior of this net by adding places. Each place is defined by its initial marking and the weights on the arcs connecting it to each transition (Figure 4).

Since the specified behavior given by the partial language L should be included in $\mathfrak{Lpo}(N, m_0)$, we only add places which do not exclude a specified behavior. Thus, we distinguish two kinds of places: In the case that there is an LPO in L which is not a run of the corresponding "one place"-net, this place restricts the behavior too much. Such a place is *non-feasible*. In the other case, the considered place is *feasible*.

Definition 6 (Feasible place). *Let L be a partial language over the finite set of labels T and let* (N_p, m_p), $N_p = (\{p\}, T, F_p, W_p)$ *be a marked p/t-net with only one place p.* (N_p, m_p) *is called associated to p. The place p is called* feasible *(w.r.t. L), if* $L \subseteq \mathfrak{Lpo}(N_p, m_p)$, *otherwise* non-feasible *(w.r.t. L).*

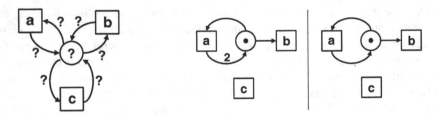

Fig. 4. An unknown place of a p/t-net   **Fig. 5.** Left: a feasible place. Right: a place which is not feasible.

Figure 5 shows on the left side a place which is feasible w.r.t. the partial language L in Figure 3. This is because, after the occurrence of a, the place is marked by two tokens. In this marking the step $a + b$ is enabled to occur (as specified by lpo_2). The place shown on the right side is non-feasible, because, after the occurrence of a, the place is again marked by only one token. In this marking the step $a + b$ is not enabled to occur. Thus, lpo_2 is not enabled w.r.t. the one-place-net shown on the right side.

Adding only feasible places to the set of transitions given by the labels of a partial language L results in a p/t-net $(N, m_0) = (P, T, F, W, m_0)$ for which holds $\mathcal{L} \subseteq \mathfrak{Lpo}(N, m_0)$ because $\mathfrak{Lpo}(N, m_0) = \bigcap_{p \in P} \mathfrak{Lpo}(N_p, m_p)$ (if $P \neq \emptyset$). If (N, m_0) includes any non-feasible place, $\mathcal{L} = \mathfrak{Lpo}(N, m_0)$ is not possible. Adding places reduces $\mathfrak{Lpo}(N, m_0)$. Therefore, considering the infinite set of all feasible places leads to a p/t-net which is a positive solution to the synthesis problem or there is no positive solution. The first problem hereby is to identify feasible places. Feasible places can be found through so called regions. A region is a function defined on the structure of the language fulfilling certain properties. Every region corresponds to a feasible place. In this paper we discuss two different types of regions of partial languages and their algorithmic applicability. The second problem is that there are always infinitely many feasible places. But we have shown in [15] that they can be represented by a finite set of basis feasible places (basis representation [17]). Instead of considering a basis, in this paper we use another approach to tackle the problem of computing an appropriate finite set of feasible places. We apply the principle called separation representation [17] to select feasible places. The idea is to represent non-specified behavior through a finite set of so called *wrong continuations*. If there exists a positive solution of the synthesis problem, it is possible to exclude all wrong continuations through a finite set of feasible places. In this case the resulting net is a solution of the synthesis problem.

Note that if all wrong continuations can be excluded through a finite set of feasible places, we still cannot conclude that there is a positive solution of the synthesis problem. In [14] it is explained that the partial language shown later on in Figure 7 is an example exhibiting this problem. For an actual decision whether there is a positive solution to the synthesis problem an additional equality test whether the synthesized net has the specified behavior as shown in [14] is necessary. This is out of scope of the paper and therefore we abstracted from this question in our formulation of the synthesis problem in Definition 5.

4 LPO Transition Regions

A naive approach to synthesize a p/t-net from a finite partial language L is to consider the set of step sequences L' generated by the LPOs in L (see Figure 6 left column). Following ideas of [10], where regions of trace languages are defined, it is possible to define regions of languages of step sequences [14].

A region of a language of step sequences L' can simply be defined as a tuple of natural numbers which represents the initial marking of a place and the number of tokens each transition consumes, respectively produces, in that place, satisfying some property which ensures that no step sequence of the given language L' is prohibited by this place. The set of regions of L' defines the set of feasible places of L' [14]. Such regions, which are directly given by the parameters of a place, are subsumed under the region type of transition regions [17]. The regions of a language of step sequences considered here are called step transition regions.

Maximal step sequences	Maximal right continuations	Wrong continuations			
bc	(∅, a)	(∅, 2a)	(b, a)	(a+b, b)	(2a+b, 2c)
aabc	(∅, b)	(∅, a+b)	(b, b)	(a+b, 2c)	(a+b+c, 2a)
abac	(a, a+b)	(∅, 2b)	(b, 2c)	(b+c, a)	(a+b+c, b)
abca	(b, c)	(∅, c)	(2a, a)	(b+c, b)	(a+b+c, c)
a(a+b)c	(a+b, a+c)	(a, 2a)	(2a, 2b)	(b+c, c)	(2a+b+c, a)
ab(a+c)		(a, 2b)	(2a, c)	(2a+b, a)	(2a+b+c, b)
		(a, c)	(a+b, 2a)	(2a+b, b)	(2a+b+c, c)

Fig. 6. Sets of maximal step sequences, maximal right continuations and wrong continuations generated by the partial language depicted in Figure 3

Definition 7 (Step transition region). *Denoting* $T = \{t_1, \ldots, t_m\}$ *the transitions occurring in* L', *a step transition region of* L' *is a tuple* $\mathbf{r} = (r_0, \ldots, r_{2m}) \in \mathbb{N}^{2m+1}$ *satisfying for every* $\sigma = \tau_1 \ldots \tau_n \in L'$ *and every* $j \in \{1, \ldots, n\}$:

$$(i) \qquad r_0 + \sum_{i=1}^{m}((\tau_1 + \ldots + \tau_{j-1})(t_i) \cdot r_i - (\tau_1 + \ldots + \tau_j)(t_i) \cdot r_{m+i}) \geq 0.$$

Every step transition region $\mathbf{r}$ *of* L' *defines a place* p_r *via* $m_0(p_r) := r_0$, $W(t_i, p_r) := r_i$ *and* $W(p_r, t_i) := r_{m+i}$ *for* $1 \leq i \leq m$. *The place* p_r *is called the* corresponding place to $\mathbf{r}$.

We use this definition as a basis of an advanced approach. Calculating all step sequences from a given LPO is very costly. Furthermore, the step transition region definition leads to a lot of needless constraints. Many of them are equal to or less restrictive than others. In particular, given a step sequence $\sigma = \tau_1 \ldots \tau_j$, a different partitioning of the steps $\tau_1 \ldots \tau_{(j-1)}$ does not change the corresponding inequality (i). Thus, it is sufficient to only count the number of transition occurrences in the steps $\tau_1 \ldots \tau_{(j-1)}$ to formulate the corresponding inequality. Given a step sequence $\tau_1 \ldots \tau_j$, we call the sum $\pi = \sum_{i=1}^{j-1} \tau_i$ a *prefix step*. Given a finite partial language L, we consider the set of all possible prefix steps of L which can be computed as the set of all Parikh vectors corresponding to prefixes of L. Furthermore, sequentializations of steps yield less restrictive and thus redundant constraints. Given a partition of a step τ_j into l steps $\tau_j = \sum_{i=1}^{l} \tau_j^i$, a solution of the inequalities arising from the step sequence $\tau_1 \ldots \tau_n$ according to Definition 7 also solves all inequalities arising from $\tau_1 \ldots \tau_{(j-1)} \tau_j^1 \ldots \tau_j^l \tau_{(j+1)} \ldots \tau_n$. That means, given a prefix step π and a step τ_j occurring after π within a partial language L, it is not necessary to consider a prefix step $\pi' \geq \pi$ together with a step τ_j' occurring after π' if $\tau_j' + (\pi' - \pi) \leq \tau_j$. For illustration of the latter inequality, we consider the net of the running example (Figure 2). In this net the step $\tau_j = a + b$ is enabled after the prefix step $\pi = a$. If $\pi' = \pi$, the inequality only states that all smaller steps, i.e. $\tau_j' = a$ or $\tau_j' = b$ (or $\tau_j' = \emptyset$), are also enabled after π. If $\pi' = 2a$ (as an example for

$\pi' > \pi$) the inequality states that $\tau'_j = b$ is enabled after π'. Note especially that it is sufficient to consider the occurrence of steps τ_j corresponding to cuts of LPOs in the partial language, but some cuts within the partial language may be redundant.

Definition 8 (Prefix step, right continuation, maximal right continuation). *Given a partial language L with set of labels T, the set $\Pi_L = \{\pi \in \mathbb{N}^T \mid \exists \, lpo \in L, lpo' = (V', <', l')$ prefix of lpo : $\pi(t) = |\{v \in V' \mid l'(v) = t\}|$ for all $t \in T\}$ contains all multi-sets corresponding to the numbers of transition occurrences in a prefix of an LPO in L. $\pi \in \Pi_L$ is called a* prefix step *of L.*

Given a prefix step π, a multi-set $\tau \in \mathbb{N}^T$ is called a step after π, *if there exists an LPO $lpo = (V, <, l) \in L$, a co-set $C \subseteq V$ and a prefix $(V', <', l')$ of C for which holds: $\forall \, t \in T \, (\tau(t) = |\{v \in C \mid l(v) = t\}| \ \wedge \ \pi(t) = |\{v' \in V' \mid l(v') = t\}|)$. If C is a cut, then τ is called* cut after π *(note that given some π there is not necessarily some cut after π).*

We denote $L_{co}^{\Pi} = \{(\pi, \tau) \mid \pi \in \Pi_L, \tau$ a step after $\pi\}$. $(\pi, \tau) \in L_{co}^{\Pi}$ is called a right continuation *of L.*

We define a partial order $\ll$ on $\mathbb{N}^T \times \mathbb{N}^T$ as follows: Given $(\pi, \tau), (\pi', \tau') \in \mathbb{N}^T \times \mathbb{N}^T$, then $(\pi, \tau) \ll (\pi', \tau')$ if and only if $(\pi, \tau) \neq (\pi', \tau')$, $\pi' \le \pi$ and $\tau + (\pi - \pi') \le \tau'$.

We denote the set of all pairs of a prefix step of L together with a cut after this prefix step as $L^{\Pi} = \{(\pi, \tau) \mid \pi \in \Pi_L, \tau$ a cut after $\pi\}$. The set $L_{max}^{\Pi} = \{(\pi, \tau) \in L^{\Pi} \mid \forall (\pi', \tau') \in L^{\Pi} : (\pi, \tau) \not\ll (\pi', \tau')\}$ is the set of all right continuations, given by cuts, which are maximal w.r.t. $\ll$. $(\pi, \tau) \in L_{max}^{\Pi}$ is called a maximal right continuation *of L.*

Note that a right continuation which is not given by a cut is never maximal w.r.t. $\ll$, i.e. $L_{max}^{\Pi} = \{(\pi, \tau) \in L_{co}^{\Pi} \mid \forall (\pi', \tau') \in L_{co}^{\Pi} : (\pi, \tau) \not\ll (\pi', \tau')\}$. In the running example (see Figure 3) we have $\Pi_L = \{\emptyset, a, b, a+b, 2a, b+c, 2a+b, a+b+c, 2a+b+c\}$, $L_{co}^{\Pi} = \{(\emptyset, \emptyset), (a, \emptyset), (b, \emptyset), (a+b, \emptyset), (2a, \emptyset), (b+c, \emptyset), (2a+b, \emptyset), (a+b+c, \emptyset), (2a+b+c, \emptyset), (\emptyset, a), (\emptyset, b), (b, c), (a, a), (a, b), (2a, b), (a+b, a), (a+b, c), (2a+b, c), (a+b+c, a), (a+b, a+c), (a, a+b)\}$ and $L^{\Pi} = L_{max}^{\Pi} = \{(\emptyset, a), (\emptyset, b), (a, a+b), (b, c), (a+b, a+c)\}$ (see also Figure 6).

Our aim is to construct feasible places. For this purpose, roughly speaking, each cut of an LPO in L has to be enabled after the occurrence of its prefix. Each cut together with its prefix is a candidate to define a maximal right continuation. The set of maximal right continuations, which is the basis of the following region definition, can easily be calculated from the cuts of the LPOs in L. In the

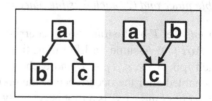

Fig. 7. Another partial language

running example, each cut of lpo_1 and lpo_2 corresponds to a maximal right continuation. Figure 7 shows an example partial language, where this is not the case. The cut in the first LPO consisting only of the a-labeled event and having an empty prefix can be neglected, because in the second LPO there is a cut containing an a-labeled and a b-labeled event and having an empty prefix $((\emptyset, a) \ll (\emptyset, a+b))$. The cut in the second LPO consisting only of the c-labeled event and defining the prefix step $a + b$ can be neglected, because in the first LPO there is a cut containing a b- and a c-labeled event and defining a prefix step containing only a $((a + b, c) \ll (a, b + c))$. Thus, although this partial language has four cuts, there are only two maximal right continuations.

Following these ideas leads to the definition of *LPO transition regions*.

Definition 9 (LPO transition region). *Let L be a partial language and denote $T = \{t_1, \ldots, t_m\}$ the transitions occurring in L. An LPO transition region of L is a tuple $\mathbf{r} = (r_0, \ldots, r_{2m}) \in \mathbb{N}^{2m+1}$ satisfying for every maximal right continuation (π, τ) of L:*

$$(ii) \qquad r_0 + \sum_{i=1}^{m} ((\pi)(t_i) \cdot r_i - (\pi + \tau)(t_i) \cdot r_{m+i}) \geq 0.$$

Like step transition regions, every LPO transition region $\mathbf{r}$ of L defines a place p_r via $m_0(p_r) := r_0$, $W(t_i, p_r) := r_i$ and $W(p_r, t_i) := r_{m+i}$ for $1 \leqslant i \leqslant m$. Again, the place p_r is called the corresponding *place to $\mathbf{r}$.*

Remark 1. As mentioned, it can be shown that if $(\pi, \tau) \ll (\pi', \tau')$, then whenever (π', τ') is enabled in some net this is also the case for (π, τ). This implication justifies that we only have to consider $\ll$-maximal right continuations in the definition of regions. The motivation for considering $\ll$ came from the idea that sequentializations and prefixes do not have to be regarded. We can show that if lpo_1 is a prefix of a sequentialization of lpo_2 and $(\pi, \tau) \in \{\text{lpo}_1\}^{\Pi}$, then $(\pi, \tau) \in \{\text{lpo}_2\}^{\Pi}$ or there is $(\pi', \tau') \in \{\text{lpo}_2\}^{\Pi}$ with $(\pi, \tau) \ll (\pi', \tau')$. Therefore, the $\ll$-relation tackles the problem that L might include sequentializations and prefixes. We allow that L contains sequentializations and prefixes because searching and removing them is too costly. But the $\ll$-relation also helps to neglect redundant constraints in many other cases as shown in the example of Figure 7. Of course, still the set of maximal right continuations may generate some redundant constraints. It is possible to apply further procedures to remove such constraints, which is a topic of further fine tuning the region definition and the resulting synthesis algorithm.

Theorem 1. *Given a partial language L, (1) each LPO transition region defines a feasible place and (2) each feasible place is defined by an LPO transition region.*

Proof. Let T be the transitions occurring in L.

Part (1): Assume an LPO transition region $\mathbf{r}$ and p_r its corresponding place. For each $\text{lpo} = (V, <, l) \in L$ and each cut C of lpo, we show that the step given by C is enabled after the occurrence of the prefix of C in the respective "one place"-net. To each cut C there exists a (not necessarily unique) maximal right continuation (π, τ) of L such that the multi-set of transitions given by the prefix of C is larger than π and the multi-set given by C together with the difference of the multi-set given by the prefix of C and π is smaller than or equal to τ. In the case the cut and its prefix correspond to a maximal right continuation (π_C, τ_C), we can choose $\pi = \pi_C$ and $\tau = \tau_C$. If this is not the case the existence of such (π, τ) can be followed from the definition of L_{max}^{Π} by the transitivity of $\ll$ and the finiteness of L^{Π}. Formally, the chosen (π, τ) fulfills: $\forall t \in T$ $(\pi(t) \leq |\{v \in V| \ l(v) = t, \ \exists c \in C : v < c\}| \ \wedge \ \tau(t) \geq |\{v \in C| \ l(v) = t\}| + |\{v \in V| \ l(v) = t, \ \exists c \in C : v < c\}| - \pi(t))$. $\mathbf{r}$ is an LPO transition region and fulfills (ii): $r_0 + \sum_{t \in T} ((\pi)(t) \cdot r_i - (\pi + \tau)(t) \cdot r_{m+i}) \geq 0$.

Since p_r corresponds to $\mathbf{r}$, we first express the left side of the inequality using the arc weights and the initial marking of p_r. Then, we replace π and τ by summing over the actual events of C and its prefix. For this purpose we use the above two inequalities:
$r_0 + \sum_{t \in T}((\pi)(t) \cdot r_i - (\pi+\tau)(t) \cdot r_{m+i}) = m_0(p_r) + \sum_{t \in T}((\pi)(t) \cdot W(t, p_r) - (\pi + \tau)(t) \cdot W(p_r, t)) \leq m_0(p_r) + \sum_{v \in V \wedge v < C} W(l(v), p_r) - \sum_{v \in V \wedge v < C} W(p_r, l(v)) - \sum_{v \in C} W(p_r, l(v))$. It holds: $m_0(p_r) + \sum_{v \in V \wedge v < C} W(l(v), p_r) - \sum_{v \in V \wedge v < C} W(p_r, l(v)) \geq \sum_{v \in C} W(p_r, l(v))$. Thus, each lpo $\in L$ is enabled w.r.t. p_r. We conclude that p_r is feasible w.r.t. L.

Part (2): Given a feasible place p, define a tuple $\mathbf{r}$ by $r_0 := m_0(p)$, $r_i := W(t_i, p)$ and $r_{m+i} := W(p, t_i)$ for $1 \leq i \leq m$. For each maximal right continuation (π, τ) there exists an LPO lpo $\in L$ and a cut C of lpo fulfilling $\forall t \in T$ $(\tau(t) = |\{v \in C|\ l(v) = t\}| \wedge \pi(t) = |\{v \in V|\ l(v) = t, \exists . c \in C : v < c\}|)$. Since lpo is enabled w.r.t. the "one place"-net containing only the place p, it holds: $m_0(p) + \sum_{v \in V \wedge v < C} W(l(v), p) - \sum_{v \in V \wedge v < C} W(p, l(v)) - \sum_{v \in C} W(p, l(v)) \geq 0$ and $m_0(p) + \sum_{v \in V \wedge v < C} W(l(v), p) - \sum_{v \in V \wedge v < C} W(p, l(v)) - \sum_{v \in C} W(p, l(v)) = r_0 + \sum_{t \in T}((\pi)(t) \cdot r_i - (\pi+\tau)(t) \cdot r_{m+i})$ such that $r_0 + \sum_{t \in T}((\pi)(t) \cdot r_i - (\pi + \tau)(t) \cdot r_{m+i}) \geq 0$, i.e. we have (ii). Thus, $\mathbf{r}$ is an LPO transition region defining p. $\qquad\square$

The set of LPO transition regions of L (resp. step transition regions of L') can be characterized as the set of non-negative integral solutions of an homogenous linear inequality system $\mathbf{A}_L^{(ii)} \cdot \mathbf{r} \geq \mathbf{0}$ (resp. $\mathbf{A}_L^{(i)} \cdot \mathbf{r} \geq \mathbf{0}$.). Every inequality (ii) (resp. (i)) as given by Definition 9 (resp. Definition 7) defines a row of the inequality system. By this approach every feasible place can be computed as a solution of the inequality system. This is shown in Theorem 1 (resp. in [14]).

The idea in [1,6] to get an effective synthesis algorithm for a finite language is to prohibit non-specified behavior by feasible places. In the original algorithm (Alg. I) [14] based on Definition 7 we try to calculate a region for each step sequence not specified in L' such that the corresponding place guarantees that this step sequence is not enabled. If $\tau_1 \ldots \tau_n$ is not enabled then also $\tau_1 \ldots \tau_n \tau_{n+1}$ and $\tau_1 \ldots \tau_n'$ with $\tau_n \leq \tau_n'$ are not enabled. Therefore, it is sufficient to consider the finite set of so called wrong continuations as defined in [14] instead of the set of all step sequences not in L'. The same principle holds if we consider prefix steps and steps after prefix steps of a finite partial language. Therefore in Alg. III, we only have to consider all possible prefix steps π and all minimal steps not being a step after π. We call this a *wrong continuation* (see Figure 6 right column):

Definition 10 (Wrong continuation). *The set of wrong continuations of L is defined by* $L_{wrong} = \{(\pi, \tau) \notin L_{co}^{\Pi} \mid \pi \in \Pi_L, \tau \in \mathbb{N}^T, \forall \tau' < \tau : (\pi, \tau') \in L_{co}^{\Pi}\}$.

In order to compute a feasible place which prohibits a wrong continuation (π, τ) of L, one defines so called *separating LPO transition regions* defining such places:

Definition 11 (Separating LPO transition region). *Let* (π, τ) *be a wrong continuation of L. An LPO transition region $\mathbf{r}$ of L is a* separating LPO transition region *w.r.t.* (π, τ) *if*

$$(iii) \qquad r_0 + \sum_{i=1}^{m}(\pi(t_i) \cdot r_i - (\pi + \tau)(t_i) \cdot r_{m+i}) < 0.$$

A separating region $\mathbf{r}$ w.r.t. (π, τ) can be calculated (if it exists) as a non-negative integer solution of an homogenous linear inequality system with integer coefficients of the form

$$\mathbf{A}_L^{(ii)} \cdot \mathbf{r} \geq 0$$
$$\mathbf{b}_{\pi\tau}^{(iii)} \cdot \mathbf{r} < 0.$$

The vector $\mathbf{b}_{\pi\tau}^{(iii)}$ is defined in such a way that $\mathbf{b}_{\pi\tau}^{(iii)} \cdot \mathbf{r} < 0 \Leftrightarrow (iii)$.

If there exists no non-negative integer solution of the system $\mathbf{A}_L^{(ii)} \cdot \mathbf{r} \geq 0$, $\mathbf{b}_{\pi\tau}^{(iii)} \cdot \mathbf{r} < 0$, there exists no separating region w.r.t. (π, τ) and thus no feasible place prohibiting (π, τ). If there exists a non-negative integer solution of the system, any such solution defines a feasible place prohibiting (π, τ). If we choose one arbitrary separating region $\mathbf{r}_{\pi\tau}$ for each wrong continuation (π, τ) for which such a region exist, then we call the finite set of all these regions a *separation representation* (of the set of all regions). A place corresponding to each separating region of the separation representation is added to the synthesized net (N, m_0). Algorithmically, the places are introduced step by step according to a fixed ordering of the wrong continuations. If a wrong continuation is already prohibited by previously introduced places, it is not searched for a respective separating region, i.e. the costly step of computing a solution of the corresponding inequality system can be skipped (as a consequence the synthesized net depends on the chosen ordering of wrong continuations). This yields Alg. III:

```
 1: WC ← L.getWrongContinuations()
 2: (N, m₀) ← (∅, L.getLabels(), ∅, ∅, ∅)
 3: for all (π, τ) ∈ WC do
 4:     if (N, m₀).enables(π, τ) then
 5:         r ← Solver.getIntegerSolution(A_L^(ii) · r ≥ 0, r ≥ 0, b_πτ^(iii) · r < 0)
 6:         if r ≠ null then
 7:             (N, m₀).addCorrespondingPlace(r)
 8:         end if
 9:     end if
10: end for
11: return (N, m₀)
```

Remark 2. Note that for performance issues, it is important to make the matrix $\mathbf{A}_L^{(ii)}$ as small as possible, because throughout the synthesis algorithm a lot of inequality systems of the form $\mathbf{A}_L^{(ii)} \cdot \mathbf{r} \geq 0$, $\mathbf{b}_{\pi\tau}^{(iii)} \cdot \mathbf{r} < 0$ have to be solved. Therefore, we tried to reduce the set of constraints by only considering right continuations which are maximal w.r.t. $\ll$ (see Remark 1). But it is not that important to have a small set WC. This is because testing whether a wrong continuation is prohibited by previously added places is fast. Only very efficient procedures to a-priori reduce the size of the set WC are candidates to improve the overall performance of the synthesis algorithm. Therefore, we here did not apply methods to reduce WC. An example for a method to reduce WC is to only consider minimal wrong continuations w.r.t. $\ll$, which in our running example (see Figure 6) removes wrong continuations such as $(\emptyset, a + b)$ being larger w.r.t. $\ll$ than (b, a). Applying and testing such methods is again a matter of further fine tuning of the synthesis algorithms.

Theorem 2. *There exists a positive solution of the synthesis problem for the finite partial language L if and only if* $\mathfrak{Lpo}(N, m_0) = \mathcal{L}$ *((N, m_0) is a net computed by Alg. III).*

Proof. It is only necessary to prove the *only if*-part. Assume there is a positive solution (N', m_0') of the synthesis problem for the given partial language L, i.e. $\mathfrak{Lpo}(N', m_0') = \mathcal{L}$, but $\mathfrak{Lpo}(N, m_0) \neq \mathcal{L}$. This implies $\mathfrak{Lpo}(N, m_0) \supsetneq \mathcal{L}$, because we already know $\mathfrak{Lpo}(N, m_0) \supseteq \mathcal{L}$ (Theorem 1).

We can distinguish two cases: either the set $\mathfrak{Lpo}(N, m_0)_{co}^{\Pi}$ of right continuations of the set of LPOs enabled in (N, m_0) coincides with L_{co}^{Π} ($= \mathcal{L}_{co}^{\Pi}$) or not. In the case that $\mathfrak{Lpo}(N, m_0)_{co}^{\Pi} \neq L_{co}^{\Pi}$ there exists $(\pi, \tau) \in \mathfrak{Lpo}(N, m_0)_{co}^{\Pi}$ fulfilling $(\pi, \tau) \notin L_{co}^{\Pi}$ (since $\mathfrak{Lpo}(N, m_0) \supseteq \mathcal{L}$). The prefix step π corresponds to an LPO lpo $= (V, <, l)$ (i.e. $\pi(t) = |\{v \in V \mid l(v) = t\}|$) given in $\mathfrak{Lpo}(N, m_0)$. If $\pi \in \Pi_L$ set $(\pi', \tau') := (\pi, \tau)$. If $\pi \notin \Pi_L$ there exists a maximal (w.r.t. the prefix relation $\prec$) prefix lpo$'$ of lpo such that the corresponding prefix step π' is in Π_L. Let τ' be a non-empty step after π' in lpo (which exists, since lpo$'$ is a strict prefix of lpo). Then $(\pi', \tau') \in \mathfrak{Lpo}(N, m_0)_{co}^{\Pi}$ fulfills $(\pi', \tau') \notin L_{co}^{\Pi}$ (by the maximality of lpo$'$). Since $(\pi', \tau') \notin L_{co}^{\Pi}$ and $\pi' \in \Pi_L$, there exists $(\pi', \tau'') \notin L_{co}^{\Pi}$, $\tau'' \leq \tau'$ and $(\pi', \tau'') \in L_{wrong}$. Since $(\pi', \tau') \in \mathfrak{Lpo}(N, m_0)_{co}^{\Pi}$ and $\tau'' \leq \tau'$, it holds $(\pi', \tau'') \in \mathfrak{Lpo}(N, m_0)_{co}^{\Pi}$. Now considering the procedure of the synthesis algorithm, since (π', τ'') is a wrong continuation, the algorithm adds a place corresponding to a separating region, if there exists a separation region w.r.t. (π', τ''). Since $(\pi', \tau'') \in \mathfrak{Lpo}(N, m_0)_{co}^{\Pi}$, the step τ'' is enabled in (N, m_0) after the occurrence of the transitions given in π'. This would not be possible if (N, m_0) contains a place corresponding to a separating region w.r.t. (π', τ''). That means, there exists no separating region excluding (π', τ'') and thus there is no respective feasible place. Consequently, τ'' is enabled after π' in any net containing the behavior given by L. It follows: $(\pi', \tau'') \in \mathfrak{Lpo}(N', m_0')_{co}^{\Pi}$, but $(\pi', \tau'') \notin L_{co}^{\Pi}$. Thus, there is an LPO enabled in (N', m_0') and not given in $\mathcal{L}$, i.e. $\mathfrak{Lpo}(N', m_0') \neq \mathcal{L}$. This is a contradiction to (N', m_0') is a solution of the synthesis problem.

Let the set $\mathfrak{Lpo}(N, m_0)_{co}^{\Pi}$ of right continuations of the set of LPOs enabled in (N, m_0) coincide with L_{co}^{Π} and let lpo $\notin \mathcal{L}$ be enabled in (N, m_0). This means each right continuation $(\pi, \tau) \in \{\text{lpo}\}_{co}^{\Pi}$ given by lpo is enabled in (N, m_0) (firing all transitions in π enables step τ), i.e. $(\pi, \tau) \in \mathfrak{Lpo}(N, m_0)_{co}^{\Pi}$. Since $\mathfrak{Lpo}(N, m_0)_{co}^{\Pi} = L_{co}^{\Pi}$, we conclude $(\pi, \tau) \in L_{co}^{\Pi}$. By assumption we have $L_{co}^{\Pi} = \mathfrak{Lpo}(N', m_0')_{co}^{\Pi}$. Therefore, we get $(\pi, \tau) \in \mathfrak{Lpo}(N', m_0')_{co}^{\Pi}$. That means, all right continuations of lpo are also enabled in (N', m_0'). Thus, also lpo $\notin \mathcal{L}$ is enabled in (N', m_0') and consequently $\mathfrak{Lpo}(N', m_0') \neq \mathcal{L}$. This is a contradiction. $\qquad\square$

As already explained in the introduction, the advanced algorithm (Alg. III) to synthesize a p/t-net from a finite partial language using LPO transition regions exhibits some major improvements compared to the version computing step sequences and using transition regions of step sequences as described in [14] (Alg. I). We implemented Alg. III in our tool VipTool [3] and compared it to Alg. I (optimized by also only considering Parikh images of $\tau_1 \dots \tau_{j-1}$ in the inequality system $\mathbf{A}_L^{(i)} \cdot \mathbf{r} \geq \mathbf{0}$) by experimental tests. As expected, the tests clearly showed that the method using directly LPO transition regions (Alg. III) is superior to the approach generating step sequences first and then using step transition regions (Alg. I). In the tests, Alg. III had a significantly faster runtime and it had a tendency to compute a slightly smaller set of places. However, we do not want to discuss this comparison in more detail in this paper.

5   Token Flow Regions

The main problem of considering LPO transition regions is the possible exponential number of cuts of a finite partial language compared to the number of events and skeleton arcs (the set of skeleton arcs $<_s$ of an LPO $(V, <, l)$ is the unique smallest binary relation on V having $<$ as its transitive closure). The number of cuts can be very huge if there is a certain degree of concurrency in the partial language. Loosely speaking, the more parallelism appears in a given partial language, the worse is the runtime of the synthesis algorithm using LPO transition regions. To tackle this problem, we use another notion of regions, so called *token flow regions* [16,15,14]. Again, every token flow region corresponds to a feasible place.

The idea of defining token flow regions of a given partial language L is as follows: Assign to every edge (v, v') of an LPO in L a natural number representing the number of tokens which are produced by the occurrence of $l(v)$ and consumed by the occurrence of $l(v')$ in the place to be defined. Then, the number of tokens consumed overall by a transition $l(v')$ in this place is given by the sum of the natural numbers assigned to ingoing edges of v'. This number can be interpreted as the weight of the arc connecting the new place with the transition $l(v')$. Similarly, the number of tokens produced overall by a transition $l(v)$ in this place is given by the sum of the natural numbers assigned to outgoing edges of v, and this number can be interpreted as the weight of the arc connecting the transition $l(v)$ with the new place. Moreover, transitions can also consume tokens from the initial marking of the new place. In order to specify the number of such tokens, we extend an LPO by an *initial event* v_0 representing a transition producing the initial marking. The sum of the natural numbers assigned to outgoing edges of the initial event v_0 can be interpreted as the initial marking of the new place. Transitions can produce tokens in the new place which remain in the final marking. In order to specify the number of such tokens, we extend an LPO by a *final event* v_m representing a transition consuming the final marking.

We call such an LPO extended by an initial and a final event a $\star$-extension of an LPO. Formally, v_0 is ordered before each other event and v_m after each other event in the $\star$-extension. Both events have labels not in T. For a given partial language L we consider a $\star$-extension of each LPO in L. The set of all these $\star$-extensions is denoted by $L^\star$. $L^\star$ is defined in such a way that all initial events have the same label while all final events have different labels. The set of edges of the $\star$-extensions in $L^\star$ is denoted by $E_L^\star$. For formal definitions see [15,14].

According to the above explanation, we can define a token flow region r by assigning a natural number $r(v, v')$ to each edge $(v, v') \in E_L^\star$. We call the value $r(v, v')$ the *token flow* between v and v'. The sum of the natural numbers assigned to ingoing edges of an event v' we call the *intoken flow* of v'. The sum of the natural numbers assigned to outgoing edges of an event v we call the *outtoken flow* of v. The sum of the natural numbers assigned to outgoing edges of the initial event of an LPO lpo we call the *initial token flow* of lpo.

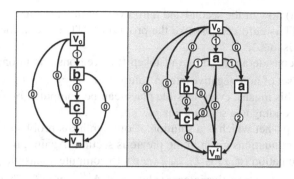

Fig. 8. A token flow region

Since equally labeled events of L formalize occurrences of the same transition, this is well-defined only if equally labeled events of L have equal intoken flow (property IN) and equal outtoken flow (property OUT). Also, all LPOs must have the same initial token flow (property INIT). Each token flow region r (fulfilling IN, OUT and INIT) defines a place p_r. $W(p_r, t)$ is given by the intoken flow of events labeled by t, $W(t, p_r)$ is given by the outtoken flow of events labeled by t and $m_0(p_r)$ is given by the outtoken flow of initial events. We call p_r the *corresponding place* of r. For the running example a token flow region r is shown in Figure 8. The corresponding place of r is depicted on the left side of Figure 5.

Definition 12 (Token flow region). *Let L be a partial language and $E_L^\star$ the set of edges of a $\star$-extension $L^\star$. A token flow region of L is a function $r : E_L^\star \to \mathbb{N}$ fulfilling*

$$(iv) \qquad IN, \ OUT \ and \ INIT.$$

It was shown in [16] that the set of places corresponding to token flow regions of a partial language equals the set of feasible places w.r.t. this partial language.

Theorem 3. *Given a partial language L, (1) each token flow region defines a feasible place and (2) each feasible place is defined by a token flow region.*

The set of token flow regions can be computed as the set of non-negative integer solutions of an homogenous linear equation system with integer coefficients $\mathbf{A}_L^{(iv)} \cdot \mathbf{r} = \mathbf{0}$. To compute a token flow region r, we need to assign a value $r(v, v')$ to every edge $e_i = (v, v')$ in the finite set of edges $E_L^\star = \{e_1, \ldots, e_k\}$ of the $\star$-extensions of the LPOs in the given partial language L. The vector $\mathbf{r}$ contains a variable r_i for each edge $e_i \in E_L^\star$ representing $r(e_i)$. We encode the properties IN, OUT and INIT such that r fulfills IN, OUT and INIT on L if and only if $\mathbf{A}_L^{(iv)} \cdot \mathbf{r} = \mathbf{0}$. This can be done by defining for pairs of equally labeled events a row of $\mathbf{A}_L^{(iv)}$ counting the token flow on ingoing edges of one event positively and of the other event negatively. Similarly, another row of $\mathbf{A}_L^{(iv)}$ counts the token flow on outgoing edges of one event positively and of the other event negatively. Concerning IN (OUT) it is enough for each label t to ensure that the intoken (outtoken) flow of the first and second event with label t are equal, that the intoken (outtoken) flow of the second and third event with label t are equal, and so on. The property INIT is enforced just like the property OUT, but considering the initial events of the LPOs in L^*.

By the above considerations the set of token flow regions r is in one-to-one correspondence to the set of non-negative integer solutions $\mathbf{r} = (r_1, \ldots, r_k)$ of $\mathbf{A}_L^{(iv)} \cdot \mathbf{r} = 0$ via $r(e_i) = r_i$. This means, every feasible place can be computed by such a solution. The place corresponding to a solution $\mathbf{r}$ is p_r.

To calculate a p/t-net which is a solution of our synthesis problem we use the concept of wrong continuations given in the previous section. Again, given a finite L, for each wrong continuation $(\pi, \tau) \in L_{wrong}$ we try to compute a feasible place p_r which prohibits (π, τ) as a non-negative integer solution of $\mathbf{A}_L^{(iv)} \cdot \mathbf{r} = 0$. Therefore, we need to translate the separating inequality (iii) into the new notion of token flow regions. For every transition t_i of the given language L we choose an example event v_{t_i} labeled by t_i. The set of all ingoing (outgoing) edges of v_{t_i} corresponds to the arc (t_i, p_r) $((p_r, t_i))$. Given a fixed indexing of the edges of the LPOs in L^* and a fixed example event v_{t_i} for each label t_i, we define sets T_i^{in} containing the indices of the ingoing edges into v_{t_i} and sets T_i^{out} containing the indices of the outgoing edges of v_{t_i}. The set T^{init} contains the indices of the outgoing edges of an initial example event.

Definition 13 (Separating token flow region). *Let (π, τ) be a wrong continuation of a given finite partial language L. A token flow region r of L is a separating token flow region w.r.t. (π, τ) if*

$$(v) \quad \sum_{j \in T^{init}} r(e_j) + \sum_{i=1}^{m} \left(\sum_{j \in T_i^{out}} \pi(t_i) \cdot r(e_j) - \sum_{j \in T_i^{in}} (\pi + \tau)(t_i) \cdot r(e_j) \right) < 0.$$

A separating token flow region r w.r.t. (π, τ) can be calculated (if it exists) as a non-negative integer solution of an homogenous linear inequality system with integer coefficients of the form

$$\mathbf{A}_L^{(iv)} \cdot \mathbf{r} = 0$$
$$\mathbf{b}_{\pi\tau}^{(v)} \cdot \mathbf{r} < 0.$$

The vector $\mathbf{b}_{\pi\tau}^{(v)}$ is defined in such a way that $\mathbf{b}_{\pi\tau}^{(v)} \cdot \mathbf{r} < 0 \Leftrightarrow (v)$.

Trying to calculate for each wrong continuation a separating token flow region leads to the second algorithm (Alg. IV) described in this paper. As in Alg. III, feasible places corresponding to separating token flow regions are introduced step by step. Again, if a wrong continuation is already prohibited by a previously added place, the algorithm does not search for a respective separating region. In pseudo-code Alg. IV can be described equally as Alg. III with the only difference that the statement "$\mathbf{r} \leftarrow Solver.getIntegerSolution(\mathbf{A}_L^{(ii)} \cdot \mathbf{r} \geq 0, \mathbf{r} \geq 0, \mathbf{b}_{\pi\tau}^{(iii)} \cdot \mathbf{r} < 0)$" has to be replaced by "$\mathbf{r} \leftarrow Solver.getIntegerSolution(\mathbf{A}_L^{(iv)} \cdot \mathbf{r} = 0, \mathbf{r} \geq 0, \mathbf{b}_{\pi\tau}^{(v)} \cdot \mathbf{r} < 0)$". Following the same arguments used in Theorem 2, we get:

Theorem 4. *There exists a positive solution of the synthesis problem for the finite partial language L if and only if $\mathfrak{L}po(N, m_0) = \mathcal{L}((N, m_0)$ is a net computed by Alg. IV).*

We have already explained in the introduction (without going into detail) that there are some general results and experiences, e.g. in [14], clearly indicating that the principle of separation representation applied in Alg. IV is usually superior to the principle of basis representation used in Alg. II (Alg. II and Alg. IV use the same notion of region). Again, we implemented the new algorithm, Alg. IV, in our tool VipTool. As expected, experimental tests showed that Alg. IV is in fact much faster than Alg. II and generates a way smaller set of places than Alg. II. This probably results from the problem that Alg. II computes a complete basis of the set of token flow regions. However, again we do not discuss this comparison in more detail in this paper.

6 LPO Transition Regions vs. Token Flow Regions

In this section we compare the new algorithms Alg. IV, using separating token flow regions, and Alg. III, using separating LPO transition regions. We compare their complexity on a theoretical level, we use the implementations of the algorithms in VipTool to conduct systematic experimental tests by varying different parameters of the input partial language and finally we apply the implementations to real-size examples.

The implementations use the Simplex algorithm of LpSolve to solve the occurring inequality systems (http://lpsolve.sourceforge.net). All tests are performed using Java SE 1.6.0 on an Intel Core2Duo 2.66 GHz (2 CPUs) machine with 2012 MB RAM running a Microsoft Windows Vista operating system. For the sake of reproducibility we provide all tested example partial languages and the running example as xml-files on the VipTool homepage viptool.ku-eichstaett.de. The order of the elements in the xml-files influences the order of processing the wrong continuations in the synthesis algorithms (note that also the order of the files which is given by the file names matters). Therefore, the results of the algorithms depend on this order which in turn depends on the order of drawing the LPOs in VipTool.

Since in both algorithms the same notion of wrong continuation is used, the presented comparison can be seen as a comparison between the concepts of transition regions and token flow regions. But most importantly, the two investigated algorithms are the most promising algorithms to synthesize a p/t-net from a finite partial language developed so far because, as explained before and confirmed by experiments, the separation representation is in general superior to the basis representation and the LPO transition regions proved to be better applicable than the step transition regions. Therefore, it is quite natural to let these two algorithms compete with each other.

6.1 Complexity Comparison

First, we compare the complexity of the two algorithms. The input for both algorithms is an arbitrary finite partial language L, i.e. a finite set of LPOs. To be realistic, we assume that each LPO $\text{lpo} = (V, <, l) \in L$ is specified by its set of events V together with a label $l(v)$ for each $v \in V$ and its set of skeleton arcs $<_s$. Therefore, the size of the input is $\Sigma_{\text{lpo}=(V,<,l)\in L}(2|V| + |<_s|)$. The number of labels $|T| \leq |V|$ can often be seen as a constant (in cases where the activities are fixed and not dependant on the number and the size of the LPOs).

The structure of both algorithms is very similar. The pseudo-code is the same except for one statement. They both start by computing the set WC (line 1). To first compute all prefix steps, all prefixes of each lpo $= (V, <, l) \in L$ have to be computed, where the number of prefixes of lpo can be exponential in $|V|$ and $|<_s|$. More precisely, ordered events only generate a linear number of prefixes, while a co-set generates prefixes for all subsets, i.e. highly concurrent partial languages are problematic. Given a prefix step π, computing all wrong continuations (π, τ) may be exponential in the largest step τ' defining a right continuation (π, τ') (in the worst case all subsets of τ' have to be extended by an event and tested whether they are wrong continuations). The overall number of wrong continuations is bounded by the product of the number of labels $|T|$ and the number of prefixes within the partial language.

In the next step (line 2) simply the labels of L are considered. Then, each element $(\pi, \tau) \in WC$ is processed (line 3). First, it is checked whether (π, τ) is enabled in the so far constructed net $(N, m_0) = (P, T, F, W, m_0)$ (line 4). This means checking whether $m_0(p) + \sum_{t \in \pi} \pi(t) \cdot (W(t, p) - W(p, t)) \geq \sum_{t \in \tau} \tau(t) \cdot W(p, t)$ for each place $p \in P$. Since (π, τ) is given by the events V of one LPO lpo $= (V, <, l) \in L$, checking this inequality is linear in the size of one LPO. But the number of checks may be exponential in the size of the input, because $|P|$ can be exponential, but usually it is very small. The reason for this is that a place for $(\pi, \tau) \in WC$ may only be added if this enabledness check is positive. In the negative case the next wrong continuation is processed (line 9-10). The enabledness check is very often negative because, although each place is added to the net to separate a certain wrong continuation, it often separates a lot of further wrong continuations leading to a negative check when they are processed later on. In each case of a positive enabledness check the next step is the costly one of solving an inequality system (line 5). Here is the only difference of the two algorithms because the regarded inequality systems are different.

Solving the inequality systems is the most interesting part for comparing the algorithms and we discuss this part in more detail. First, the inequality systems have to be built. The build time is not that important because in both algorithms the inequality system only has to be built once because the systems for different $(\pi, \tau) \in WC$ only differ in one inequality (which is directly given by (π, τ)). In Alg. IV, first respective pairs of events have to be found which requires at most quadratic complexity and then the actual build time directly corresponds to the size of the respective inequality system (discussed in the following). In Alg. III, the build time is roughly quadratic in the size of the respective inequality system (discussed in the following) due to comparisons of inequalities to find redundant ones. We use a quite simple comparison of inequalities as described above (by the $\ll$-relation), but also more advanced and possibly more costly methods can be applied.

Second, the crucial part is now to solve the respective inequality systems. In both algorithms we have to find an integer solution of an homogeneous inequality system. It is important here that the homogeneity of the system enables the use of solvers searching for rational solutions, since in this case multiplying with the common denominator of the entries of a rational solution yields an integer solution. Therefore, it is not necessary to consider integer linear programming solvers [19] such as branch and bound algorithms or the cutting-plane method (Gomory), which have exponential runtime in the size of the inequality system. But in order to decide the solvability of the inequality system and to compute a solution in the positive case rational linear programming solvers, such as the Simplex method, the method of Khachyan or the method of Karmarkar, can be applied [19]. The methods of Khachiyan (ellipsoid method) and Karmarkar (interior point method) require only polynomial runtime [19] in the size of the inequality system. The Simplex algorithm is exponential in the worst case, but probabilistic and experimental results [19] show that it has a fast average runtime. Also incremental linear programming approaches are applicable, because the inequality systems for the different wrong continuations only differ in one inequality.

Although we can apply the same linear programming techniques in both algorithms, the sizes of the systems to be solved are different depending on the input partial language. These sizes heavily influence the runtime of the solvers. In Alg. III, the number of rows of $\mathbf{A}_L^{(ii)}$ equals the number of maximal right continuations. Each cut is a candidate to define a maximal right continuation, i.e. in the worst case the number of rows of $\mathbf{A}_L^{(ii)}$ is equal to the number of cuts in L, but often some cuts can be neglected because they define the same right continuation as another cut or the respective right continuation is not maximal w.r.t. $\ll$. In this way, the number of rows of $\mathbf{A}_L^{(ii)}$ can be small compared to the number of cuts in L. The number of cuts of one LPO lpo $= (V, <, l) \in L$ can in the worst case be exponential in $|V|$ and $|<_s|$. But if there is few concurrency in lpo, the number is small, e.g. it equals $|V|$ if lpo is totally ordered. Also in the unrealistic case that there is almost no ordering in lpo, the number of cuts is small, namely the number of cuts is bounded by $2^{|<_s|}$. The number of columns of $\mathbf{A}_L^{(ii)}$, i.e. the number of variables of the inequality system, is $2|T| + 1$. This is a very small value and can sometimes even be seen as a constant.

In Alg. IV, the number of rows of $\mathbf{A}_L^{(iv)}$ in all cases roughly coincides with the number of events in L. For each label we can fix one reference event. For each further event referring to a label, we get two equations for IN and OUT. Moreover, we can fix one LPO as a reference for the initial node. For each further LPO we get one equation for INIT. Altogether, the number of rows is $2\Sigma_{\text{lpo}=(V,<,l)\in L}|V| - 2|T| + |L| - 1$. In the worst case, there are few labels and many LPOs, but then the number of rows is still linear in the number of events in L with a maximal factor of three (in terms of figures the factor is three in the case all events have the same label and each LPO has only one event). The number of columns of $\mathbf{A}_L^{(iv)}$ is equal to the number of edges $|E_L^\star|$. The number of edges $|<|$ of one LPO lpo $= (V, <, l) \in L$ may be quadratic in $|<_s|$. In the worst case lpo is a total order, then $|<_s| = |V| - 1$, but $|<| = (((|V| - 1)|V|)/2$. The set $E_L^\star$ also contains the additional edges of lpo*. This number is always equal to $2|V|$. Altogether, the number of columns of $\mathbf{A}_L^{(iv)}$, i.e. the number of variables of the inequality system, can be quadratic in the size of the input if there is few concurrency in L. But if there is a certain degree of concurrency in L the number is often nearly linear in the size of the input. If L contains prefixes and sequentializations this might unnecessarily increase the number of variables that have to be considered in $\mathbf{A}_L^{(iv)}$ (due to the additional arcs). Therefore, for further fine tuning it might be reasonable to delete prefixes and sequentializations within L (but this is also costly) before running Alg. IV.

In a nutshell, we can say that $\mathbf{A}_L^{(ii)}$ has a small number of columns (variables) while $\mathbf{A}_L^{(iv)}$ has a small number of rows (constraints). The number of rows (constraints) of $\mathbf{A}_L^{(ii)}$ can be exponential if there is a lot of concurrency in L. The number of columns (variables) of $\mathbf{A}_L^{(iv)}$ can be quadratic if there is few concurrency in L. Comparing Alg. III and Alg. IV we can conclude that their performance depends on the structure of the input partial language L. Usually, Alg. III seems to be more efficient if L exhibits few concurrency and Alg. IV seems to be more efficient if there is a lot of concurrency. If the set of maximal right continuations is by far smaller than the set of cuts of L this significantly improves the performance of Alg. III.

The number of places of the nets synthesized by the two algorithms in both cases depends on how often the check whether a wrong continuation is enabled in the so far constructed net is positive. For each positive check one place is added or the wrong continuation cannot be excluded. The number of positive checks depends on the places computed so far and therefore it depends on the order of processing the wrong continuations and on the computed solutions of respective previous inequality systems. A system has many solutions and one arbitrary one is chosen by a solver. The chosen solution depends on the inequality system, which is different for the two algorithms, and also on the used solver. It is not possible to estimate which kinds of places are computed here. Thus, estimating the overall number of places of the synthesized nets is not possible for both algorithms. Nevertheless, for both algorithms in the worst case the number of places can be exponential, but usually, as explained before, there are only few positive checks and thus the number is small.

For both algorithms it is also possible to use objective functions in order to guide the computation of solutions of the respective inequality systems in such a way that places separating many wrong continuations are computed. Such a procedure reduces the overall number of places. But to correctly regard an objective function, integer linear programming solver have to be applied which have exponential runtime. Another possibility to reduce the number of places is to (heuristically) find an appropriate order of processing the wrong continuations. Finally, it is of course always possible to afterwards search for implicit places in the synthesized nets and to delete such places. These topics are interesting for further fine tuning the algorithms.

6.2   Systematic Experimental Tests

To get more detailed insights, we examined the two algorithms Alg. III and Alg. IV by systematic experimental tests. In particular, we tried to better understand the dependency of the performance and the resulting nets of the algorithms from the size and the structure of the input partial language. In this subsection, we present the main results of our tests by means of three test series. In the first series the number of sequentially ordered events in three LPOs a_n, b_n, c_n is increased (see Figure 9 left), in the second one the number of events in three concurrent sequences of events in one LPO abc_n (see Figure 9 middle) is increased, and in the third one the number of considered LPOs $l_1 \ldots l_n$ (here all LPOs only differ in the last event, see Figure 9 right) is increased. For illustration, we show three p/t-nets in Figure 10 having those respective partial languages of runs. In all three examples the synthesized nets exactly have the specified behaviors.

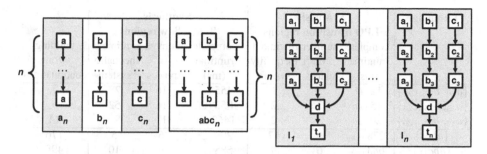

Fig. 9. Example LPOs. Left: Three LPOs modeling three alternative sequences (a_n, b_n, c_n). Middle: One LPO modeling three concurrent sequences (abc_n). Right: n LPOs ($l_1 \ldots l_n$) modeling an alternative between n transitions at the end.

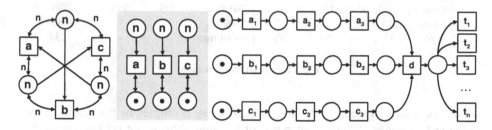

Fig. 10. Left: P/t-net having the behavior given by the three LPOs a_n, b_n, c_n. Middle: P/t-net having the behavior given by the LPO abc_n. Right: P/t-net having the behavior given by the n LPOs $l_1 \ldots l_n$.

The results of the tests are shown in the table in Figure 11. The first column describes the partial language given in each test. The columns two to four contain the mean overall runtime of 50 test runs together with the deviations in ms of a 95% confidence interval for the expectation of the overall runtime (given in brackets), the number of calculated places and the average runtime for the calculation of one separating region in the case of Alg. III. The columns five to seven contain the same for Alg. IV. The last column contains the mean runtime used to calculate the wrong continuations of the given language which is equal in both algorithms.

In the first test series, given in the first three rows of the table, Alg. III performs significantly better than Alg. IV. The partial language contains no concurrency. Then, as usual, the number of cuts is not too large and the number of arcs is relatively high. This results in a moderate number of rows in the inequality systems solved by Alg. III and a large number of columns in the inequality systems solved by Alg. IV.

In the second test series it is the other way round. There is a very high degree of concurrency within the partial language. Here, Alg. IV is faster than Alg. III. However, although Alg. IV has a significantly better performance in solving the inequality systems, the difference in the overall runtime is not that large. The reason for this is that in examples with a high degree of concurrency the time consumption for building the

Language	Synthesis LPO transition regions			Synthesis Token flow regions			building
	complete runtime (ms)	number of places	runtime p. inequality system	complete runtime (ms)	number of places	runtime p. inequality system	wrong continuations
$\{a_{20}, b_{20}, c_{20}\}$	12 (±1)	3	1	455 (±3)	16	28	7
$\{a_{25}, b_{25}, c_{25}\}$	15 (±1)	3	2	853 (±5)	16	52	9
$\{a_{30}, b_{30}, c_{30}\}$	20 (±1)	3	2	1462 (±7)	17	85	14
$\{abc_9\}$	278 (±6)	6	17	209 (±5)	6	5	176
$\{abc_{11}\}$	683 (±8)	6	30	588 (±7)	9	10	498
$\{abc_{13}\}$	1499 (±9)	6	54	1301 (±9)	7	18	1171
$\{l_1 \ldots l_{20}\}$	288 (±6)	31	4	1238 (±8)	31	35	140
$\{l_1 \ldots l_{30}\}$	501 (±7)	31	6	2782 (±11)	31	80	297
$\{l_1 \ldots l_{40}\}$	834 (±8)	31	8	3194 (±12)	23	114	559
Apoptosis	9145 (±29)	22	97	8898 (±18)	29	65	7006
Transit	5563 (±22)	168	30	120189 (±107)	399	300	380
Prod./cons.	56370 (±78)	23	65	55024 (±75)	23	7	54854
Log file	5253 (±19)	59	68	out of mem	-	-	1233

Fig. 11. Experimental results

wrong continuations seems to be large (which is also confirmed by the considerations in the last section) and this common part of the algorithms is not yet implemented in a performance optimized way (in contrast to solving the inequality systems where the very fast methods of LpSolve are used). Improving this situation is a task of further tuning the algorithms.

In the third test series, partial languages consisting of a lot of relatively small LPOs are considered. The single LPOs are all similar such that they have many common maximal right continuations yielding a small overall number of maximal right continuations. As a consequence, the number of rows in the inequality systems solved by Alg. III is small. Due to the large overall number of arcs in the partial language, which results from the large number of LPOs, the number of columns in the inequality systems solved by Alg. IV is large. Therefore, although the considered LPOs contain a certain degree of concurrency, Alg. III performs better than Alg. IV.

It is difficult to make a general statement about the numbers of places generated by the algorithms. While in the second and third test series the numbers are similar, Alg. IV generates significantly more places than Alg. III in the first test series. In the first and second series Alg. III even generates the minimal number of places necessary to reproduce the specified behavior. This tendency, that in most cases the two algorithms generate similar numbers of places and in the remaining cases Alg. IV generates more places than Alg. III, was also supported by further tests (of course there are exceptions). An illustration on which kinds of places are computed by the algorithms is shown for our running example in the next section, in Figure 12.

6.3 Experimental Tests in Realistic Settings

In this section we apply the algorithms in realistic settings. We consider two Petri nets of real case studies and one classical Petri net example and compute the runs in the form of LPOs of these Petri nets (by an unfolding algorithm implemented in VipTool [3]). As a fourth example we consider LPOs generated from a log file. Then, we apply the two synthesis algorithms to each of these four sets of LPOs and measure the performance and the number of places generated.

The first example is a Petri net designed by the DSSZ group from Cottbus (see http://www-dssz.informatik.tu-cottbus.de) to model the biological process of apoptosis [9] (different versions of this net are presented on the webpage in different examples, talks and the paper [9]; we refer to the version which is used by them for analysis purposes shown e.g. in the slides of the talk "Biopathways and Petri Nets - Demonstrated for Apoptosis" as well as in more recent talks such as "Petri Nets for Systems and Synthetic Biology"). We added tokens instead of the input transitions. Thereby, we provided all substances twice to make the scenarios more manifold. This example consists of seven LPOs with altogether 167 events and 13 different labels. The LPOs exhibit a certain degree of concurrency due to the non-safe initial marking.

The second example is the second version of the workflow of a transit case study [22] (transit2 process, can also be found at http://www.petriweb.org). To make the example finite, we added three places to restrict the number of possible repetitions of the message transfer due to erroneous messages, allowed by the three central (bordered) communication channels, to one, i.e. we restricted each of the three transitions related to sending a fun-nck message ($"s\_fun\_nck"$) by a place having an arc to the transition and containing one token. Then, this example consists of eight LPOs with altogether 288 events and 36 different labels. In this example almost the whole communication is ordered such that the LPOs exhibit very few concurrency.

To get an example with a high degree of concurrency we thirdly considered a net consisting of three concurrently operating instances of the classical producer/consumer net shown in most Petri net textbooks (see also http://www.petriweb.org), where we restricted each producer to be able to produce two times to get finite behavior. This example consists of one LPO with altogether 21 events and 12 different labels.

The fourth example is a log file of a business process distributed with the tool ProM [20] (http://prom.win.tue.nl/tools/prom/), which is used to test the different process mining algorithms of ProM. The name of the file is $grouped\_a22f0n00.xml$. Note that our synthesis algorithms try to exactly reproduce the numbers of repetitions of each transition depending on repetitions and occurrences of other transitions. Since this is very difficult, we first filtered the log by removing all cases containing more than one repetition of the loop transitions m and i. Then, to be able to take into account real partial order behaviors we applied the partial order generator of ProM to translate the log file into a set of LPOs. This example consists of 58 LPOs with altogether 841 events and 22 different labels. The LPOs have a typical degree of concurrency for business processes. The results of the four tests are shown in the above table in Figure 11 (note that Alg. IV failed for the last example). They basically confirm our interpretations explained so far. Altogether, our tests showed that for some examples it is better to use Alg. III, while for other examples it is better to use Alg. IV. In the first case, sometimes there are significant differences between the algorithms in both the runtime and the numbers of generated places. In the latter case, the advantages of Alg. IV seem small for the current implementation.

7   Tool Support and Applicability

Finally, we give an idea about the tool support of the new synthesis algorithms (Alg. III, Alg. IV) and shortly sketch possible application fields. Both algorithms are integrated as plug-ins of the current version 0.5.1 of VipTool [3].

Figure 12 shows a screenshot containing the partial language of our running example edited in VipTool and the two resulting nets synthesized by the two new algorithms. The example language can be represented by a very compact net (see Figure 2) which contains a complex place having a selfloop to one transition and a further outgoing arc to another transition. This is a typical example where both algorithms calculate redundant places. But such redundant places can easily be deleted afterwards by plug-ins of VipTool searching for implicit places (often applications require a possibly small number of places). VipTool contains a very fast method to delete places dominated by one other place and also different more costly methods searching more generally for implicit places.

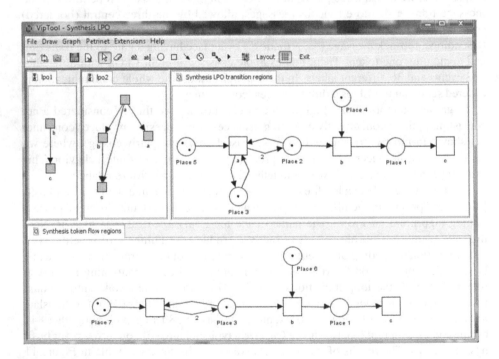

Fig. 12. Screenshot of VipTool

The implemented synthesis algorithms can be applied to support the modeling of systems by Petri nets. In early stages of system modeling, scenarios or use cases are often the most intuitive and appropriate modeling concept. In VipTool scenarios can be specified by LPOs which allow to model concurrency within scenarios in a natural way.

However, for the final purposes of modeling, namely the follow-up with documentation, analysis, simulation, optimization, design or implementation of a system, usually integrated state-based system models are desired. To bridge the gap between the scenario view of a system and a final system model, automatic construction of a system model from a specification of the system behavior in terms of single scenarios is an important challenge in many application areas. In particular, in the field of software engineering the step of coming from a user oriented scenario specification to an implementation oriented state based model of a software system received much attention in the last years and offers great potential for automation [13]. Similar problems occur in the domains of business process design (not restricted to the well-known field of process mining [21]), hardware design and controller synthesis. In all these areas a popular choice for a final system model, especially if concurrency is involved, are Petri nets and domain specific dialects of Petri nets. Here, the synthesis algorithms come into play to automatically generate a p/t-net from a finite set of finite scenarios given by LPOs.

In practice, many system specifications are given by such finite sets of finite scenarios, e.g. sets of sequence diagrams without loop fragments to specify scenarios modeling the external behavior of a software system, instance event driven process chains to specify scenarios of a business process or scenario specifications automatically generated from log files. Thus, for typical applications of synthesis from partial languages the two algorithms dealing with finite partial languages are often sufficient.

For specifications including infinite behavior these algorithms also build a good starting point, because it should be possible to extend the algorithms such that they can deal with certain kinds of infinite languages. For instance, in [2] we show some general principles how to accomplish synthesis from so called LPO-terms which allow iteration of certain behavior and thus represent infinite partial languages. But in general infinite languages are more difficult to handle effectively. In particular, for algorithmic purposes they first have to be finitely represented.

8 Conclusion

Both presented algorithms (Alg. III, Alg. IV) performed better than their previous versions (Alg. I, Alg. II) described in [14,15]. The first algorithm (Alg. III) using LPO transition regions does not have to translate the given partial language into a language of step sequences any more. The second algorithm (Alg. IV) using token flow regions benefits from applying the newly developed wrong continuations approach. The two algorithms developed in this paper use the two most promising combinations of a region definition together with a finite representation approach for the set of all feasible places. The algorithm using token flow regions performs fast if there is a lot of concurrency given in the partial language, the algorithm using LPO transition regions performs fast in the converse case and in the case that the number of maximal right continuations is by far smaller than the number of cuts in the partial language.

From the theoretical viewpoint our main interest for future research is to extend the algorithms to certain representations of infinite partial languages. While this is also interesting from the practical viewpoint, there are some more urgent topics of further research concerning practical applicability of the algorithms. Here, it is important to fine tune the algorithms (along the lines mentioned throughout the paper) and it would be nice to automatically choose the better performing algorithm from the characteristics of the specified language. Moreover, we are interested in considering certain domain specific problems of different application fields, e.g. synthesis of workflow nets or even sound workflow nets (such restrictions on the resulting nets can for instance be regarded by additional constraints in the inequality systems).

Acknowledgements. We thank the anonymous referees for their fruitful comments. We are particularly grateful to one of the referees for extensively checking and improving the technical parts.

References

1. Badouel, E., Darondeau, P.: On the Synthesis of General Petri Nets. Technical Report 3025, Inria (1996)
2. Bergenthum, R., Desel, J., Lorenz, R., Mauser, S.: Synthesis of Petri Nets from Infinite Partial Languages. In: ACSD 2008, pp. 170–179. IEEE, Los Alamitos (2008)
3. Bergenthum, R., Desel, J., Lorenz, R., Mauser, S.: VipTool-Homepage (2009), http://viptool.ku-eichstaett.de
4. Cortadella, J., Kishinevsky, M., Kondratyev, A., Lavagno, L., Yakovlev, A.: Petrify: A Tool for Manipulating Concurrent Specifications and Synthesis of Asynchronous Controllers. IEICE Trans. of Informations and Systems E80-D(3), 315–325 (1997)
5. Cortadella, J., Kishinevsky, M., Kondratyev, A., Lavagno, L., Yakovlev, A.: Hardware and Petri Nets: Application to Asynchronous Circuit Design. In: Nielsen, M., Simpson, D. (eds.) ICATPN 2000. LNCS, vol. 1825, pp. 1–15. Springer, Heidelberg (2000)
6. Darondeau, P.: Deriving Unbounded Petri Nets from Formal Languages. In: Sangiorgi, D., de Simone, R. (eds.) CONCUR 1998. LNCS, vol. 1466, pp. 533–548. Springer, Heidelberg (1998)
7. Ehrenfeucht, A., Rozenberg, G.: Partial (Set) 2-Structures. Part i: Basic Notions and the Representation Problem. Part ii: State Spaces of Concurrent Systems. Acta Inf. 27(4), 315–368 (1989)
8. Grabowski, J.: On Partial Languages. Fundamenta Informaticae 4(2), 428–498 (1981)
9. Heiner, M., Koch, I., Will, J.: Model Validation of Biological Pathways Using Petri Nets - Demonstrated for Apoptosis. Journal BioSystems 75(1-3), 15–28 (2004)
10. Hoogers, P., Kleijn, H., Thiagarajan, P.: A Trace Semantics for Petri Nets. Information and Computation 117(1), 98–114 (1995)
11. Josephs, M.B., Furey, D.P.: A Programming Approach to the Design of Asynchronous Logic Blocks. In: Cortadella, J., Yakovlev, A., Rozenberg, G. (eds.) Concurrency and Hardware Design. LNCS, vol. 2549, pp. 34–60. Springer, Heidelberg (2002)
12. Kiehn, A.: On the Interrelation Between Synchronized and Non-Synchronized Behaviour of Petri Nets. Elektronische Informationsverarbeitung und Kybernetik 24(1/2), 3–18 (1988)
13. Liang, H., Dingel, J., Diskin, Z.: A Comparative Survey of Scenario-Based to State-Based Model Synthesis Approaches. In: SCESM 2006, pp. 5–12. ACM, New York (2006)
14. Lorenz, R., Bergenthum, R., Desel, J., Mauser, S.: Synthesis of Petri Nets from Finite Partial Languages. Fundamenta Informaticae 88(4), 437–468 (2009)

15. Lorenz, R., Bergenthum, R., Mauser, S., Desel, J.: Synthesis of Petri Nets from Finite Partial Languages. In: ACSD, pp. 157–166. IEEE, Los Alamitos (2007)
16. Lorenz, R., Juhás, G.: Towards Synthesis of Petri Nets from Scenarios. In: Donatelli, S., Thiagarajan, P.S. (eds.) ICATPN 2006. LNCS, vol. 4024, pp. 302–321. Springer, Heidelberg (2006)
17. Lorenz, R., Juhás, G., Mauser, S.: How to Synthesize Nets from Languages - a Survey. In: WSC, pp. 637–647. IEEE, Los Alamitos (2007)
18. Pratt, V.: Modelling Concurrency with Partial Orders. Int. Journal of Parallel Programming 15, 33–71 (1986)
19. Schrijver, A.: Theory of Linear and Integer Programming. Wiley, Chichester (1986)
20. van der Aalst, W.M.P., van Dongen, B.F., Günther, C.W., Mans, R.S., de Medeiros, A.K.A., Rozinat, A., Rubin, V., Song, M., Verbeek, E., Weijters, A.J.M.M.: ProM 4.0: Comprehensive Support for Real Process Analysis. In: Kleijn, J., Yakovlev, A. (eds.) ICATPN 2007. LNCS, vol. 4546, pp. 484–494. Springer, Heidelberg (2007)
21. van der Aalst, W.M.P., Weijters, T., Maruster, L.: Workflow Mining: Discovering Process Models from Event Logs. IEEE Trans. Knowl. Data Eng. 16(9), 1128–1142 (2004)
22. Verbeek, E., van der Toorn, R.A.: Transit Case Study. In: Cortadella, J., Reisig, W. (eds.) ICATPN 2004. LNCS, vol. 3099, pp. 391–410. Springer, Heidelberg (2004)
23. Vogler, W.: Modular Construction and Partial Order Semantics of Petri Nets. LNCS, vol. 625. Springer, Heidelberg (1992)

On Bisimulation Theory in Linear Higher-Order π-Calculus

Xian Xu[1,2,*]

[1] Department of Computer Science and Technology,
East China University of Science and Technology,
130 Mei Long Road, Shanghai, China P.R. (200237)
[2] BASICS, Department of Computer Science and Engineering,
Shanghai Jiao Tong University,
800 Dong Chuan Road, Shanghai, China P.R.(200240)
xuxian@ecust.edu.cn, xuxian@sjtu.edu.cn, xuxian2004@gmail.com

Abstract. Higher-order process calculi are an important branch of process model for its significance in both theory and practice. In this paper, we establish new results on bisimulation theory in linear higher-order π-calculus. By exploiting the properties of linear higher-order processes, we work out two simpler variants than local bisimulation, which is an intuitive observational equivalence. We prove that they both coincide with local bisimilarity. The first variant, called *local linear bisimulation*, simplifies the matching of higher-order input and higher-order output based on the feature of checking equivalence with some special processes (in input or output) instead of general ones. The second variant, called *local linear variant bisimulation*, rewrites the first-order bound output clause in local bisimulation by harnessing the congruence properties.

Keywords: Bisimulation, Linear, Higher-order, π-Calculus, Process calculi.

1 Introduction

Compared with first-order calculi [1] [2] [3], higher-order process calculi excel in that they provide the ability to communicate an integral program, rather than a simple value or a reference to a program. The importance are twofold. Firstly, theoretically higher-order features offer a broader spectrum of communication capability that can possibly compute more conveniently and efficiently, and the (technical) framework of process calculi is widened. Secondly, practically distributed and mobile computing are increasingly expanding in various forms, in which communication involving higher-order elements can be witnessed in nearly anywhere, such as sending a small application through network or packaging and transmitting a script over several computing nodes in the network and

* The author is supported by The National 973 Project (2003CB317005), The National Nature Science Foundation of China (60473006,60573002,60773094), Plan of Aurora of Shanghai (07SG32).

K. Jensen, J. Billington, and M. Koutny (Eds.): ToPNoC III, LNCS 5800, pp. 244–274, 2009.

then configuring and running in a remote computer. To our knowledge, up-to-date research on bisimulation and related topics in higher-order process calculi mainly includes the following parts.

CHOCS and plain CHOCS. Thomsen studies two kinds of higher-order CCS, which extends CCS by higher-order communication with processes, that is CHOCS [4] [5] and Plain CHOCS [6] [7], with dynamic and static binding of restriction respectively. The higher-order bisimulation (in CHOCS) and applicative higher-order bisimulation (in Plain CHOCS) are examined. They are quite structural but not intuitive in that communicated and continual processes are separated in comparison [8]. The bisimulations use delayed approach, which prohibits internal moves after an observable action, to counter the technical obstacle in equivalence proving.

Higher-order π-calculus. Sangiorgi studies higher-order π-calculus [9] [8] that extends first-order π-calculus [2] [10] with communication of processes. The bisimulation he puts forth, context bisimulation [8], improves that in Plain CHOCS by considering residual and transmitted processes in the meanwhile. However the style is still delayed. Yet it is proven that the early and late versions of context bisimulations coincide. Triggered bisimulation, normal bisimulation are proposed to simplify context bisimulation on its heavy usage of universal quantification. And barbed bisimulation [11] is also considered in higher-order paradigm.

Nevertheless, the work mentioned above is not perfect in several points. For example,

- Delayed bisimulations. The bisimulations are all delayed versions. That is weak transitions have no trailing silent moves.
- Computational power. The computational power of higher-order process calculi is equal to that of Turing machines. This leads to the status that no axiom system is available for higher-order process calculi for a long period.

Fu attempts to settle these problems with linear higher-order π-calculus (LHOPi for short) [12], which demands that a same process variable shall never appear simultaneously in concurrent positions, typically the parallel composition and higher-order output. LHOPi takes idea from the practical scene that one program can be used only once in one network application on the client side, such as on-line games and video on demand. Fu shows that linearity can effectively downscale the computational power, and entail a sound and complete axiom system as well as an algorithm checking the equivalence of two processes. Besides, the bisimulation he uses, called local bisimulation, is a general one, not delayed. The technical difficulty is solved by the bisimulation lemma [12]. And most of the results in [12] hold in general higher-order π-calculus. Another related work comes from Cao, who establishes a equation system for a variant of higher-order π-calculus [13], and the idea is similar to Fu's. Linear type systems are another research interest closer to linear process calculi [14] [15]. They focus more on practical programming languages and live as a theoretical foundation.

Table 1. On the expressiveness

Calculus	HOcore			LHOPi		...
input	+	+	+	+	+	...
output	+	+	+	+	+	...
Composition	+	+	+	+	+	...
Restriction	-	+	-	+	+	...
Recursion	-	-	+	-	+	...
Linearity	-	-	-	+	+	...
TC	YES	YES	YES	NO	?YES	...
DSB	YES	NO	?	YES	?	...
DWB	?	NO	?	YES	?	...
ASSB	YES	NO	?	YES	?	...
ASWB	?	NO	?	YES	?	...

A very recent contribution is due to Sangiorgi et al.. In [16], Sangiorgi et al. study a variant (HOcore) of HOPi that has only the following operators: input prefix, process output, and parallel composition. This work arouses interest because it presents somewhat a counterintuitive result: HOcore is Turing complete, and hence the terminating property is not decidable; But, surprisingly, the strong bisimilarities (several instances in higher-order calculi) are decidable, and this entails a complete axiom system. Moreover, it is shown in [16] that the addition of restriction operator results in undecidability of strong bisimilarities. In contrast, LHOPi constraints the occurrence of process variables instead of removing restriction operator. This renders LHOPi not Turing complete in absence of recursion and entails a sound and complete axiom system for (weak) context bisimilarity [12]. To summary, the ways of restricting HOPi and their effect can be shown in the following (partial) table (Table 1). For those calculi without explicit names, we omit them. In Table 1, '+' means the inclusion of the corresponding operator or applying of the corresponding condition, while '-' means the opposite. 'YES' or 'NO' indicate whether the listed properties hold or not. 'TC' stands for Turing Complete, 'DSB' for Decidability of Strong Bisimilarity, 'DWB' for Decidability of Weak Bisimilarity, 'ASSB' for Axiom System for Strong bisimilarity, and 'ASWB' for Axiom System for Weak bisimilarity. Moreover, '?' reads Unknown, and '?YES'or '?NO" says inclined conjecture that still lacks of rigorous proofs.

The table is not completely filled in. Both the combination of the operators and conditions and the corresponding properties remain to be investigated. Some of the holes for property (i.e. TC, DSB and etc.) may be derived from the available ones, while the others may need a thorough new round of research.

Something worth noting is that in presence of recursion, the expressiveness of LHOPi would be (strictly) extended [12], but to our belief it would not invalidate the main results of this paper.

Albeit HOcore well models Minsky Machines theoretically and thus is Turing complete, it could be very indirect for HOcore to model various (computational) phenomena, from practical point of view. The point is that the set of operators of HOcore is too small to describe complicated procedures, such as a distributed protocol or a biological process, in a straightforward way, while an indirect description, which is usually not favorable, is often possible. This conveys the intuition that various computational models with the same level of computational power can have distinct modeling capacity, in the sense of directness and succinctness.

For linearity, as a matter of fact, the related work on higher-order process calculi and the examples, as will be seen, encourages to think in the following way: in a wide range of theoretical work and applications, linear processes can be up to the requirement. For a theoretical example, in [17], Homer, a variant of Plain CHOCS, is shown to be able to encode first-order pi-calculus, but actually only a subcalculus of Homer, which is closer to linear processes, is needed in the encoding.

To make it more explicit, we will exemplify the expressiveness and application potential of LHOPi. To present the examples without damaging the structure of the paper, we put the discussion in the appendix (Appendix B).

Contribution

Our endeavor in this paper is focused on the bisimulation theory of LHOPi. The work can be regarded as a successive work of Fu's [12][1]. We seek for a simpler expression of *local bisimulation*. This is achieved by making full use of the properties of linear processes, that is the *Abstraction Theorem* for higher-order input and the *Concretion Theorem* for higher-order output. These theorems are firstly used by Fu to lay the foundation for an axiom system, and we harness them to simplify local bisimulation, by loosening the requirements of bisimulation.

Higher-order input. We only demand the receiving of a special process in the comparison of higher-order input rather than a general process. So ($\mathcal{R}$ is a certain bisimulation, and we omit some relatively unimportant details).

$$\text{If } P \xrightarrow{a(A)} P', \text{ then } Q \xrightarrow{a(A)} Q' \text{ for some } Q' \text{ and } P'\mathcal{R}Q'$$

is simplified to

$$\text{If } P \xrightarrow{a(\mathbf{c})} P', \text{ where } c \text{ is a fresh name, then } Q \xrightarrow{a(\mathbf{c})} Q' \text{ for some } Q', \text{ and } P' \, \mathcal{R} \, Q'$$

(where $\mathbf{c}$ abbreviates $c(x).0$)

The simplification on input resembles the idea of triggers, except that the special process is different. The nature of it is that it can detect any subtle distinction between two receiving environments, with its freshness and sensitiveness enough to test the stableness of the receiving processes. In fact, this property can be extended to general higher-order π-calculus, as pointed out in [12].

---

[1] LHOPi was first presented in [18] of workshop CHINA2008.

Higher-order output. In higher-order output, the demand that the transmitted process should be considered in an arbitrary environment with the residual process is weakened to the holding of the similar property on a special environment. Such special processes can be infinite, but we merely need one of them for checking bisimulation. Thus we usually say 'a' instead of '$every/any$' in a simulating clause. On the other hand, since we do not have recursion in the calculi and fresh names are infinite, we can imply that 'any' such processes satisfy the bisimulation clause. So

If $P \xrightarrow{(\widetilde{x})\bar{a}A} P'$ then $Q \xLongrightarrow{(\widetilde{y})\bar{a}B} Q'$ for some $\widetilde{y}, B, Q'$, and for every process $E[X]$ it holds that $(\widetilde{x})(E[A]|P') \mathcal{R} (\widetilde{y})(E[B]|Q')$

is simplified to

If $P \xrightarrow{(\widetilde{x})\bar{a}A} P'$, then $Q \xLongrightarrow{(\widetilde{y})\bar{a}B} Q'$ for some $\widetilde{y}, B, Q'$. And for a process $E[X] \equiv \bar{c}.(X+d)$, where c, d are fresh names, it holds that $(\widetilde{x})(E[A]|P') \mathcal{R} (\widetilde{y})(E[B]|Q')$

The upshot of the simplification on output is similar to normal bisimulation in [8], which also sets up some kind of particular environment, except that here the special environment is not necessarily a repertoire of the transmitted process, and simply one instance suffices to observe the compositional behavior of the residual process and the sent process in the new environment, thanks to linearity of processes. The crux is that in LHOPi one has to ensure that the contexts as well as the general processes are linear.

The obtained bisimulation after the two modifications (on higher-order input and output) above is called *local linear bisimulation* ($\approx_{ll}$ for the corresponding bisimilarity). We show that the *Abstraction Theorem* and the *Concretion Theorem* also hold on local linear bisimilarity, which leads to the coincidence between local bisimilarity and local linear bisimilarity. The proofs are non-trivial. An example is also given in Section 3.2 to show how the simplification works.

First-order bound output. We also examine the first-order bound output, whose corresponding clause in local bisimulation is

If $P \xrightarrow{\bar{a}(x)} P'$, then $Q \xLongrightarrow{\bar{a}(x)} Q'$ for some Q', and for every process O, $(x)(O|P') \mathcal{R} (x)(O|Q')$

We argue in an informal way that this cannot be simplified. An explanation is that first-order bound output has little room for versatility of the form of the process acting as the environment. The name-sending mechanism is more fine-grained in form and hardly any special process can be figured out. Neither can we applying the idea of "triggers" because the target to be triggered can be anywhere and in various form. This fact also adds to the differentiation between name-sending and process-sending.

To gain a handy manipulation, we define a variant on first-order bound output clause by using the congruence properties, and obtained the following clause.

If $P\xrightarrow{\overline{a}(x)}P'$, then $Q\xRightarrow{\overline{a}(x)}Q'$ for some Q', and for all processes O_1 and O_2 such that $O_1 \mathcal{R} O_2$, $(x)(O_1|P') \mathcal{R} (x)(O_2|Q')$

The bisimulation gained after this adjustment on local linear bisimulation is called *local linear variant bisimulation* ($\approx_{ll}^v$ for the corresponding bisimilarity). We show that local linear variant bisimilarity coincides with local linear bisimilarity.

To conclude and make it more explicit, the contribution of this paper mainly includes the following points: to offer a handy way to handle bisimulation in LHOPi; to pave the way for a possible logical framework and proof system for LHOPi; to add merits to LHOPi which effectively cuts down the expressiveness of general HOPi, i.e. in a sense restriction to linearity weighs more than removing restriction operator with respect to expressiveness cutting down. Specifically, HOcore [16] is not linear, Turing Complete, but strong bisimilarly decidable; while LHOPi is not Turing Complete so its semantic properties are decidable, including strong and weak bisimilarities. Thus further work on expressiveness of (L)HOPi can be developed on the basis of a relatively 'light' bisimulation theory.

Recursion. We are in faith that the main results still hold in presence of recursion, since the results are basically on the transition semantical analysis, and the definition and intermediate lemmas can be extended accordingly. A rough intuition is that recursion brings around infinite behavior (transitions), however, any finite fragment of the transition can be scaled down using the similar simplification strategy as in this paper, though the details, including certain constraint (to maintain linearity of processes) and proofs, may need further studying and adjustment. For example, the concept "isolated" is needed in defining recursion [12] to ensure linearity when unfolding a recursive expression. And the proof concerning recursion, as another example, should be shifted from structural induction to transition induction (or induction on derivation height) since the structure of a process may grow more complex during evolving. However, we have the intuition that all obstacles can be overcome so that the main results of this paper remain true.

Organization. The paper is organized as follows. Section 2 gives an introduction to linear higher-order π-calculus defined by Fu. We recall the syntax, operational semantics, local bisimulation, and equivalence and congruence properties of local bisimilarity. Critical results, which describe the relationship between the equivalence of the prefixed processes and the equivalence of the continuations and lay foundation for axiom system for linear higher-order π-calculus, are also presented. In section 3, we define several variants, local linear bisimulation and local linear variant bisimulation, which simplify but coincide with the original bisimulation, that is local bisimulation (in the largest sense). The detailed proofs are given. We also discuss possibility of simplifying first-order bound output clause in local bisimulation. Section 4 concludes our work before some future work is discussed.

2   Linear Higher-Order π-Calculus

In this section, we give an introduction to LHOPi [12].

2.1   Syntax

Linear higher-order π (LHOPi) processes (or simply processes) are denoted by $A, B, E, F, P,\ Q, \ldots$. We denote names by small letters $a, b, c, \ldots$ and process variables by X, Y, Z. LHOPi processes are defined by the following abstract grammar. Most of them have their usual meanings, such as composition, restriction, choice and match. The prefixes worth noting are higher-order input $(a(X).P)$, which means receiving a process on port a to instantiate variable X and continuing as P, and higher-order output $(\bar{a}Q.P)$, which means sending process Q on port a and progressing as P.

$$P := 0 \mid X \mid a(x).P \mid \bar{a}x.P \mid a(X).P \mid \bar{a}Q.P \mid$$
$$P|Q \mid (x)P \mid [x{=}y]P \mid P{+}Q$$

The linearity is guaranteed by demanding $fv(P) \cap fv(Q) = \emptyset$ in $\bar{a}Q.P$ and $P|Q$, where $fv(\cdot)$ denotes free (process) variables. For clarity, we may write $\bar{a}[A]$ for $\bar{a}A$. An important fact is that linearity rules out the possibility to define replication (or recursion), which renders the LHOPi processes finite. This is crucially related to the non-Turing-completeness of LHOPi, as mentioned in Section 1.

We have derived prefixes: $\tau.P \triangleq (m)(m(x)|\bar{m}m.P), m$ fresh and $\bar{a}(x).P \triangleq (x)\bar{a}x.P$. Notice $\triangleq$ means definition. The first is silent action and the latter is first-order bound output. Most conventions are standard. $fn(P_1, ..., P_n), bn(P_1, . .., P_n), n(P_1, ..., P_n); fv(P_1, ..., P_n), bv(P_1, ..., P_n), v(P_1, ..., P_n)$ denote free names, bound names, names; free variables, bound variables and variables in processes $P_1, ..., P_n$, respectively. We generally focus on closed processes, which contain no free variables. Sometimes we use "(closed)" before processes to indicate this. "Process expressions" indicate general processes. Name substitution on processes $P\{y/x\}$ and higher-order (process or variable) substitution on processes $P\{Q/X\}$ are defined structurally on processes as usual. We demand that higher-order substitutions do not produce a non-linear processes. σ denotes name substitution. $\tilde{x}$ denotes a finite set of names, that is $x_1, x_2, ..., x_n$. $E[X_1, X_2, ..., X_n]$ stands for the process with (at most) a series of process variables occurring in it. We write $E[A_1, A_2, ..., A_n]$ for $E[X_1, X_2, ..., X_n]\{A_1/X_1, A_2/X_2, ..., A_n/X_n\}$. We usually omit the process 0. When the object (or content) of an input or output is irrelevant to the context, it is omitted. For example, some abbreviations are: a for $a(x).0$ (this shall not be confused with first-order name with the help of context); $\bar{a}$ for $\bar{a}x.0$ when x would never be used (w.l.o.g. we name such a name as x); τ for $\tau.0$. I_a is defined as $a(X).X$. A fresh name (resp. variable) is a name (resp. variable) that does not occur in the processes under consideration. We assume structural congruence, denoted by $\equiv_s$, on processes, like that in [19]. To be more specifically, apart from the law for α-conversion, a typically assumed structural congruence includes the associative and commutative laws for choice, composition and restriction, which forms monoids on each operators, and the distribution law on the distributivity of restriction with respect to choice and composition, in which the one worth noting is $(x)(P|Q) \equiv_s (x)P|Q$ if $x \notin fn(Q)$.

Contexts are processes with holes for processes. It is important that process behavior has some invariance under various contexts. We define three kinds of contexts below. Note context $C[\cdot]$ is different from $E[X]$ in that the latter should not let name capturing occur whereas the former does not take care of this.

Contexts: $[\cdot]$ is a context; If $C[\cdot]$ is a context, then $a(x).C[\cdot]$, $\overline{a}x.C[\cdot]$, $\tau.C[\cdot]$, $a(X).C[\cdot]$, $\overline{a}A.C[\cdot]$, $(x)C[\cdot]$, $P|C[\cdot]$ and $[x=y]C[\cdot]$ are contexts.

Full contexts: A context is a full context; If $C[\cdot]$ is a full context, then $\overline{a}[C[\cdot]].P$ and $C[\cdot] + P$ are full contexts.

Local contexts: A full context of the form $(\widetilde{x})([\cdot]||O)$. The more usual form of a local context is ($\widetilde{x}, \widetilde{c}$ are all pairwise distinct)$(x_1)\cdots(x_n)(\overline{c_1}x_1|\cdots|\overline{c_n}x_n|[\cdot])$ (or $(\widetilde{x})(\widetilde{\overline{c}x}|[\cdot]))$.

2.2 Semantics

For the labelled transition system, we need a concept $cp(P, X)$ of a process variable, indicating the process variables locating at the concurrent positions of X, such as parallel composition and higher-order output. It is due to Fu [12] and defined as follows.

$$
\begin{aligned}
cp(0, X) &\triangleq \emptyset \\
cp(X, X) &\triangleq \emptyset \\
cp(Y, X) &\triangleq \{Y\}, Y \neq X \\
cp(\tau.P, X) &\triangleq cp(P, X) \\
cp(a(x).P, X) &\triangleq cp(P, X) \\
cp(\overline{a}x.P, X) &\triangleq cp(P, X) \\
cp(a(Y).P, X) &\triangleq cp(P, X)
\end{aligned}
$$

$$
\left. \begin{aligned} cp(\overline{a}Q.P, X) \\ cp(P|Q, X) \end{aligned} \right\} \triangleq \begin{cases} fv(Q) \cup cp(P, X) & \text{if } X \in (fv(P) - fv(Q)) \\ fv(P) \cup cp(Q, X) & \text{if } X \in (fv(Q) - fv(P)) \\ fv(Q) \cup fv(P) \cup cp(P, X) \cup cp(Q, X) & \text{if } X \in (fv(Q) \cap fv(P)) \\ \emptyset & \text{if } X \notin (fv(Q) \cup fv(P)) \end{cases}
$$

$$
\begin{aligned}
cp((x)P, X) &\triangleq cp(P, X) \\
cp([x=y]P, X) &\triangleq cp(P, X) \\
cp(P+Q, X) &\triangleq cp(P, X) \cup cp(Q, X)
\end{aligned}
$$

The semantics of LHOPi is given in Figure 1. Symmetric rules are omitted. We use $\alpha, \beta, \lambda, \ldots$ for actions, which can be: τ (internal action), $a(x)$ (first-order input), $\overline{a}x$ (first-order output), $\overline{a}(x)$ (first-order bound output), $a(A)$ (higher-order input), and $(\widetilde{x})\overline{a}A$ (higher-order output). In higher-order output, $\widetilde{x}$ can be empty, and we often write $(\widetilde{x})\overline{a}[A]$ to make it clearer.

The transition rules are mostly self-explained. In higher-order input, the received process shall not break the linearity of processes, which is why we demand $fv(A) \cap cp(P, X) = \emptyset$. $\Longrightarrow$ is the reflexive transitive closure of silent actions, and

$$\frac{}{a(x).P\xrightarrow{a(y)}P\{y/x\}} \quad \frac{}{\overline{a}x.P\xrightarrow{\overline{a}x}P} \quad \frac{fv(A)\cap cp(P,X)=\emptyset}{a(X).P\xrightarrow{a(A)}P\{A/X\}} \quad \frac{}{\overline{a}A.P\xrightarrow{\overline{a}A}P}$$

$$\frac{P\xrightarrow{\lambda}P'}{P+Q\xrightarrow{\lambda}P'} \quad \frac{P\xrightarrow{\lambda}P'}{[x{=}x]P\xrightarrow{\lambda}P'} \quad \frac{P\xrightarrow{\lambda}P'}{P|Q\xrightarrow{\lambda}P'|Q}bn(\lambda)\cap fn(Q)=\emptyset$$

$$\frac{P\xrightarrow{a(x)}P',Q\xrightarrow{\overline{a}x}Q'}{P|Q\xrightarrow{\tau}P'|Q'} \quad \frac{P\xrightarrow{a(x)}P',Q\xrightarrow{\overline{a}(x)}Q'}{P|Q\xrightarrow{\tau}(x)(P'|Q')} \quad \frac{P\xrightarrow{a(A)}P',Q\xrightarrow{(\widetilde{x})\overline{a}[A]}Q'}{P|Q\xrightarrow{\tau}(\widetilde{x})(P'|Q')}$$

$$\frac{P\xrightarrow{\lambda}P'}{(x)P\xrightarrow{\lambda}(x)P'}x\notin n(\lambda) \quad \frac{P\xrightarrow{\overline{a}x}P'}{(x)P\xrightarrow{\overline{a}(x)}P'}x\neq a \quad \frac{P\xrightarrow{(\widetilde{x})\overline{a}[A]}P'}{(y)P\xrightarrow{(y)(\widetilde{x})\overline{a}[A]}P'}y\in fn(A)-\{\widetilde{x},a\}$$

Fig. 1. LTS of LHOPi

$\xRightarrow{}$ is $\Longrightarrow\xrightarrow{\lambda}\Longrightarrow$. $\xRightarrow{\hat{\lambda}}$ is $\Longrightarrow$ when λ is τ and $\xRightarrow{\lambda}$ otherwise. It is not difficult to show that the LTS preserves linearity of processes through transition induction [12].

The following lemmas state the properties of LTS concerning substitutions. Their proofs are basically transition inductions. Lemma 1 presents the updating of the action λ by name substitution. Lemma 2 and Lemma 3 are mutually inverse, and they state that the action actually comes from the context.

Lemma 1. *If P is a LHOPi process and $P\xrightarrow{\lambda}P'$ then $P\sigma\xrightarrow{\lambda\sigma}P'\sigma$.*

Lemma 2. *If P is a LHOPi process and $fv(P)=\{X_1,X_2,...,X_n\}$. And $P_1,P_2,...,P_n$ are LHOPi process. If $P\xrightarrow{\lambda}P'$ then $P\{P_1/X_1,P_2/X_2,...,P_n/X_n\}$ $\xrightarrow{\lambda\{P_1/X_1,P_2/X_2,...,P_n/X_n\}}P'\{P_1/X_1,P_2/X_2,...,P_n/X_n\}$.*

Lemma 3. *If P is a LHOPi process and $fv(P)=\{X_1,X_2,...,X_n\}$. And $b_1,b_2,...,b_n$ are fresh names. If $P\{b_1/X_1,b_2/X_2,...,b_n/X_n\}$ $\xrightarrow{\lambda\{b_1/X_1,b_2/X_2,...,b_n/X_n\}}P'\{b_1/X_1,b_2/X_2,...,b_n/X_n\}$, then $P\xrightarrow{\lambda}P'$.*

2.3 Bisimulation

A binary relation $\mathcal{R}$ on processes is closed under substitution of names if for each substitution σ, $(P\sigma,Q\sigma)\in\mathcal{R}$ whenever $(P,Q)\in\mathcal{R}$. A relation closed under substitution on process variables can be defined similarly. We first give the higher-order structural equivalence [12], which is essentially from Thomsen's applicative higher-order bisimilarity [6].

Definition 1 (Structural equivalence). *A symmetric binary relation $\mathcal{R}$ on (closed) processes is a structural bisimulation if it is closed under substitution of names and whenever $P\mathcal{R}Q$ the following holds:*

- *If $P\xrightarrow{\lambda}P'$, where λ is a silent action, first-order input, first-order output, first-order bound output, or higher-order input. Then $Q\xrightarrow{\lambda}Q'$ for some Q', and $P'\mathcal{R}Q'$;*

– If $P\xrightarrow{(\tilde{x})\bar{a}A}P'$, then some B,Q' exist s.t. $Q\xrightarrow{(\tilde{x})\bar{a}B}Q'$, $P'\mathcal{R}Q'$, and $A\mathcal{R}B$.

Two processes P,Q are structural equivalent, written $P \sim Q$, if there exists a structural bisimulation $\mathcal{R}$ s.t. $P\mathcal{R}Q$.

$\sim$ is a congruence.

Below we give the definition of local bisimulation, which is due to Fu [12]. It is the bisimulation we try to simplify.

Definition 2 (Local bisimulation). *A symmetric binary relation $\mathcal{R}$ on (closed) processes is a local bisimulation, if it is closed under substitution of names, and whenever $P\mathcal{R}Q$, the following properties hold:*

1. *If $P\xrightarrow{\tau}P'$, then $Q\Longrightarrow Q'$ for some Q' and $P'\mathcal{R}Q'$;*
2. *If $P\xrightarrow{a(x)}P'$, then $Q\overset{a(x)}{\Longrightarrow}Q'$ for some Q' and $P'\mathcal{R}Q'$;*
3. *If $P\xrightarrow{\bar{a}x}P'$, then $Q\overset{\bar{a}x}{\Longrightarrow}Q'$ for some Q' and $P'\mathcal{R}Q'$;*
4. *If $P\xrightarrow{\bar{a}(x)}P'$, then $Q\overset{\bar{a}(x)}{\Longrightarrow}Q'$ for some Q', and for every process O, $(x)(O|P')\ \mathcal{R}\ (x)(O|Q')$;*
5. *If $P\xrightarrow{a(A)}P'$, then $Q\overset{a(A)}{\Longrightarrow}Q'$ for some Q' and $P'\mathcal{R}Q'$;*
6. *If $P\xrightarrow{(\tilde{x})\bar{a}A}P'$ then $Q\overset{(\tilde{y})\bar{a}B}{\Longrightarrow}Q'$ for some $\tilde{y},B,Q'$, and for every process $E[X]$ s.t. $\tilde{x}\tilde{y}\cap fn(E)=\emptyset$ it holds that $(\tilde{x})(E[A]|P')\ \mathcal{R}\ (\tilde{y})(E[B]|Q')$.*

We say P is local bisimilar to Q, written $P \approx_l Q$ ($\approx_l$ is local bisimilarity), if there exists a local bisimulation $\mathcal{R}$ s.t. $P\mathcal{R}Q$.

The bisimulation is non-delayed and on closed processes, and can be extended to open processes in the standard way. Clauses $4,6$ are in a late style, whereas 5 is in an early style. Corresponding early cases for $4,6$ and late case for 5 can be defined. Moreover, it is proven that the corresponding early (or late) case is equivalent to the late (or early) case [12]. Local bisimilarity is an equivalence and congruence relation (excluding choice operator), and what's more, as showed by Fu, it is an observed bisimulation [12] that is a general bisimulation satisfying the least requirements to be qualified for an observational equivalence and somewhat like barbed bisimulation [20] [11]. Up-to technique can be defined on local bisimilarity. For example local bisimulation up-to $\sim$ can be defined by replacing $\mathcal{R}$ with $\sim \mathcal{R} \sim$ in the clauses.

The next three theorems clarify the relationship between the equivalence of prefixed processes and the equivalence of continual processes. Intuitively, these theorems provide us with alternative ways to express, analyze, or prove the equivalence of three kinds of prefixed processes: first-order bound output, higher-order input, and higher-order output, respectively. In Theorem 2 and Theorem 3 some special process can play the role of universal ones, as required in the definition of local bisimulation. For example, process I_b and environment $\bar{d}.(X+e)$ are taken advantage of in the Abstraction theorem and the Concretion theorem respectively. In the latter, the use of e is to set some kind of *guard*, out of technical reason. It is designed to make the proof more smooth and can be safely eliminated without invalidating the results. Below by 'fresh' we mean the related names do not appear in the processes under consideration, which typically are the processes filling in the hole in the local context $(\widetilde{x})(\overline{c}\widetilde{x}|[\cdot])$. In most other cases, the processes with respect to which the names are fresh are clear from the context.

Theorem 1 (Localization). *Suppose $\widetilde{c}(fresh)$ are pairwise distinct, so are $\widetilde{z}$, and $a \notin \widetilde{c}, x \notin \widetilde{z}$. Then the following equations are equivalent:*
(i) $(\widetilde{z})(\overline{c}z|\overline{a}(x).P) \approx_l (\widetilde{z})(\overline{c}z|\overline{a}(x).Q)$;
(ii) $(\widetilde{z})(\overline{c}z|(x)(\overline{b}x|P)) \approx_l (\widetilde{z})(\overline{c}z|(x)(\overline{b}x|Q))$ *for a fresh name b;*
(iii) $(\widetilde{z})(\overline{c}z|(x)(E|P)) \approx_l (\widetilde{z})(\overline{c}z|(x)(E|Q))$ *for every process E.*

Theorem 2 (Abstraction). *Suppose $\widetilde{c}(fresh)$ are pairwise distinct, so are $\widetilde{z}$, and $a \notin \widetilde{c}$. Then the following equations are equivalent:*
(i) $(\widetilde{z})(\overline{c}z|a(X).P) \approx_l (\widetilde{z})(\overline{c}z|a(X).Q)$;
(ii) $(\widetilde{z})(\overline{c}z|P\{I_b/X\}) \approx_l (\widetilde{z})(\overline{c}z|Q\{I_b/X\})$ *for a fresh name b;*
(iii) $(\widetilde{z})(\overline{c}z|P\{b/X\}) \approx_l (\widetilde{z})(\overline{c}z|Q\{b/X\})$ *for a fresh name b;*
(iv) $(\widetilde{z})(\overline{c}z|P\{E/X\}) \approx_l (\widetilde{z})(\overline{c}z|Q\{E/X\})$ *for every process E.*

Theorem 3 (Concretion). *Suppose $\widetilde{c}(fresh)$ are pairwise distinct, so are $\widetilde{z}$, and $a \notin (\widetilde{c} \cup \widetilde{x} \cup \widetilde{y}), \widetilde{x} \cap \widetilde{z} = \emptyset, \widetilde{y} \cap \widetilde{z} = \emptyset$. Then the following equations are equivalent:*
(i) $(\widetilde{z})(\overline{c}z|(\widetilde{x})\overline{a}[A].P) \approx_l (\widetilde{z})(\overline{c}z|(\widetilde{y})\overline{a}[B].Q)$;
(ii) $(\widetilde{z})(\overline{c}z|(\widetilde{x})(\overline{b}[A]|P)) \approx_l (\widetilde{z})(\overline{c}z|(\widetilde{y})(\overline{b}[B]|Q))$ *for a fresh name b;*
(iii) $(\widetilde{z})(\overline{c}z|(\widetilde{x})(\overline{d}.(A+e)|P)) \approx_l (\widetilde{z})(\overline{c}z|(\widetilde{y})(\overline{d}.(B+e)|Q))$ *for fresh names d, e;*
(iv) $(\widetilde{z})(\overline{c}z|(\widetilde{x})(E[A]|P)) \approx_l (\widetilde{z})(\overline{c}z|(\widetilde{y})(E[B]|Q))$ *for every process $E[X]$.*

3   Local Linear Bisimulation

We will put forth a variant of local bisimulation, called local linear bisimulation, which simplifies the former by harnessing the special properties of linear processes. The proof of the coincidence between the variant and the original bisimilarity is basically an exploiting of two theorems, Theorem 2 and Theorem 3.

Note we write a for $a(x).0$ in higher-order input and higher-order substitution concerning such a simple process. Occasionally we use bold font $\mathbf{a}$ to denote it is an abbreviation and not a first-order name. Yet in most other cases we do not since it shall be clear from the context.

Definition 3 (Local linear bisimulation). *A symmetric binary relation $\mathcal{R}$ on (closed) processes is a local linear bisimulation, if it is closed under substitution of names, and whenever $P \mathcal{R} Q$, the following properties hold:*

1. *If $P \xrightarrow{\tau} P'$, then $Q \Longrightarrow Q'$ for some Q', and $P' \mathcal{R} Q'$;*
2. *If $P \xrightarrow{a(x)} P'$, then $Q \stackrel{a(x)}{\Longrightarrow} Q'$ for some Q', and $P' \mathcal{R} Q'$;*
3. *If $P \xrightarrow{\overline{a}x} P'$, then $Q \stackrel{\overline{a}x}{\Longrightarrow} Q'$ for some Q', and $P' \mathcal{R} Q'$;*
4. *If $P \xrightarrow{\overline{a}(x)} P'$, then $Q \stackrel{\overline{a}(x)}{\Longrightarrow} Q'$ for some Q', and for every process O, $(x)(O|P') \mathcal{R} (x)(O|Q')$.*

5. If $P \xrightarrow{a(c)} P'$, where c is a fresh name, then $Q \xRightarrow{a(c)} Q'$ for some Q', and $P' \mathcal{R} Q'$;

6. If $P \xrightarrow{(\widetilde{x})\overline{a}A} P'$, then $Q \xRightarrow{(\widetilde{y})\overline{a}B} Q'$ for some $\widetilde{y}, B, Q'$. And for a process $E[X]$ of the form $\overline{c}.(X+d)$, where c, d are fresh names, it holds that $(\widetilde{x})(E[A]|P') \mathcal{R} (\widetilde{y})(E[B]|Q')$, that is $(\widetilde{x})(\overline{c}.(A+d)|P') \mathcal{R} (\widetilde{y})(\overline{c}.(B+d)|Q')$.

We say P is local linear bisimilar to Q, written $P \approx_{ll} Q$, if there exists some local linear bisimulation $\mathcal{R}$ such that $P\mathcal{R}Q$. Hence $\approx_{ll}$ is the largest local linear bisimulation.

Note the difference from local bisimulation in higher-order input and higher-order output, which borrows some insight into the bisimulation feature on linear higher-order processes and down-scales general higher-order analysis. Though Theorem 1 looks like Theorem 2 and Theorem 3 in a sense, it differs in that it has no essential down-scaling effect, as will be seen. It is clear that $\sim \subseteq \approx_l \subseteq \approx_{ll}$. Local linear bisimulation up-to $\sim$ can be defined in the standard way and will be used later on.

3.1   Characterizing Local Linear Bisimulation

In this section, we examine the properties of local linear bisimulation. The lemma below is called Bisimulation Lemma and serves as a basis for the lemmas henceforth. It plays crucial role in proving the equivalence properties of local bisimilarity, where non-delayed style of weak transition is applied. The proof is straightforward.

Lemma 4 (Bisimulation Lemma). Suppose P, Q are processes. If $P \Longrightarrow \cdot \approx_{ll} Q$ and $Q \Longrightarrow \cdot \approx_{ll} P$, then $P \approx_{ll} Q$.

Next are two lemmas forming the basis of equivalence property of $\approx_{ll}$, and the second also contributes to the Concretion Theorem (Theorem 6). Their proofs make essential use of Bisimulation Lemma.

Lemma 5. Suppose R, P, Q are processes. If $(x)(a.R|P) \approx_{ll} (x)(a.R|Q)$ for a fresh name a, then $(x)(R|P) \approx_{ll} (x)(R|Q)$.

Proof. As a is fresh, $(x)(P|a.R) \xrightarrow{a} (x)(P|R)$ must be simulated by $(x)(Q|a.R) \Longrightarrow (x)(Q_1|a.R) \xrightarrow{a} (x)(Q_1|R) \Longrightarrow Q' \approx_{ll} (x)(P|R)$, which can be rewritten as $(x)(Q|a.R) \xrightarrow{a} (x)(Q|R) \Longrightarrow Q' \approx_{ll} (x)(P|R)$. Similarly we have that $(x)(P|R) \Longrightarrow P' \approx_{ll} (x)(Q|R)$ for some P'. So by Bisimulation Lemma (Lemma 4), we have $(x)(P|R) \approx_{ll} (x)(Q|R)$. $\square$

Lemma 6. Suppose A, B, P, Q are processes, and a, c, d are fresh names. If $(\widetilde{x})(\overline{a}[A]|P) \approx_{ll} (\widetilde{y})(\overline{a}[B]|Q)$, then $(\widetilde{x})(\overline{c}.(A+d)|P) \approx_{ll} (\widetilde{y})(\overline{c}.(B+d)|Q)$.

Proof. Suppose $(\widetilde{x})(\overline{a}[A]|P) \approx_{ll} (\widetilde{y})(\overline{a}[B]|Q)$ for a fresh name a. Then $(\widetilde{x})(\overline{a}[A]|P) \xrightarrow{(\widetilde{x_1})\overline{a}[A]} (\widetilde{x_2})P$, where $\widetilde{x_1} \cup \widetilde{x_2}$ is $\widetilde{x}$. As a is fresh, this must be simulated by (for some Q_1, Q')

$$(\widetilde{y})(\overline{a}[B]|Q) \Longrightarrow (\widetilde{y})(\overline{a}[B]|Q_1) \xrightarrow{(\widetilde{y_1})\overline{a}[B]} (\widetilde{y_2})Q_1 \Longrightarrow (\widetilde{y_2})Q',$$

where $\widetilde{y_1} \cup \widetilde{y_2}$ is $\widetilde{y}$, such that $(\widetilde{y_1})(G[B]\|(\widetilde{y_2})Q') \approx_{ll} (\widetilde{x_1})(G[A]\|(\widetilde{x_2})P)$, for a process $G[X] \triangleq \overline{c}.(X+d)$ $(c, d$ are fresh). By α-conversion and structure equivalence, it can be rewritten as $(\widetilde{y})(G[B]\|Q') \approx_{ll} (\widetilde{x})(G[A]\|P)$.

It follows from the simulating transition sequence that $Q \Longrightarrow Q_1 \Longrightarrow Q'$, so we have $(\widetilde{y})(G[B]\|Q) \Longrightarrow \cdot \approx_{ll} (\widetilde{x})(G[A]\|P)$, where the dot is the process $(\widetilde{y})(G[B]\|Q')$. Similarly we know that $(\widetilde{x})(G[A]\|P) \Longrightarrow \cdot \approx_{ll} (\widetilde{y})(G[B]\|Q)$. By Bisimulation Lemma (Lemma 4) we have $(\widetilde{x})(G[A]\|P) \approx_{ll} (\widetilde{y})(G[B]\|Q)$, which is exactly $(\widetilde{x})(\overline{c}.(A+d)\|P) \approx_{ll} (\widetilde{y})(\overline{c}.(B+d)\|Q)$. $\qquad\square$

Lemma 7. $\approx_{ll}$ *is an equivalence relation.*

Proof. We need to take advantage of Lemma 5, Lemma 6 and Bisimulation Lemma (Lemma 4). The proof is quite routine after taking this into consideration. $\qquad\square$

As a necessary step in making a bisimilarity useful, the congruence property is stated below.

Theorem 4. $\approx_{ll}$ *is a congruence relation on all the calculus operators except the choice operator.*

Proof. We use a similar approach to that in [12] [8]. That is, define the transition closure of a designed relation saying the desired properties of the bisimilarity, as $S_0 \triangleq \approx_{ll}$,
$$S_{i+1} \triangleq \left\{ \begin{array}{l} (\tau.P, \tau.Q), (a(x).P, a(x).Q), (\overline{a}x.P, \overline{a}x.Q), (a(X).P, a(X).Q), \\ (\overline{a}A.P, \overline{a}A.Q), (P\mid R, Q\mid R), ((x)P, (x)Q) \end{array} \middle| P S_i Q \right\}.$$
And $S \triangleq \bigcup_{i \in \omega} S_i$.

And we then show that S is a local linear bisimulation up-to $\sim$. The details are routine and we skip them here. The pattern of the proof is like that in [12], [8] or even [6] and [21]. $\qquad\square$

Lemma 8. *Suppose* $E[X], A$ *are processes and* $E[A] \xrightarrow{\alpha} E'[A']$, *where* A *contributes in the action. Then* $E[I_a] \xrightarrow{a(A)} E''[A] \xrightarrow{\alpha} E'[A']$, *for a fresh name* a *and some* E''.

Proof. Routine by induction on the derivation height of $E[A] \xrightarrow{\alpha} E'[A']$. Note the linearity plays an important part in the proof, since in general a possibly non-linear $E[I_a]$ has to make several input actions to reach the same state as from $E[A]$ because X can appear in several (concurrent) positions. Also note the choice operator that results in the necessity of E'' because it may differ from E' if distinct choice is made on which subprocess does an action. $\qquad\square$

Now we present and prove the critical theorem called Abstraction. The message it is conveying is that one can appreciate the equivalence of two higher-order input prefixed processes in several different ways. From technical viewpoint, to handle the higher-order input, at lease two kinds of processes can be utilized to simplify the discussion, that is $I_b \equiv b(X).X$ and $b \equiv b(x).0$, the latter of which is internalized in the definition of local linear bisimulation. To make it concise, we omit the local contexts in the Abstraction theorem (like that in Theorem 2), even though extending to the general scenario involving local contexts is not hard. After all, the concise form suffices to support further theory.

Theorem 5 (Abstraction). *The following equations are equivalent:*
(i) $a(X).P \approx_{ll} a(X).Q$ for some name a;
(ii) $P\{I_b/X\} \approx_{ll} Q\{I_b/X\}$ for a fresh name b;
(iii) $P\{b/X\} \approx_{ll} Q\{b/X\}$ for a fresh name b;
(iv) $P\{R/X\} \approx_{ll} Q\{R/X\}$ for every process R.

Proof. The proof can be found in Appendix A.

Below we consider the Concretion theorem, before whose proof we give some auxiliary lemmas. The first lemma (Lemma 9) says that one can safely replace a process in a context with some trigger that can later activate, through a synchronization, the replaced process (A) which is now put parallel outside the context (sometimes we informally recognize the resulting process as "trigger form"). What is more important is that we merely need one instance of the process A, because we are coping with linear processes.

Lemma 9. *Suppose a, b are fresh names, $E[X]$ is an arbitrary process with at most process variable X, and A is a process. We have the following properties:*

1. *If $(\widetilde{x})E[A] \xrightarrow{\lambda} P$, where A takes part in the action, then $(\widetilde{x})(E[a]|\overline{a}.(A+b)) \xrightarrow{\lambda} P'$ for some P', and $P \sim P'$;*
2. *The converse. If $(\widetilde{x})(E[a]|\overline{a}.(A+b)) \xrightarrow{\lambda} P'$ and a, b do not appear in P' and λ, then we have $(\widetilde{x})E[A] \xrightarrow{\lambda} P$ for some P, and $P' \sim P$.*

Proof. The proof is essentially the same as Lemma 22 in [12] and note linearity plays an essential part, so we omit the detail.   □

Another lemma contributing to the Concretion theorem is Lemma 10. It needs Lemma 9 as the implicit premise. As a matter of fact, it is exactly part of the Concretion theorem. It states that the equivalence between "trigger forms" $(\overline{a}.(A+b)|E[a])$ can lead us to the equivalence between canonical contextual form $(E[A])$.

Lemma 10. *Suppose a, b are all fresh names, and $E[X], F[X]$ are processes. If $(\widetilde{y})(\overline{a}.(A+b)|E[a]) \approx_{ll} (\widetilde{z})(\overline{a}.(B+b)|F[a])$, then $(\widetilde{y})E[A] \approx_{ll} (\widetilde{z})F[B]$.*

Proof. We define a relation $\mathcal{R}$ as follows:

$$\mathcal{R} \triangleq \left\{ ((\widetilde{y})E[A], (\widetilde{z})F[B]) \,\middle|\, \begin{array}{l} (\widetilde{y})(\overline{a}.(A+b)|E[a]) \approx_{ll} (\widetilde{z})(\overline{a}.(B+b)|F[a]), \\ a, b \text{ are fresh} \end{array} \right\} \cup \approx_{ll}$$

We show that $\mathcal{R}$ is a local linear bisimulation up-to $\sim$. It is easy to show that $\mathcal{R}$ is closed under substitution of names. Suppose $(\widetilde{y})E[A]\mathcal{R}(\widetilde{z})F[B]$, and $(\widetilde{y})E[A] \xrightarrow{\lambda} P$. We have the following analysis.

- A does not take part in the action λ. Then we know that there exists $E_1[X]$ such that $P \equiv (\widetilde{y})E_1[A]$. By this we have $(\widetilde{y})(\overline{a}.(A+b)|E[a]) \xrightarrow{\lambda} (\widetilde{y})(\overline{a}.(A+b)|E_1[a])$. From the premise, we have the next simulation for some $F_1[X]$: $(\widetilde{z})(\overline{a}.(B+b)|F[a]) \xRightarrow{\lambda} (\widetilde{z})(\overline{a}.(B+b)|F_1[a])$, which is the only possibility because a, b are fresh. Therefore, we know that $(\widetilde{z})F[B] \xRightarrow{\lambda} (\widetilde{z})F_1[B]$.

- λ is a silent action, first-order input, output or higher-order input. This case is direct. We have $(\tilde{z})(\overline{a}.(B+b)|F_1[a]) \approx_{ll} (\tilde{y})(\overline{a}.(A+b)|E_1[a])$. Then $(\tilde{y})E_1[A] \; \mathcal{R} \; (\tilde{z})F_1[B]$.

- λ is a first-order bound output or higher-order output. This case is a little complicated. We take the first-order bound output as example, the higher-order output case is similar. Suppose λ is $\overline{u}(v)$. We have for every process O, $(v)(O|(\tilde{y})(\overline{a}.(A+b)|E_1[a])) \approx_{ll} (v)(O|(\tilde{z})(\overline{a}.(B+b)|F_1[a]))$, which results in $(\tilde{y}v)(\overline{a}.(A+b)|(E_1[a]|O)) \approx_{ll} (\tilde{z}v)(\overline{a}.(B+b)|(F_1[a]|O))$, thanks to α-conversion. Define $E_1'[X] \triangleq E_1[X]|O$, $F_1'[X] \triangleq F_1[X]|O$. So we have $(\tilde{y}v)(\overline{a}.(A+b)|E_1'[a]) \approx_{ll} (\tilde{z}v)(\overline{a}.(B+b)|F_1'[a])$, and $(v)(O|(\tilde{y})E_1[A]) \sim (\tilde{y}v)(E_1[A]|O) \equiv (\tilde{y}v)(E_1'[A])$, also $(v)(O|(\tilde{z})F_1[B]) \sim (\tilde{z}v)(F_1[B]|O) \equiv (\tilde{z}v)(F_1'[B])$. Now we know $(\tilde{y}v)(E_1'[A]) \; \mathcal{R} \; (\tilde{z}v)(F_1'[B])$.

- A is involved in the action λ. Then by Lemma 9, $(\tilde{y})(\overline{a}.(A+b)|E[a]) \overset{\lambda}{\Longrightarrow} P_1$, for some P_1, and $P_1 \sim P$. From the premise, we know that $(\tilde{z})(\overline{a}.(B+b)|F[a]) \overset{\lambda'}{\Longrightarrow} Q_1$, for some Q_1. A simple analysis can tell us that neither the fresh name a nor b shall appear in Q_1. So again by Lemma 9, we have $(\tilde{z})F[B] \overset{\lambda'}{\Longrightarrow} Q$, for some Q, and $Q \sim Q_1$.

  - λ is a silent action, first-order input, output or higher-order input. In this case, λ' is just λ. We have immediately $P \sim P_1 \approx_{ll} Q_1 \sim Q$.

  - λ is a first-order bound output or higher-order output. We take the higher-order output as example, the first-order bound output case is similar. Suppose λ is $(\tilde{v})\overline{u}[H]$ and λ' is $(\tilde{v'})\overline{u}[K]$. We have for a process $G[X] \equiv \overline{d}.(X+e)$ (d, e are fresh): $(\tilde{v})(G[H]|P_1) \approx_{ll} (\tilde{v'})(G[K]|Q_1)$. Since $(\tilde{v})(G[H]|P) \sim (\tilde{v})(G[H]|P_1), (\tilde{v'})(G[K]|Q_1) \sim (\tilde{v'})(G[K]|Q)$, we are finished.

Hence $\mathcal{R}$ is a local linear bisimulation up-to $\sim$. $\square$

With the help the intermediate lemmas above, we can now prove the Concretion theorem, which, in duality with the Abstraction theorem (Theorem 5), tells us how to handle higher-order output. For example, it can be handled in some asynchronous fashion (the second clause), or it can be simplified from ranging over the variety of receiving environment by observing with a special context (the third clause). Like that in the Abstraction theorem, we omit the local contexts in the Concretion theorem to make it concise.

Theorem 6 (Concretion). *The following equations are equivalent:*
(i) $(\tilde{x})\overline{a}[A].P \approx_{ll} (\tilde{y})\overline{a}[B].Q$ *for some name* $a \notin (\tilde{x} \cup \tilde{y})$;
(ii) $(\tilde{x})(\overline{b}[A]|P) \approx_{ll} (\tilde{y})(\overline{b}[B]|Q)$ *for a fresh name* b;
(iii) $(\tilde{x})(\overline{c}.(A+d)|P) \approx_{ll} (\tilde{y})(\overline{c}.(B+d)|Q)$ *for fresh names* c, d;
(iv) $(\tilde{x})(E[A]|P) \approx_{ll} (\tilde{y})(E[B]|Q)$ *for every process* $E[X]$.

Proof. We prove this theorem in the following strategy:

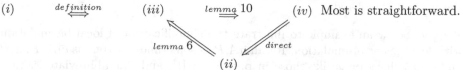

(i) $\overset{definition}{\Longleftrightarrow}$ (iii) $\overset{lemma\ 10}{\Longrightarrow}$ (iv) Most is straightforward.

Something worth noting is the application of Lemma 10 in $(iii) \Rightarrow (iv)$. Since $\approx_{ll}$ is closed under parallel composition, we have, for every process $E[X]$: $(\widetilde{x})(\overline{c}.(A+d)|P)|E[c] \approx_{ll} (\widetilde{y})(\overline{c}.(B+d)|Q)|E[c]$, which can be equivalently transformed to $(\widetilde{x})(\overline{c}.(A+d)|(P|E[c])) \approx_{ll} (\widetilde{y})(\overline{c}.(B+d)|(Q|E[c]))$. By defining $E'[X] \triangleq P|E[X], E''[X] \triangleq Q|E[X]$, we have $(\widetilde{x})(\overline{c}.(A+d)|E'[c]) \approx_{ll} (\widetilde{y})(\overline{c}.(B+d)|E''[c])$. Then by Lemma 10, $(\widetilde{x})(E'[A]) \approx_{ll} (\widetilde{y})(E''[B])$, which is exactly $(\widetilde{x})(P|E[A]) \approx_{ll} (\widetilde{y})(Q|E[B])$. To summary (by commutativity of $\approx_{ll}$) we have $(\widetilde{x})(E[A]|P) \approx_{ll} (\widetilde{y})(E[B]|Q)$, for every process E. We are done.   □

3.2 Coincidence with Local Bisimilarity

Below is the important theorem for local linear bisimilarity. It constitutes the main result of this paper.

Theorem 7. *Local linear bisimilarity coincides with local bisimilarity, that is* $\approx_l = \approx_{ll}$.

Proof. The proof is based on the definition of the two bisimulations, by taking the following two theorems into consideration: (i) The Abstraction Theorem (Theorem 5); (ii) The Concretion Theorem (Theorem 6). We prove $\approx_l = \approx_{ll}$ in two steps.

"$\subseteq$"". This direction is straightforward.

"$\supseteq$"". We show that $\mathcal{R} \triangleq \{(P,Q) \mid P \approx_{ll} Q, P, Q \text{ are LHOPi processes}\} \cup \approx_l$ is a local bisimulation. One has to examine the clauses in the definition of local bisimulation one by one. The most difficult cases are higher-order input and output. Below we analyze each of them. Suppose $P\mathcal{R}Q$ because $P \approx_{ll} Q$.

$\left. \begin{array}{l} (i).\ P\overset{\tau}{\to}P';\ (ii).\ P\overset{a(x)}{\longrightarrow}P' \\ (iii).\ P\overset{\overline{a}x}{\longrightarrow}P';\ (iv).\ P\overset{\overline{a}(x)}{\longrightarrow}P' \end{array} \right\}$ These cases are straightforward.

$(v).\ P\overset{a(A)}{\longrightarrow}P'$. Clearly we can define $P''[X]$ so that $P''\{A/X\} \equiv P'$. Then $P\overset{a(b)}{\longrightarrow}P''\{b/X\}$, for a fresh name b. Since $P \approx_{ll} Q$, we have $Q\overset{a(b)}{\Longrightarrow}Q''\{b/X\}$ for some Q'', and thus $Q\overset{a(A)}{\Longrightarrow}Q''\{A/X\} \triangleq Q'$, meanwhile $P''\{b/X\} \approx_{ll} Q''\{b/X\}$. Then by the Abstraction Theorem on $\approx_{ll}$ (Theorem 5), $P' \equiv P''\{A/X\} \approx_{ll} Q''\{A/X\} \equiv Q'$, which leads to $P' \mathcal{R} Q'$.

$(vi).\ P\overset{(\widetilde{x})\overline{a}[A]}{\longrightarrow}P'$. Because $P \approx_{ll} Q$, we know that there exist some $\widetilde{y}, B, Q'$ such that $Q\overset{(\widetilde{y})\overline{a}[B]}{\Longrightarrow}Q'$, and for a process $E[X] \triangleq \overline{c}.(X+d)$ (c, d are fresh), $(\widetilde{x})(E[A]|P') \approx_{ll} (\widetilde{y})(E[B]|Q')$. That is $(\widetilde{x})(\overline{c}.(A+d)|P') \approx_{ll} (\widetilde{y})(\overline{c}.(B+d)|Q')$. Then by the Concretion Theorem on $\approx_{ll}$ (Theorem 6, $(iii)\to(iv)$), we have, for every process $G[X]$ (with no name collision) $(\widetilde{x})(G[A]|P') \approx_{ll} (\widetilde{y})(G[B]|Q')$, which is exactly what we need to close the simulation, that is $(\tilde{x})(G[A]|P')\ \mathcal{R}\ (\tilde{y})(G[B]|Q')$. Now the proof is completed.   □

We give below an example to illustrate the simplification of local bisimulation with local linear bisimulation. Recall $\overline{a}A.P$ is sometimes written as $\overline{a}[A].P$, $\overline{x}.P$ and $x.Q$ behave on x like those in pure CCS [1], and $\overline{x}, x$ abbreviate $\overline{x}.0, x.0$ respectively. Consider the following two processes:

$$P \triangleq (x)\overline{a}[\overline{x}].(x.d\,|\,x.d), \qquad Q \triangleq (x)\overline{a}[x].(\overline{x}.d)$$

P, Q are LHOPi processes, since they have no process variables. They are intuitively local bisimilar, because $P\xrightarrow{(x)\overline{a}[\overline{x}]}x.d\,|\,x.d$ can be matched by $Q\xRightarrow{(x)\overline{a}[x]}\overline{x}.d$, (and vice versa) and for every LHOPi process $E[X]$ s.t. $x \notin fn(E[X])$,

$$(x)(E[\overline{x}]\,|\,x.d\,|\,x.d) \text{ and } (x)(E[x]\,|\,\overline{x}.d) \qquad\qquad (***)$$

are local bisimilar anyhow. Notice this is critically due to the linearity of $E[X]$, that is X would not appear free more than once on concurrent positions. However, to strictly prove this fact, one has to go through all possible forms of $E[X]$, which is really a heavy burden.

Now applying the simplification of local bisimulation, in the spirit of local linear bisimulation, we simply need to check whether the two processes in $(***)$ are equivalent when choosing $E[X] \triangleq \overline{f}.(X+g)$, where f, g are fresh names, i.e. names not appearing free in P, Q. Thus the question is reduced to checking whether the following processes are equivalent:

$$(x)(\overline{f}.(\overline{x}+g)\,|\,x.d\,|\,x.d) \text{ and } (x)(\overline{f}.(x+g)\,|\,\overline{x}.d)$$

Their equivalence is relatively easier to check.

The approach of local linear bisimulation is related to the idea of triggers [8] [9]. However the simplification is more general in several respects: local (linear) bisimulation is a more general and not delayed bisimulation; local (linear) bisimulation is also of open style; LHOPi enjoys a complete axiom system.

3.3   On First-Order Bound Output

What if we try simplifying the clause of first-order bound output in local bisimulation using the Localization Theorem (Theorem 1), in the way like what we have done for higher-order actions? Will it be effective as in the higher-order output? Considering the characteristic of linear higher-order processes and the essence of the Localization Theorem, our answer is NO. We have the following points.

- If we try to use (ii) in the Localization theorem (Theorem 1) to simplify the local bisimulation, that is, in the simulation step a special process $\overline{b}x$ (b is fresh) rather than an arbitrary process O is required, the obtained bisimulation (local linear bisimulation with this modification on first-order

bound output clause, we denote it by "LLN bisimulation") may not even
have the corresponding Localization theorem, because the (*iii*) cannot be
reached under a simplified simulation condition. So one cannot recover the
original local bisimilarity.

- In the "LLN bisimulation" , the first-order bound output cannot be eliminated in simulation, because the simulation result says the same thing as
  before the simulation, which may cause loop definition. One shall avoid this
  anytime.
- Apart from the special process $\bar{b}x$, which contributes nothing to simplification, no other special process is known to exist to replace the arbitrary
  process O without loss of generality. We tend to believe no such process
  exist.

Although the first-order bound output clause in local bisimulation cannot be
simplified, it can be rewritten in a form that eases discussion. In other words,
the 'simplification' here is in the sense that it can provide some simple means in
tackling local bisimulation in the case of first-order bound output.

As a matter of fact, the motivation of such simplification is initially technical,
yet later we become aware of more than that. The technical reason is related
to another contribution of ours on logical characterization of local bisimilarity,
whose form is not well suited for a direct characterization. Instead we turn to
local linear bisimilarity, which turns out not satisfactory either because it does
not match the logical equivalence perfectly. In this situation, we design local
linear variant bisimilarity, which rewrites the first-order bound output clause
using two universal processes O_1 and O_2 instead of only one such process (see
Definition 4). This technical manipulation works fine. Shifting from techcality to
theoretical viewpoint, we contend it shows that in a bisimulation clause defined
with respect to contextual processes like that in first-order bound output, the
universality of one process can be somewhat relaxed to those of more than one,
which does not alter the nature of the original bisimulation.

Below is the definition of the bisimulation obtained from local linear bisimulation by 'simplifying' the clause of first-order bound output. We name it local
linear variant bisimulation.

Definition 4. *A symmetric binary relation $\mathcal{R}$ on (closed) processes is a local
linear variant bisimulation, if it is closed under substitution of names, and whenever $P \mathcal{R} Q$, the following properties hold:*

1. *If $P \xrightarrow{\tau} P'$, then $Q \Longrightarrow Q'$ for some Q', and $P' \mathcal{R} Q'$;*
2. *If $P \xrightarrow{a(x)} P'$, then $Q \stackrel{a(x)}{\Longrightarrow} Q'$ for some Q', and $P' \mathcal{R} Q'$;*
3. *If $P \xrightarrow{\bar{a}x} P'$, then $Q \stackrel{\bar{a}x}{\Longrightarrow} Q'$ for some Q', and $P' \mathcal{R} Q'$;*
4. *If $P \xrightarrow{\bar{a}(x)} P'$, then $Q \stackrel{\bar{a}(x)}{\Longrightarrow} Q'$ for some Q', and for all processes O_1 and O_2 such
   that $O_1 \mathcal{R} O_2$, it holds that $(x)(O_1|P') \mathcal{R} (x)(O_2|Q')$;*
5. *If $P \xrightarrow{a(c)} P'$, where c is a fresh name, then $Q \stackrel{a(c)}{\Longrightarrow} Q'$ for some Q', and $P' \mathcal{R} Q'$;*
6. *If $P \xrightarrow{(\tilde{x})\bar{a}A} P'$, then $Q \stackrel{(\tilde{y})\bar{a}B}{\Longrightarrow} Q'$ for some $\tilde{y}, B, Q'$. And for a process $E[X]$ of
   the form $\bar{c}.(X+d)$, where c, d are fresh names, it holds that
   $(\tilde{x})(E[A]|P') \mathcal{R} (\tilde{y})(E[B]|Q')$.*

We say P is local linear variant bisimilar to Q, written $P \approx_{ll}^{v} Q$, if there exists some local linear variant bisimulation $\mathcal{R}$ such that $P \mathcal{R} Q$.

It can be shown, in a fashion similar to that of $\approx_{ll}$, that $\approx_{ll}^{v}$ is an equivalence relation and a congruence.

Theorem 8. *$\approx_{ll}^{v}$ is an equivalence relation and a congruence relation on all the calculus operators except the choice operator.*

We can now give another main result, that is local linear variant bisimilarity is coincident to the local linear bisimilarity.

Theorem 9. *Local linear variant bisimilarity coincides with local linear bisimilarity, that is $\approx_{ll} = \approx_{ll}^{v}$.*

Proof. We focus on the first-order bound output case in two bisimulations, since that is where the difference in the definitions lies. We recall the two clauses in each definition.

1. In $\approx_{ll}$: If $P \xrightarrow{\overline{a}(x)} P'$, then $Q \xRightarrow{\overline{a}(x)} Q'$ for some Q', and for every process O, $(x)(O|P') \approx_{ll} (x)(O|Q')$.
2. In $\approx_{ll}^{v}$: If $P \xrightarrow{\overline{a}(x)} P'$, then $Q \xRightarrow{\overline{a}(x)} Q'$ for some Q', and for all processes O_1 and O_2 such that $O_1 \approx_{ll}^{v} O_2$, $(x)(O_1|P') \approx_{ll}^{v} (x)(O_2|Q')$.

" $\subseteq$ ". This case is straightforward.
" $\supseteq$ ". This case is a little complex, in that it needs to exploit the congruence properties of $\approx_{ll}^{v}$ (Theorem 8), specifically the closure under parallel composition and restriction. We can define a relation: $\mathcal{R} \triangleq \{(P, Q) \mid P \approx_{ll}^{v} Q\} \cup \approx_{ll}$, and show $\mathcal{R}$ is a local linear bisimulation. In the first-order bound output case, we have $(x)(O_1|P') \approx_{ll}^{v} (x)(O_2|Q') \approx_{ll}^{v} (x)(O_1|Q')$, thanks to the congruence properties. Now we conclude that $\approx_{ll} = \approx_{ll}^{v}$. $\qquad \square$

Remark. Notice that all the bisimulations we define are of late style in first-order bound output and higher-order output. We can define (respectively) the early ones on these clauses accordingly. Using a similar approach to that in [12], one can readily prove that the early versions coincide with the corresponding late versions. We will take advantage of this fact in the future work on logical characterization, as will be mentioned in the conclusion.

4   Conclusion

Starting from previous work on bisimulation theory of higher-order process calculi, we arrive at some new results on bisimulation theory in linear higher-order π-calculus. Local bisimulation is an intuitively reasonable observational equivalence enjoying such characteristics as closure under substitution, equivalence, and congruence. By exploiting the properties of linear processes, we design two simpler variants of local bisimulation, which we prove to coincide with the original definition. The first variant, called local linear bisimulation, simplifies the higher-order input and higher-order output simulation steps in local bisimulation through examining the essence in the Abstraction theorem and the Concretion theorem. The coincidence proof is non-trivial and new. The second variant, called local linear variant bisimulation, adjusts the first-order bound output in local bisimulation to make it more appropriate for some analysis like axiomatization and logical characterization, by making use of the congruence properties.

Future Work

Some future work based on the result in this paper can be addressed. We mention several of them below.

- Recursion. Our calculus here is free of recursion operator or fix-point operator. This can provide us a complete axiom system, as shown in [12]. Albeit the inclusion of recursion would grant the calculus the power of Turing machines, it can enrich the diversity of the behavior of processes and the description capability, especially in cooperation with restriction and possibly relabelling which we do not include here either. We think that the inclusion of recursion would not shatter the main results in this paper, that is the simplification can still be obtained through a similar technical routine, as mentioned in the section Introduction. The difference worth noticing is in the proof concerning process structures, where one shall not use induction on process structure any more, but induction on derivation height instead, because the recursion can increase the complexity of a process during transitions.
- Logical characterization. Another immediate yet important task starting off from the bisimulation theory in this paper is to achieve a logical characterization of local bisimulation, which can complement the algebraic theory and enable practical modeling and verification using LHOPi. Related work on logically characterizing bisimulations in higher-order process calculi is [22], where strong context bisimulation in higher-order π-calculus is characterized, and [23], where weak context bisimulation in higher-order π-calculus is characterized. The framework is likewise. Below we summarize the rough pattern of logical characterization.
  1. The first step is to examine the target bisimulation to be characterized whether it is immediately suitable for a characterization. If the answer is positive, then one can proceed with step 3; otherwise goto step 2;
  2. In this step, one tries to find for the original bisimulation some variant tailored for logical characterization. And show that the variant coincides with the original bisimulation;
  3. In this step, one uses proper technique to obtain the approximation of the (variant) bisimilarity using a chain of "bisimulations". The approach can be standard;
  4. In this step, one designs the (modal) logic for characterizing the (variant) bisimulation. The method varies in different categories of calculi;
  5. The last step is to prove the characterization theorem, which establishes the coincidence between the target bisimilarity and designed logical equivalence. Techniques like characteristic formulas may be needed in the proof procedure.

A direct logical characterization of local bisimulation is possible by the results in [22] [23]. However since we are dealing with linear processes, the bisimulation is expected to enjoy a simpler form of logical characterization. The work in this paper indeed provides the simplification of local bisimulation. So we can characterize local linear variant bisimulation instead. Recall that the simplification resides in higher-order input and output, but first-order bound output clause cannot be simplified, though some more desirable form is available. The existence of first-order bound output results in the necessity of using constructive implication in the logic.

In summary, we can see that the results in this paper have settled several parts in the pattern above of logical characterization. The next task is to exercise the design of logic, for which we utilize related results and technique in [22] [23] [24], where the main contribution is the constructive implication operator ($\Rightarrow$) that is used to specify the property of a function process, which is a process with process variables appearing in it. For example, $\models P[X] : \phi \Rightarrow \phi'$ means that when inputted with a process R satisfying ϕ, the obtained process $P\{R/X\}$ shall satisfy ϕ'. Moreover, to accomplish the logical characterization, one has to reformulate the calculus under a new framework to tailor the processes to be suitable for a logical description. And as another preparation, the traditional approach should be adopted to approximate local bisimilarity (this result is given in the appendix [25]). We work on the task on logical characterization of local bisimilarity in [26].

Acknowledgements. We would like to thank the anonymous referees for their helpful comments and suggestions. We also thank Maciej Koutny for his effort in helping the author with the revision of this paper.

References

1. Milner, R.: Communication and Concurrency. Prentice Hall, Englewood Cliffs (1989)
2. Milner, R., Parrow, J., Walker, D.: A calculus of mobile processes (parts i and ii). Information and Computation 100(1), 1–77 (1992)
3. Sangiorgi, D., Walker, D.: The Pi-calculus: a Theory of Mobile Processes. Cambridge Universtity Press, Cambridge (2001)
4. Thomsen, B.: A calculus of higher order communication systems. In: Proceedings of POPL 1989, Austin, Texas, United States, pp. 143–154 (1989)
5. Thomsen, B.: Calculi for Higher Order Communicating Systems. Phd thesis, Department of Computing, Imperial College (1990)
6. Thomsen, B.: Plain chocs, a second generation calculus for higher-order processes. Acta Informatica 30(1), 1–59 (1993)
7. Thomsen, B.: A theory of higher order communication systems. Information and Computation 116, 38–57 (1995)
8. Sangiorgi, D.: Bisimulation for higher-order process calculi. Information and Computation 131(2), 141–178 (1996)
9. Sangiorgi, D.: Expressing Mobility in Process Algebras: First-order and Higher-order Paradigms. Phd thesis, University of Edinburgh (1992)

10. Sangiorgi, D.: A theory of bisimulation for π-calculus. Acta Informatica 33(1), 69–97 (1996)

11. Sangiorgi, D., Walker, D.: On barbed equivalences in pi-calculus. In: Larsen, K.G., Nielsen, M. (eds.) CONCUR 2001. LNCS, vol. 2154, pp. 292–304. Springer, Heidelberg (2001)

12. Fu, Y.: Checking equivalence for higher order processes. SJTU BASICS, http://basics.sjtu.edu.cn/people (2005) (Find 'Yuxi Fu')

13. Cao, Z.: Equivalence checking for a finite higher order pi-calculus. In: Beckert, B., Hähnle, R. (eds.) TAP 2008. LNCS, vol. 4966, pp. 30–47. Springer, Heidelberg (2008)

14. Wadler, P.: Linear types can change the world! In: Broy, M., Jones, C. (eds.) IFIP TC 2 Working Conference on Programming Concepts and Methods, Sea of Galilee, Israel, pp. 347–359 (1990)

15. VII, T.M.: Linear type systems for communication (2001), http://www.cs.cmu.edu/~tom7/papers/

16. Lanese, I., Perez, J., Sangiorgi, D., Schmitt, A.: On the expressiveness and decidability of higher-order process calculi. In: Proceedings of the 23rd Annual IEEE Symposium on Logic in Computer Science (LICS 2008), pp. 145–155 (2008)

17. Bundgaard, M., Hildebrandt, T., Godskesen, J.C.: A cps encoding of name-passing in higher-order mobile embedded resources. Theoretical Computer Science 356, 422–439 (2006)

18. Xu, X.: On bisimulation theory in linear higher-order π-calculus. In: The proceedings of Concurrency Methods: Issues and Applications (CHINA 2008), a satellite workshop of PETRI NETS 2008, Xi'an, China (2008) Technical report CS-TR-1102, School of Computing Science, Newcastle University, UK, http://basics.sjtu.edu.cn/~xuxian/OnBTinLHO.pdf

19. Milner, R.: Functions as processes. Journal of Mathematical Structures in Computer Science 2(2), 119–141 (1992)

20. Milner, R., Sangiorgi, D.: Barbed bisimulation. In: Kuich, W. (ed.) ICALP 1992. LNCS, vol. 623, pp. 685–695. Springer, Heidelberg (1992)

21. Baldamus, M., Frauenstein, T.: Congruence proofs for weak bisimulation equivalences on higher-order process calculi. Technical Report Report 95-21, Berlin University of Technology, Computer Science Department (1995)

22. Amadio, R., Dam, M.: Reasoning about higher-order processes. In: Mosses, P.D., Schwartzbach, M.I., Nielsen, M. (eds.) CAAP 1995, FASE 1995, and TAPSOFT 1995. LNCS, vol. 915, pp. 202–216. Springer, Heidelberg (1995)

23. Baldamus, M., Dingel, J.: Modal characterization of weak bisimulation for higher-order processes. In: Bidoit, M., Dauchet, M. (eds.) CAAP 1997, FASE 1997, and TAPSOFT 1997. LNCS, vol. 1214, pp. 285–296. Springer, Heidelberg (1997)

24. Milner, R., Parrow, J., Walker, D.: Modal logics for mobile processes. Theoretical Computer Science 114(1), 149–171 (1993)

25. Xu, X.: An appendix to on bisimulation theory in linear higher-order pi-calculus (2009), http://basics.sjtu.edu.cn/~xuxian/

26. Xu, X.: A logical characterization of local bisimulation in linear higher-order π-calculus. Technical report, BASICS Lab, SJTU (2007)

27. Nusse, R.: Wnt signaling in disease and in development. Cell research 15(1), 28–32 (2005)

28. Gross, L.: One signal, multiple pathways: Diversity comes from the receptor. PLoS Biol. 4(4), 115–131 (2006)

Appendix

A   Proof of Abstraction Theorem for $\approx_{ll}$

In this section, we give the detailed proof of the Abstraction Theorem for $\approx_{ll}$. We reproduce the theorem below.

Theorem 5 [Abstraction]. The following equations are equivalent:
(i) $a(X).P \approx_{ll} a(X).Q$ for some name a; (ii) $P\{I_b/X\} \approx_{ll} Q\{I_b/X\}$ for a fresh name b; (iii) $P\{b/X\} \approx_{ll} Q\{b/X\}$ for a fresh name b; (iv) $P\{R/X\} \approx_{ll} Q\{R/X\}$ for every process R.

Now we present the proof.

Proof. $(i) \Leftrightarrow (iii)$ is easy by definition.

$$\begin{matrix}(ii) \Leftrightarrow (iv) \\ (iii) \Leftrightarrow (iv)\end{matrix} \Bigg\} \quad \text{These two cases are similar in style, and note} \\ (a)(E\{I_a/X\}|\overline{a}[A]) \sim (a)(E\{a/X\}|\overline{a}.A).$$

So we simply consider $(iii) \Leftrightarrow (iv)$ here. Since $(iv) \Rightarrow (iii)$ is obvious, we cope with $(iii) \Rightarrow (iv)$. We define the following relation:

$$\mathcal{R} \triangleq \{(P\{R/X\}, Q\{R/X\}) \mid \begin{matrix} P\{b/X\} \approx_{ll} Q\{b/X\} \\ b \text{ is fresh, } R \text{ is a process} \end{matrix}\} \cup \approx_{ll}$$

First we show that $\mathcal{R}$ is closed under name substitution. Suppose $P\{R/X\} \mathcal{R}$ $Q\{R/X\}$ for $P\{b/X\} \approx_{ll} Q\{b/X\}$. Let σ be a substitution on names and d be a fresh name (not in $n(P, Q)$ and σ). We have $P\sigma\{d/X\} \approx_{ll} Q\sigma\{d/X\}$, since $\approx_{ll}$ is closed under substitution of names. Then we know
$P\sigma\{R\sigma/X\} \mathcal{R} Q\sigma\{R\sigma/X\})$.

Secondly we show $\mathcal{R}$ is a local linear bisimulation up-to $\sim$. Suppose $P\{R/X\} \overset{\alpha}{\rightarrow} P'$. Note α is of the form $a(\mathbf{d})$ in higher-order input. There are several cases to analyze.

– The action α is caused by a copy of R, that is $R \overset{\alpha}{\rightarrow} R'$ for some R'. Note substitution on process variables should avoid name capturing. So $P\{R/X\} \overset{\alpha}{\rightarrow} P_1\{R'/X\} \equiv P'$. Now we have the following reasoning (for some P_1, Q_1):

$$P\{R/X\} \overset{\alpha}{\longrightarrow} P_1\{R'/X\}$$

$$P\{b/X\} \overset{b(x)}{\longrightarrow} P_1\{0/X\}$$

$$\approx_{ll} \vdots \qquad\qquad\qquad \vdots \approx_{ll}$$

$$Q\{b/X\} \overset{b(x)}{\Longrightarrow} Q_1\{0/X\}$$

$$Q\{R/X\} \overset{\alpha}{\Longrightarrow} Q_1\{R'/X\}$$

And obviously, one can get, by a simple reasoning, $P_1\{d/X\} \approx_{ll} Q_1\{d/X\}$, for some fresh name d. Hence $P_1\{R'/X\} \mathcal{R} Q_1\{R'/X\}$.

- The action α is caused by P, that is $P\{R/X\} \xrightarrow{\alpha} P_1\{R/X\} \equiv P'$. Now we have the following reasoning (for some P_1, Q_1):

$$P\{R/X\} \xrightarrow{\quad\alpha\quad} P_1\{R/X\}$$

$$P\{b/X\} \xrightarrow{\quad\alpha\quad} P_1\{b/X\}$$

$$\approx_{ll} \qquad\qquad\qquad\vdots$$

$$Q\{b/X\} \xLongrightarrow{\quad\hat{\alpha}\quad} Q_1\{b/X\}$$

$$Q\{R/X\} \xLongrightarrow{\quad\hat{\alpha}\quad} Q_1\{R/X\}$$

If α is first-order input, output or higher-order input, the result is straightforward. If α is first-order bound output or higher-order output, some (similar) minor manipulation is needed. We take first-order bound output as the example. Suppose α is $\overline{u}(v)$, in this case, we have, for every process O, and some $P_1'[X] \triangleq (v)(O|P_1[X]), Q_1'[X] \triangleq (v)(O|Q_1[X])$:

$$(v)(O|P_1\{b/X\}) \cdots\!\overset{\approx_{ll}}{\cdots}\!\cdots (v)(O|Q_1\{b/X\})$$

$$\equiv \qquad\qquad\qquad\qquad \equiv$$

$$P_1'\{b/X\} \qquad\qquad\qquad Q_1'\{b/X\}$$

Then we know $P_1'\{R/X\} \mathcal{R} Q_1'\{R/X\}$. In summary, $P\{R/X\} \xrightarrow{\alpha} P_1\{R/X\}$ can be simulated by
$Q\{R/X\} \xLongrightarrow{\hat{\alpha}} Q_1\{R/X\}$.

- The action α is τ, and is caused by a communication between P and R. This is the most involved case and has totally six sub-cases. We examine them below.

  • $P \xrightarrow{u(v)} P'$, $R \xrightarrow{\overline{u}(v)} R'$, and $P\{R/X\} \xrightarrow{\tau} (v)(P'\{R'/X\})$.

    We have the following reasoning (for some P'', Q'', Q'). Basically P'', Q'' may differ from P, Q respectively, because even in linear paradigm P'' can differ (from P) in some sub-process originally in P. In specially cases, they can indeed be identical or structural congruent. Here, however, their existence, which is from the operational semantics, suffices to serve the proceeding of the proof. We do not make similar comment in the following subcases. Note in the following (chasing diagram [1]) the upper row is simulated by the lower row, and b is fresh.

$$P\{b/X\} \xrightarrow{b(x)} P''\{0/X\} \xrightarrow{u(v)} P'\{0/X\}$$

$$\approx_{ll} \qquad\qquad \approx_{ll} \qquad\qquad \approx_{ll}$$

$$Q\{b/X\} \Longrightarrow^{b(x)} Q''\{0/X\} \Longrightarrow^{u(v)} Q'\{0/X\}$$

Moreover, from the premise we also have

$$P\{b/X\} \xrightarrow{u(v)} P'\{b/X\}$$

$$\approx_{ll} \qquad\qquad \approx_{ll}$$

$$Q\{b/X\} \Longrightarrow^{u(v)} Q'\{b/X\}$$

So to summarize a little, we can have (because $\overset{u(v)}{\Longrightarrow}$ comes from Q)

$$Q\{R/X\}\overset{u(v)}{\Longrightarrow}Q'\{R/X\}, \quad Q\{R/X\}\overset{\overline{u}(v)}{\Longrightarrow}Q''\{R'/X\}\overset{u(v)}{\Longrightarrow}Q'\{R'/X\},$$
$$\text{and it holds that } P'\{b/X\} \approx_{ll} Q'\{b/X\}.$$

It follows from this, (congruence) property of $\approx_{ll}$ and a routine reasoning that

$$Q\{R/X\}\overset{\tau}{\Longrightarrow}(v)(Q'\{R'/X\}),$$
$$\text{and } (v)P'\{b/X\} \approx_{ll} (v)Q'\{b/X\}.$$

Now define $P'''[X] \triangleq (v)P'[X], Q'''[X] \triangleq (v)Q'[X]$ so that

$$P'''\{R'/X\} \triangleq (v)(P'\{R'/X\}), \quad Q'''\{R'/X\} \triangleq (v)(Q'\{R'/X\}).$$

Thus we have in summary

$$P\{R/X\}\overset{\tau}{\to}P'''\{R'/X\}, \text{ is simulated by } Q\{R/X\}\overset{\tau}{\Longrightarrow}Q'''\{R'/X\},$$
$$\text{and } P'''\{b/X\} \approx_{ll} Q'''\{b/X\}.$$

Hence it follows that

$$P'''\{R'/X\} \; \mathcal{R} \; Q'''\{R'/X\}.$$

- $P\xrightarrow{u(v)}P'$, $R\xrightarrow{\overline{u}v}R'$, and $P\{R/X\}\overset{\tau}{\to}(P'\{R'/X\})$. This case is similar to the last case.
- $P\xrightarrow{\overline{u}(v)}P'$, $R\xrightarrow{u(v)}R'$, and $P\{R/X\}\overset{\tau}{\to}(v)(P'\{R'/X\})$.
  We have the following reasoning (for some P'', Q'', Q'). Note the upper row is simulated by the lower row, and b is fresh.

$$P\{b/X\} \xrightarrow{b(x)} P''\{0/X\} \xrightarrow{\overline{u}(v)} P'\{0/X\}$$

$$\approx_{ll} \qquad\qquad \approx_{ll}$$

$$Q\{b/X\} \Longrightarrow^{b(x)} Q''\{0/X\} \Longrightarrow^{\overline{u}(v)} Q'\{0/X\}$$

And for every process O, $(v)(O|P'\{0/X\}) \approx_{ll} (v)(O|Q'\{0/X\})$. Moreover, from the premise we also have

$$P\{b/X\} \xrightarrow{\overline{u}(v)} P'\{b/X\}$$

$$\approx_{ll}$$

$$Q\{b/X\} \overset{\overline{u}(v)}{\Longrightarrow} Q'\{b/X\}$$

And for every process O, $(v)(O|P'\{b/X\}) \approx_{ll} (v)(O|Q'\{b/X\})$. So to summarize a little, we can have

$$Q\{R/X\} \overset{\overline{u}(v)}{\Longrightarrow} Q'\{R/X\}, \quad Q\{R/X\} \overset{u(v)}{\Longrightarrow} Q''\{R'/X\} \overset{\overline{u}(v)}{\Longrightarrow} Q'\{R'/X\},$$
and it holds that $(v)(O|P'\{b/X\}) \approx_{ll} (v)(O|Q'\{b/X\})$, for every O.

It follows from this, taking O as 0 (null process) and a routine reasoning that

$$Q\{R/X\} \Longrightarrow (v)(Q'\{R'/X\}),$$
and $(v)P'\{b/X\} \approx_{ll} (v)Q'\{b/X\}$.

Now define $P'''[X] \triangleq (v)P'[X], Q'''[X] \triangleq (v)Q'[X]$ so that

$$P'''\{R'/X\} \triangleq (v)(P'\{R'/X\}), \quad Q'''\{R'/X\} \triangleq (v)(Q'\{R'/X\}).$$

Thus we have in summary

$$P\{R/X\} \overset{\tau}{\to} P'''\{R'/X\}, \text{ is simulated by } Q\{R/X\} \overset{\tau}{\Longrightarrow} Q'''\{R'/X\},$$
and $P'''\{b/X\} \approx_{ll} Q'''\{b/X\}$.

Hence it follows that $P'''\{R'/X\} \mathcal{R} Q'''\{R'/X\}$.

- $P \overset{\overline{u}v}{\to} P'$, $R \overset{u(v)}{\to} R'$, and $P\{R/X\} \overset{\tau}{\to} (P'\{R'/X\})$. This case is similar to the last case.
- $P \overset{u(A)}{\to} P'$, $R \overset{(\tilde{z})\overline{u}[A]}{\to} R'$, and $P\{R/X\} \overset{\tau}{\to} (\tilde{z})(P'\{R'/X\})$. This case is somewhat similar to the first case. We have the following reasoning (for some P'', Q'', Q_1, P_1). Note the upper row is simulated by the lower row, and b is fresh. Suppose c is fresh.

$$P\{b/X\} \xrightarrow{b(x)} P''\{0/X\} \xrightarrow{u(c)} P_1\{0/X\}$$

$$\approx_{ll} \qquad\qquad \approx_{ll} \qquad\qquad \approx_{ll}$$

$$Q\{b/X\} \overset{b(x)}{\Longrightarrow} Q''\{0/X\} \overset{u(c)}{\Longrightarrow} Q_1\{0/X\}$$

where $P_1 \equiv P'\{c/A\}$. Moreover, from the premise we also have

$$P\{b/X\} \xrightarrow{u(c)} P_1\{b/X\}$$

$$\approx_{ll} \qquad\qquad \approx_{ll}$$

$$Q\{b/X\} \overset{u(c)}{\Longrightarrow} Q_1\{b/X\}$$

So to summarize a little, we can have

$$Q\{R/X\}\overset{u(\mathbf{c})}{\Longrightarrow}Q_1\{R/X\}, \quad Q\{R/X\}\overset{(\widetilde{z})\overline{u}[A]}{\Longrightarrow}Q''\{R'/X\}\overset{u(\mathbf{c})}{\Longrightarrow}Q_1\{R'/X\},$$
$$\text{and it holds that } P_1\{b/X\}\approx_{ll} Q_1\{b/X\}.$$

Then we have (for some Q')

$$Q\{R/X\}\overset{u(A)}{\Longrightarrow}Q'\{R/X\}, \quad Q\{R/X\}\overset{(\widetilde{z})\overline{u}[A]}{\Longrightarrow}Q''\{R'/X\}\overset{u(A)}{\Longrightarrow}Q'\{R'/X\},$$
$$\text{and it also holds that } P'\{\mathbf{c}/A\}\{b/X\}\approx_{ll} Q'\{\mathbf{c}/A\}\{b/X\},$$

where $Q_1 \equiv Q'\{\mathbf{c}/A\}$. It follows from this and a routine reasoning

$$Q\{R/X\}\overset{\tau}{\Longrightarrow}(\widetilde{z})(Q'\{R'/X\}). \tag{1}$$

We define some $P_1'[Y]$ and $Q_1'[Y]$ (Y is different from X and fresh) such that
$$P_1'\{\mathbf{c}/Y\} \equiv P_1, P_1'\{A/Y\} \equiv P';$$
$$Q_1'\{\mathbf{c}/Y\} \equiv Q_1, Q_1'\{A/Y\} \equiv Q'.$$

Then $P_1'\{\mathbf{c}/Y\}\{b/X\}\approx_{ll} Q_1'\{\mathbf{c}/Y\}\{b/X\}$. Since Y and X are different, we know that $P_1'\{b/X\}\{\mathbf{c}/Y\}\approx_{ll} Q_1'\{b/X\}\{\mathbf{c}/Y\}$. Now by the definition (structure of $\mathcal{R}$) we know $P_1'\{b/X\}\{A/Y\} \mathcal{R} Q_1'\{b/X\}\{A/Y\}$. That is $P_1'\{A/Y\}\{b/X\} \mathcal{R} Q_1'\{A/Y\}\{b/X\}$. And this is exactly $P'\{b/X\} \mathcal{R} Q'\{b/X\}$. Again by the definition (of $\mathcal{R}$) we have in any case $P'\{b'/X\}\approx_{ll} Q'\{b'/X\}$, for a fresh name b'. By the (congruence) property of $\approx_{ll}$ it follows that

$$(\widetilde{z})P'\{b'/X\}\approx_{ll} (\widetilde{z})Q'\{b'/X\}. \tag{2}$$

Now define $P'''[X] \triangleq (\widetilde{z})P'[X], Q'''[X] \triangleq (\widetilde{z})Q'[X]$ so that

$$P'''\{R'/X\} \triangleq (\widetilde{z})(P'\{R'/X\}), \quad Q'''\{R'/X\} \triangleq (\widetilde{z})(Q'\{R'/X\}).$$

Thus we have in summary

$$P\{R/X\}\overset{\tau}{\to}P'''\{R'/X\}, \text{ is simulated by}$$
$$Q\{R/X\}\overset{\tau}{\Longrightarrow}Q'''\{R'/X\}, \quad \text{(by (1))},$$
$$\text{and } P'''\{b'/X\}\approx_{ll} Q'''\{b'/X\}, \quad \text{(by (2))}.$$

Hence it follows that $P'''\{R'/X\} \mathcal{R} Q'''\{R'/X\}$.

- $P\overset{(\widetilde{z})\overline{u}[A]}{\longrightarrow}P'$, $R\overset{u(A)}{\longrightarrow}R'$, and $P\{R/X\}\overset{\tau}{\to}(\widetilde{z})(P'\{R'/X\})$.
  We have the following reasoning (for some $\widetilde{z}', B, P'', Q'', Q'$). Note the upper row is simulated by the lower row, and b is fresh.

$$
\begin{array}{ccccc}
P\{b/X\} & \overset{b(x)}{\longrightarrow} & P''\{0/X\} & \overset{(\widetilde{z})\overline{u}[A]}{\longrightarrow} & P'\{0/X\} \\
\approx_{ll} \vdots & & \approx_{ll} \vdots & & \vdots \\
Q\{b/X\} & \overset{b(x)}{\Longrightarrow} & Q''\{0/X\} & \overset{(\widetilde{z}')\overline{u}[B]}{\Longrightarrow} & Q'\{0/X\}
\end{array}
$$

And for a process $E[X] \equiv \bar{c}.(X+d)$ (c, d are fresh),
$(\tilde{z})(E[A]\|P'\{0/X\}) \approx_{ll} (\tilde{z}')(E[B]\|Q'\{0/X\})$. Moreover, from the premise we also have:

$$P\{b/X\} \xrightarrow{(\tilde{z})\overline{u}[A]} P'\{b/X\}$$

$$\approx_{ll}$$

$$Q\{b/X\} \xRightarrow{(\tilde{z}')\overline{u}[B]} Q'\{b/X\}$$

And for a process $E[X] \equiv \bar{c}.(X+d)$ (c, d are fresh),
$(\tilde{z})(E[A]\|P'\{b/X\}) \approx_{ll} (\tilde{z}')(E[B]\|Q'\{b/X\})$. So to summarize a little, we can have (for some R'' such that $R'' \equiv R'\{B/A\}$)

$$Q\{R/X\} \xRightarrow{(\tilde{z}')\overline{u}[B]} Q'\{R/X\}, \tag{3}$$

$$Q\{R/X\} \xRightarrow{u(B)} Q''\{R''/X\} \xRightarrow{(\tilde{z}')\overline{u}[B]} Q'\{R''/X\}, \tag{4}$$

and it holds that

$$(\tilde{z})(\bar{c}.(A+d)\|P'\{b/X\}) \approx_{ll} (\tilde{z}')(\bar{c}.(B+d)\|Q'\{b/X\}). \tag{5}$$

It follows from (3), (4), and a routine reasoning that

$$Q\{R/X\} \xRightarrow{\tau} (\tilde{z}')(Q'\{R''/X\}). \tag{6}$$

By Concretion Theorem (Theorem 6) and (5), $(\tilde{z})(G[A]\|P'\{b/X\}) \approx_{ll} (\tilde{z}')(G[B]\|Q'\{b/X\})$, for every $G[Y]$. It is easy to know that there exists some $R'''[Y]$ such that

$$R'''\{A/Y\} \equiv R', \quad R'''\{B/Y\} \equiv R''.$$

Now choose $G[Y] \equiv \bar{b}.(R'''[Y]+d)$, then we get

$$(\tilde{z})(\bar{b}.(R'''\{A/Y\}+d)\|P'\{b/X\}) \approx_{ll} (\tilde{z}')(\bar{b}.(R'''\{B/Y\}+d)\|Q'\{b/X\}),$$

that is $(\tilde{z})(\bar{b}.(R'+d)\|P'\{b/X\}) \approx_{ll} (\tilde{z}')(\bar{b}.(R''+d)\|Q'\{b/X\})$. Thus by Lemma 10, we have
$(\tilde{z})(P'\{R'/X\}) \approx_{ll} (\tilde{z}')(Q'\{R''/X\})$. Hence
$(\tilde{z})(P'\{R'/X\}) \, \mathcal{R} \, (\tilde{z}')(Q'\{R''/X\})$. Taking (6) (the simulating action step) into consideration, this closes the simulation.

By here the proof is completed.   □

B   Examples of Modeling with LHOPi

In this section, we present (in two subsections below) two examples in their prototypes using LHOPi. One is on distributed computing, where server process responding to request from clients merely allows one instance. The other is about biological signaling transduction, where the signaling pathway starts and continues with one message molecule. All these phenomena can be modeled within the capability of LHOPi.

Notice the examples are not full-scaled for the sake of space limit and not deviating from the topic of this paper. We intend to expand the study on applying (L)HOPi to various concurrent systems.

B.1   Singleton Design Pattern

In software development, Singleton is a well-known design pattern. It is the solution to occasions when an application requires a single instance of a given class together with a global point of access to the same class. It is important to ensure there is one instance of certain class (process), for example there ought to be only one instance of the Connection class to operate a database system or a printer, only one File system and windows manager, and only one accounting system for a certain company. Singleton is more accurate than global variables in meeting such requirement.

As another evidence, in an important component (Remoting) of the .NET framework, an object can be activated in two ways: server-activating or client-activating. In the former, the object activated by the server also have two forms: Singleton object and SingleCall object. In the first form, all requests from clients are handled by the singleton object, which remains one instance throughout the running time. In the second form, the system creates a new serving object each time there comes a call from some client, even though the same client makes different calls, and moreover once the call is over, such a serving object will be destroyed. Both of the forms: Singleton object and SingleCall object can be modeled by LHOPi. Note LHOPi (without recursion) can effectively control the number of instance, by imposing on the occurrence of process variables. With recursion one can model the singleton that continuously offers service, though the recursion should be limited so as not to cause non-linearity [12].

For example, In recursion free LHOPi, the case of SingleCall can be described as below:

$$Server \triangleq (m)(c(X).P[X] \mid m(Y).\bar{r}[Ack])$$

c is the requesting port, whereas r is the responding port. $P[X]$ is the process doing the transaction (for example a database accessing). When finished, $P[X]$ signals by local channel m to send the result back to the client. Ack plays the role of acknowledging and returning result. In LHOPi with recursion, the case of Singleton can be described as below:

$$Server \triangleq (m)(c(X).P[X] \mid m(Y).\bar{r}[Ack].Server)$$

The only difference lies in that the Server is now circularly defined. This process can continuously provide its service for calls from clients, but it can not spawn a clone process using non-linear mechanism, such as $a(X).(X|X)$.

In some other commercial scenarios, a process taking the role of a product key or a disposable cellphone, or such kind of service that can be received only once like those in charged television, LHOPi can be utilized to provide the formal model. Though there is still more to do to examine the degree of faithfulness of modeling and related questions like analysis and verification, we believe its interest is worthy of further effort.

B.2 WNT Signaling Pathway

WNT is a conserved family of signaling proteins that plays a pivotal part in organ development [27]. The malfunction of them is connected with diseases. Known processes regulated by the signaling (transduction) pathways triggered by WNT signaling protein include cell survival and growth, differentiation, proliferation, migration, mutation, and numerous other respects. The cascades of the initial signaling consists of many steps and more than one branches [28], each of which involves many interacting objects. The most-studied and best-understood branch is the "canonical" pathway, which can be described concisely as follows: Firstly the WNT signaling molecule (recognized as ligand) binds to the receptor Frizzled (FRZ) residing on the cell surface (cell membrane); Then FRZ interacts with a transmembrane protein called LRP, resulting in a complex composed of WNT, FRZ and LRP; Secondly, this complex cuts out the proteolytic degradation of β-catenin (BCAT), an intracellular signaling molecule, by inhibiting the composition of the destruction complex (aiming at BCAT) that we simply denote as BCDC (β-Catenin Destruction Complex), which actually also plays an interesting role with the help of its elements like Axin, APC, and GSK3. Then BCAT protein stabilizes in cytoplasm. Thirdly, the surviving BCAT is able to enter the nucleus and associate with TCF/LEF (TL) complex, which belongs to a family of transcription factors, to promote specific gene expression (such as Myc, Cyclin D1 and etc.) contributing to certain stage of cell development. As reported, singularity in the WNT signaling pathway is linked to a few types of cancer.

Here we abstract the signal as "single" one that runs through the signaling pathway. The model is qualitative. To study further more factors should be taken into account, such as authentication mechanism and competition on grasping β-catenin.

As a whole, the steps of the signaling transduction can be modelled using LHOPi in Table 2. We accompany the elements (labelled by different number of $\star$'s) of the model with necessary explanations.

   $\star$ WNT binds FRZ, and then LRP binds FRZ. WNT', FRZ' and processes with one prime are left for more activities we do not go further into. m is the port where signal starts cascading. b is for establishing the connection between WNT and FRZ.

   t is for triggering the expression of the target genes.

Table 2. Model of WNT signaling pathway

Location	Objects	Process in LHOPi	
Extracellular	WNT	$WNT \triangleq (b)(\overline{m}b.\overline{b}[SIG].WNT')$ $SIG \triangleq \overline{t}.0$	$\star$
Membrane	Frizzled (FRZ) LRP	$FRZ \triangleq (l)(m(x).x(Y).(\overline{l}.\overline{c}[Y].FRZ') \mid LRP)$ $LRP \triangleq l.LRP'$	$\star\star$
Cytoplasm	β-catenin (BCAT) BCDC	$BCAT \triangleq v(Z).\overline{w}[ON].\overline{n}[Z].BCAT'$ $BCDC \triangleq (u)(c(X).\overline{u}[OFF].\overline{v}[X].BCDC')$	$\star\star\star$
Nucleus	TCF/LEF (TL) Myc (MYC) Cyclin D1 (CD1) ...	$TL \triangleq n(X).X$ $MYC \triangleq t.MYC''$ ...	$\star\star\star\star$

$\star\star$ (The synchronization on) l indicates the binding of LRP and FRZ. c is the channel that relays the signal into cytoplasm.

$\star\star\star$ v is the channel for activating β-catenin (after releasing it from degradation). BCAT sends through channel w the process ON to activate related molecules, which is omitted. BCDC sends through channel u the process OFF to turn off the destruction complex that destroys β-catenin. Process ON and OFF can be simple processes acting as the triggers to another round of reactions.

$\star\star\star\star$ n is the channel that relays the signal into the nucleus. MYC$''$ may behaves as expressing related genes into corresponding proteins. Other target genes can be modeled similarly.

The whole canonical WNT signaling system is then defined as:

$$WNT-SIG-SYSTEM \triangleq$$
$$(m)(c)(t)(WNT \mid FRZ \mid (v)(w)(n)(BCAT \mid BCDC \mid TL \mid MYC))$$

Author Index